Industrial Applications of Microbial Enzymes

Microbial enzymes are important because they can be used for a wide variety of industrial purposes. There is dispersed and scanty information available with respect to microbial enzymes and their industrial applications. In this edited book, leading scientists have covered the various aspects of microbial enzymes and their industrial applications. Using microbial enzymes can help expedite various manufacturing processes and contribute to sustainable development, which is a priority worldwide. Research gaps in the entrainment of microbial enzymes with their direct application in product development are a major focus of this volume.

KEY FEATURES

- Covers microbial enzymes with comprehensive and in-depth information
- Benefits students by describing recent advancements in microbial enzymology
- Provides updates regarding microbial enzymes for researchers and industrial scientists
- Includes findings on the microbial actions for a better life

RELATED TITLES

Thatoi, H., et al., eds. *Microbial Fermentation and Enzyme Technology* (ISBN 978-0-3671-8384-4).

Svendsen, A., ed. *Understanding Enzymes: Function, Design, Engineering, and Analysis* (ISBN 978-9-8146-6932-0).

Seneviratne, C. J., ed. *Microbial Biofilms: Omics Biology, Antimicrobials and Clinical Implications* (ISBN 978-0-3676-5799-4).

Suzuki, H. *How Enzymes Work: From Structure to Function* (ISBN 978-9-8148-0066-2).

de Lourdes, M., et al., eds. *Fungal Enzymes* (ISBN 978-1-4665-9454-8).

Industrial Applications of Microbial Enzymes

Edited by
Pankaj Bhatt

CRC Press
Taylor & Francis Group
Boca Raton   London

CRC Press is an imprint of the
Taylor & Francis Group, an **informa** business

First edition published 2023
by CRC Press
6000 Broken Sound Parkway NW, Suite 300, Boca Raton, FL 33487–2742

and by CRC Press
4 Park Square, Milton Park, Abingdon, Oxon, OX14 4RN

CRC Press is an imprint of Taylor & Francis Group, LLC

Library of Congress Cataloging-in-Publication Data
Names: Bhatt, Pankaj, editor.
Title: Industrial applications of microbial enzymes / edited by Pankaj Bhatt.
Description: First edition. | Boca Raton : CRC Press, 2022. | Includes
    bibliographical references and index.
Identifiers: LCCN 2021060977 (print) | LCCN 2021060978 (ebook) |
    ISBN 9781032065137 (hardback) | ISBN 9781032065984 (paperback) |
    ISBN 9781003202998 (ebook)
Subjects: LCSH: Microbial enzymes—Industrial applications. | Industrial
    microbiology.
Classification: LCC QR90 .I518 2022 (print) | LCC QR90 (ebook) |
    DDC 660.6/2—dc23/eng/20220330
LC record available at https://lccn.loc.gov/2021060977
LC ebook record available at https://lccn.loc.gov/2021060978

ISBN: 9781032065137 (hbk)
ISBN: 9781032065984 (pbk)
ISBN: 9781003202998 (ebk)

DOI: 10.1201/9781003202998

Typeset in Times
by Apex CoVantage, LLC

Contents

Editor ix
Contributors xi

Chapter 1   Recent Advancement in Microbial Enzymes and Their
Industrial Applications 1

*Pankaj Bhatt, Sajjad Ahmad, Samiksha Joshi,
and Kalpana Bhatt*

Chapter 2   Production, Purification, and Application of the
Microbial Enzymes 19

Anupam Pandey, Ankita H. Tripathi, and Priyanka H. Tripathi

Chapter 3   Recent Advancements in Microbial Enzymes and Their
Application in Bioremediation of Xenobiotic Compounds................ 41

*Saurabh Gangola, Pankaj Bhatt, Samiksha Joshi, Saurabh
Kumar, Narendra Singh Bhandari, Samarth Terwari,
Om Prakash, and Amit Kumar Mittal*

Chapter 4   Industrial Applications of Bacterial Enzymes.................................... 59

Md. Shahbaz Anwar

Chapter 5   A Quick Look-Around of Microbial Enzymes in
Modern Food Industries and Dietary Research................................. 91

Vineet Singh, Anjali Pande, and Jae-Ho Shin

Chapter 6   Fungal Enzymes in Organic Pollutants Bioremediation 101

*Adam Grzywaczyk, Wojciech Smułek, Jakub Zdarta,
and Ewa Kaczorek*

Chapter 7   Enzymes Involved in the Bioremediation of Pesticides 133

*Sajjad Ahmad, Pankaj Bhatt, Hafiz Waqas Ahmad,
Dongming Cui, Jiatai Guo, Guohua Zhong, and Jie Liu*

Chapter 8 Esterases and Their Industrial Applications 169

Hamza Rafeeq, Asim Hussain, Ayesha Safdar,
Sumaira Shabbir, Muhammad Bilal, Farooq Sher,
Marcelo Franco, and Hafiz M. N. Iqbal

Chapter 9 Soil Microbial Enzymes and Their Importance, Significance,
and Industrial Applications... 191

Hemant Dasila, Sarita Joshi, and Sudipta Ramola

Chapter 10 Application of Microbial Enzymes in Industry and Antibiotic
Production 207

Rishendra Kumar, Lokesh Tripathi, and Pankaj Bhatt

Chapter 11 ACC-Deaminase-Producing Bacteria: From Alleviating Plant
Stress to their Commercial Application for Sustainable
Agriculture.. 221

Anjali Pande, Vineet Singh, and Byung Wook Yun

Chapter 12 Phytases and Their Characteristic Features and Biotechnological
Applications in Animal Feed.. 231

Syed Zakir Hussain Shah, Mahroze Fatima, Mehwish Khan,
and Muhammad Bilal

Chapter 13 Applications of Immobilized Ligninolytic Enzymes in the
Degradation of Industrial Pollutants... 249

Muhammad Bilal, Hamza Rafiq, Sarmad Ahmad Qamar,
Asim Hussain, Pankaj Bhatt, and Hafiz M. N. Iqbal

Chapter 14 Role of Streptokinase as a Thrombolytic Agent for Medical
Applications 271

Hamza Rafeeq, Muhammad Anjum Zia, Asim Hussain,
Ayesha Safdar, Muhammad Bilal and Hafiz M. N. Iqbal

Chapter 15 Laccase-Assisted Biocatalytic Removal of Lignin from
Lignocellulosic Biomass 295

Sadia Noreen, Sara Rehman, Memoona Asif,
Muhammad Bilal, and Hafiz M. N. Iqbal

Chapter 16 Omics Approaches for the Production of the Microbial Enzymes
and Applications 317

Heena Parveen, Anuj Chaudhary, Parul Chaudhary,
Rabiya Sultana, Govind Kumar, Priyanka Khati,
Meenakshi Rana, and Pankaj Bhatt

Index.. 333

Editor

Dr. Pankaj Bhatt earned his PhD in microbiology from G.B Pant University of Agriculture and Technology, Pantnagar, U.S Nagar, India. His PhD work was on the molecular and proteomic basis of biodegradation of pesticides. He has published 60 research and review articles in reputable journals. He has authored 20 book chapters. Dr. Bhatt edited five books with Elsevier, Springer, and IGI Global. Previously, he was an assistant professor at the Department of Microbiology, Dolphin (PG) College of Biomedical and Natural Sciences, Dehradun, India. Dr. Bhatt worked at Integrative Microbiology Research center, South China Agriculture University, Guangzhou, China for three years. In addition, he have associated with the Department of Environmental Engineering, Kyungpook National University, Daegu, South Korea. Presently, he is working as Visiting researcher at Department of Agriculture and Biological Engineering, Purdue University, Indiana, United States of America.

Contributors

Sajjad Ahmad
Key Laboratory of Integrated Pest
  Management of Crop in South China
Ministry of Agriculture and Rural Affairs
Key Laboratory of Natural Pesticide
  and Chemical Biology
Ministry of Education
South China Agricultural University
Guangzhou P. R. China

Department of Agricultural & Biological
  Engineering Purdue University
West Lafayette, IN, USA

Hafiz Waqas Ahmad
Department of Food Engineering
  Faculty of Agricultural Engineering
  and Technology
University of Agriculture
Faisalabad, Pakistan

Md. Shahbaz Anwar
Department of Microbiology
Dum Dum Motijheel College
WBSU, Kolkata

Memoona Asif
Department of Biochemistry
University of Agriculture
Faisalabad, Pakistan

Narendra Singh Bhandari
School of Agriculture
Graphic Era Hill University
Bhimtal, Uttarakhand, India

Kalpana Bhatt
Department of Botany and
  Microbiology

Gurukul Kangri University
Haridwar, Uttarakhand, India

Pankaj Bhatt
Department of Agricultural &
  Biological Engineering
Purdue University
West Lafayette, Indiana, USA

Department of Integrative Microbiology
  Research Centre
South Agricultural University China
State Key Laboratory for Conservation
  and Utilization of Subtropical Agro-
  bioresources
Guangdong Laboratory for
  Lingnan Modern Agriculture
  Integrative Microbiology
  Research Centre
South China Agricultural
  University
Guangzhou, China

Muhammad Bilal
School of Life Science and Food
  Engineering
Huaiyin Institute of Technology
Huai'an China

Anuj Chaudhary
School of Agriculture and
  Environmental Sciences
Shobhit University
Gangoh, Uttar Pradesh, India

Parul Chaudhary
Department of Animal Biotechnology
NDRI, Karnal
Haryana, India

Dongming Cui
Key Laboratory of Integrated Pest
    Management of Crop in South China
Ministry of Agriculture and Rural
    Affairs
Key Laboratory of Natural Pesticide and
    Chemical Biology
Ministry of Education
South China Agricultural University
Guangzhou, P. R. China

Hemant Dasila
Department of Microbiology
College of Basic Sciences and Humanities
G. B. Pant University of Agriculture and
    Technology
Pantnagar, India

Mahroze Fatima
Department of Fisheries and
    Aquaculture
University of Veterinary and
    Animal Sciences
Lahore, Pakistan

Marcelo Franco
Department of Exact Sciences and
    Technology
State University of Santa Cruz
Ilhéus, Brazil

Saurabh Gangola
School of Agriculture
Graphic Era Hill University
Bhimtal, India

Adam Grzywaczyk
Institute of Chemical Technology and
    Engineering
Poznan University of Technology
Berdychowo Poznan, Poland

Jiatai Guo
Key Laboratory of Integrated Pest
    Management of Crop in South China
Ministry of Agriculture and Rural Affairs

Key Laboratory of Natural Pesticide and
    Chemical Biology
Ministry of Education
South China Agricultural University
Guangzhou P. R. China

Asim Hussain
Department of Biochemistry
Riphah International University
Faisalabad, Pakistan

Hafiz M. N. Iqbal
Tecnologico de Monterrey
School of Engineering and Sciences
Monterrey, Mexico

Samiksha Joshi
School of Agriculture
Graphic Era Hill University
Bhimtal, Uttarakhand, India

Sarita Joshi
Department of Environmental Science
College of Basic Sciences and
    Humanities
G. B. Pant University of Agriculture and
    Technology
Pantnagar, India

Ewa Kaczorek
Institute of Chemical Technology and
    Engineering
Poznan University of Technology
Berdychowo Poznan, Poland

Mehwish Khan
Department of Fisheries and
    Aquaculture
University of Veterinary and Animal
    Sciences
Lahore, Pakistan

Priyanka Khati
Crop Production Division
ICAR-VPKAS, Almora
Uttarakhand, India

Govind Kumar
Department of Crop Production
ICAR-Central Institute for Subtropical
   Horticulture
Lucknow, India

Rishendra Kumar
Department of Biotechnology
Sir J. C. Bose Technical Campus Bhimtal
Kumaun University
Nainital, Uttarakhand, India

Saurabh Kumar
ICAR Research Complex for Eastern
   Region
Patna, India

Jie Liu
Key Laboratory of Integrated Pest
   Management of Crop in South China
Ministry of Agriculture and Rural Affairs
Key Laboratory of Natural Pesticide and
   Chemical Biology
Ministry of Education
South China Agricultural University
Guangzhou, P. R. China

Amit Kumar Mittal
Department of Allied Sciences
Graphic Era Hill University
Bhimtal, India

Sadia Noreen
Department of Biochemistry
Government College Women University
Faisalabad, Pakistan

Anjali Pande
Department of Applied Biosciences,
Kyungpook National University
   South Korea

Department of Agriculture and Life
   Sciences and Department of Applied
   Life Sciences
Kyungpook National University
South Korea

Anupam Pandey
Sir J. C. Bose Technical Campus, Bhimtal
Kumaun University
Nainital Uttarakhand, India
and
ICAR Directorate of Coldwater
   Fisheries Research
Bhimtal, Uttarakhand, India

Heena Parveen
Department of Dairy Microbiology
NDRI, Karnal
Haryana, India

Om Prakash
Department of Allied Sciences
Graphic Era Hill University
Bhimtal, India

Sarmad Ahmad Qamar
State Key Laboratory of Bioreactor
   Engineering and School of
   Biotechnology
East China University of Science and
   Technology
Shanghai, China

Hamza Rafeeq
Department of Biochemistry
Riphah International University
Faisalabad, Pakistan

Sudipta Ramola
College of Chemical Engineering
Zhejiang University of Technology
Hangzhou, China

Meenakshi Rana
School of Agriculture
Lovely Professional University
Phagwara, Punjab, India

Sara Rehman
Department of Biochemistry
Government College Women University
Faisalabad, Pakistan

Ayesha Safdar
Department of Biochemistry
University of Agriculture
Faisalabad, Pakistan

Sumaira Shabbir
Department of Zoology, Wildlife, and
    Fisheries
University of Agriculture
Faisalabad, Pakistan

Farooq Sher
Department of Engineering
School of Science and Technology
Nottingham Trent University
Nottingham NG11 8NS, UK

Jae-Ho Shin
Department of Applied Biosciences
Department of Integrative
    Biotechnology
Kyungpook National University
South Korea

Vineet Singh
Department of Applied Biosciences
Department of Agriculture and
    Life Sciences
Department of Applied Life
    Sciences
Kyungpook National University
South Korea

Syed Zakir Hussain Shah
Department of Zoology
University of Gujrat
Gujrat, Pakistan

Wojciech Smułek
Institute of Chemical Technology
    and Engineering
Poznan University of Technology
Berdychowo Poznan, Poland

Rabiya Sultana
Department of Biotechnology
Gauhati University
Assam, India

Samarth Terwari
Department of Agricultural and
    Biological Engineering

Purdue University
West Lafayette, Indiana, USA

Ankita H. Tripathi
Sir J. C. Bose Technical Campus,
    Bhimtal
Kumaun University
Nainital Uttarakhand, India

Priyanka H. Tripathi
Sir J. C. Bose Technical Campus,
    Bhimtal
Kumaun University Nainital
Uttarakhand, India
and
ICAR Directorate of Coldwater
    Fisheries Research
Bhimtal, Uttarakhand, India

Lokesh Tripathi
Department of Biotechnology
Sir J. C. Bose Technical Campus
Bhimtal, Kumaun University
Nainital, Uttarakhand, India

Byung Wook Yun
Department of Agriculture and Life
    Sciences
Department of Applied Life Sciences
Kyungpook National University
South Korea

Jakub Zdarta
Institute of Chemical Technology and
    Engineering
Poznan University of Technology
Berdychowo Poznan, Poland

Guohua Zhong
Key Laboratory of Integrated Pest
    Management of Crop in South China
Ministry of Agriculture and Rural
    Affairs
Key Laboratory of Natural Pesticide and
    Chemical Biology
Ministry of Education
South China Agricultural University
Guangzhou, P. R. China

Muhammad Anjum Zia
Department of Biochemistry
University of Agriculture
Faisalabad, Pakistan

1 Recent Advancement in Microbial Enzymes and Their Industrial Applications

Pankaj Bhatt, Sajjad Ahmad, Samiksha Joshi, and Kalpana Bhatt

ABSTRACT

Microbes are an inexhaustible source of enzymes having biocatalytic potential. The huge diversity of microbial enzymes makes them essential for application in different industries such as textiles, agriculture, chemical, food and beverages, pharmaceuticals, leather, and paper and pulp. Enzymes reduce environmental pollution utilizing strategies including biodegradation and bioremediation. Consequently, it is not astonishing to see the burgeoning enzyme market at a global level as they are greener, more eco-friendly, stable, highly catalytic, and easier to modify than animal and plant enzymes. Alternative strategies such as recombinant DNA technology and protein engineering are used to modify enzymes to obtain novel products, as well as a high quantity of microbial enzymes with enhanced substrate specificity and stability. The varied applications of microbial enzymes make them a prominent candidate for industries. This chapter highlights and discusses the various microbial enzymes, their applications in industrial sectors, and their present status in the worldwide enzyme market.

Keywords: microbes, enzymes, industry, catalysis, metagenomics

CONTENTS

1.1   Introduction 2
1.2   Pharmaceutical and Analytical Industry 5
1.3   Food Industry 8
1.4   Dairy Industry 9
1.5   Feed Industry 10
1.6   Paper and Pulp Industry 10
1.7   Leather Industry 11
1.8   Textile Industry 11
1.9   Detergent Industry 12

DOI: 10.1201/9781003202998-1

1.10 Conclusion 13
References... 13

1.1 INTRODUCTION

Microbial enzymes have acquired worldwide attention as a result of their broad use in industrial bioprocesses in a variety of industries, including food, pharmaceutical, textiles, and leather (Adrio and Demain, 2014) (Figure 1.1).

Factors associated with increased attention toward industrial usage of microbial enzymes include their great efficiency, economic value, catalytic activity, specificity, stability, nontoxicity, cost-effectiveness, ease of production, and low risk to the environment (Choi et al., 2015). Different types of fungi, actinomycetes, and bacteria are studied worldwide for the production of enzymes that can be exploited for industrial and commercial applications. A variety of enzymes produced intracellularly or extracellularly by different bacteria, actinomycetes, fungi, and yeasts exhibit a wide range of commercial applications. Microbial enzymes, such as proteases, amylases, pectinases, cellulases, xylanases, laccases, and lipases, are extracellular, whereas a few, like catalases, are intracellular (Fiedurek and Gromada, 2000). In comparison to plant and animal enzymes, the microbial enzymes have proven more active, stable, and beneficial in industries. Among all, 50% of industrial enzymes are obtained from yeast and fungi, 35% from bacteria, and 15% are from plants (Saranraj and Naidu, 2014). Microbial enzymes have shown superior performances and work efficiently under a wide range of varied chemical and physical conditions. Microorganisms are favored sources of enzymes due to their fast growth rate, easy availability, and high yield, and they are easy to modify and optimize, making them more susceptible to gene manipulation with biochemical diversity. Moreover, they can easily convert toxic compounds (containing amines, nitriles, phenolic groups, and carboxylic groups) into nontoxic forms, either by bioconversion or biodegradation processes. The method of fermentation by which microbial enzymes are synthesized at a large scale is not interrupted by seasonal variations and hence do not affect the standardized supply of enzyme. Extremoenzymes are obtained from microbes surviving in extreme environments, like polar regions, volcanic springs, the sea, and areas of very high salt concentrations, possess an amazing array of enzymes catalyzing biochemical reactions and perform well under extreme pH and temperature. Thus, they have direct applications in industrial processes that occur under extreme conditions. Thermophilic microbes are well known for thermostable enzymes. These thermostable enzymes have a low risk of microbial contamination during the fermentation process for applications in industries on commercial scales (Adrio and Demain, 2014). Major groups of enzymes used in industries are hydrolases, like proteases for dairy and detergent industries, and carbohydrases, like cellulases and amylases for detergents, baking, and the textile industry (Gurung et al., 2013). Some of the examples include bacterial species belonging to genera *Pseudomonas*, *Bacillus*, and *Clostridium* for the synthesis of alkaline proteases and fungi like *Trichoderma*, *Penicillium*, and *Aspergillus* for the synthesis of xylanases with significant applications in bio-industries (Nigam, 2013). Currently, out of 4,000 known microbial enzymes, only 200 are used commercially and 20 are produced on a truly industrial scale. Top three companies contributing 75%

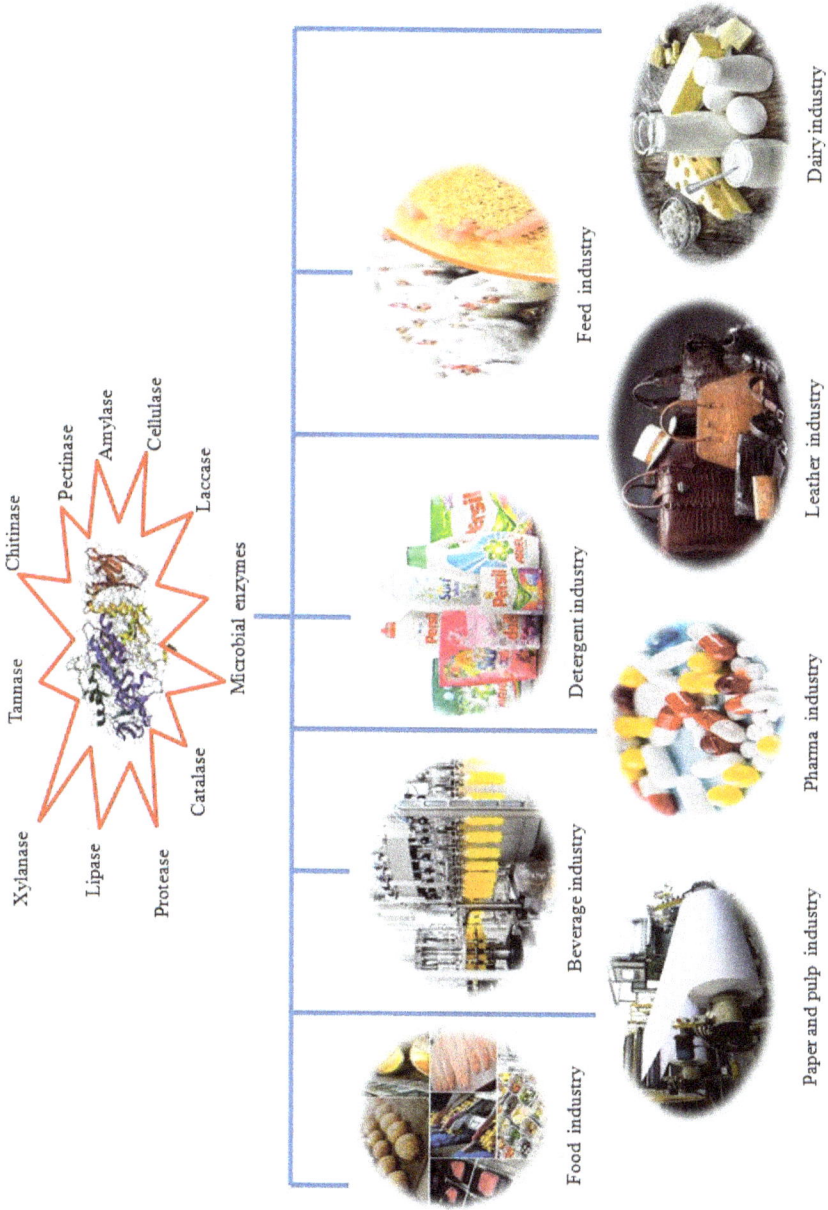

FIGURE 1.1 Applications of microbial enzymes in different areas.

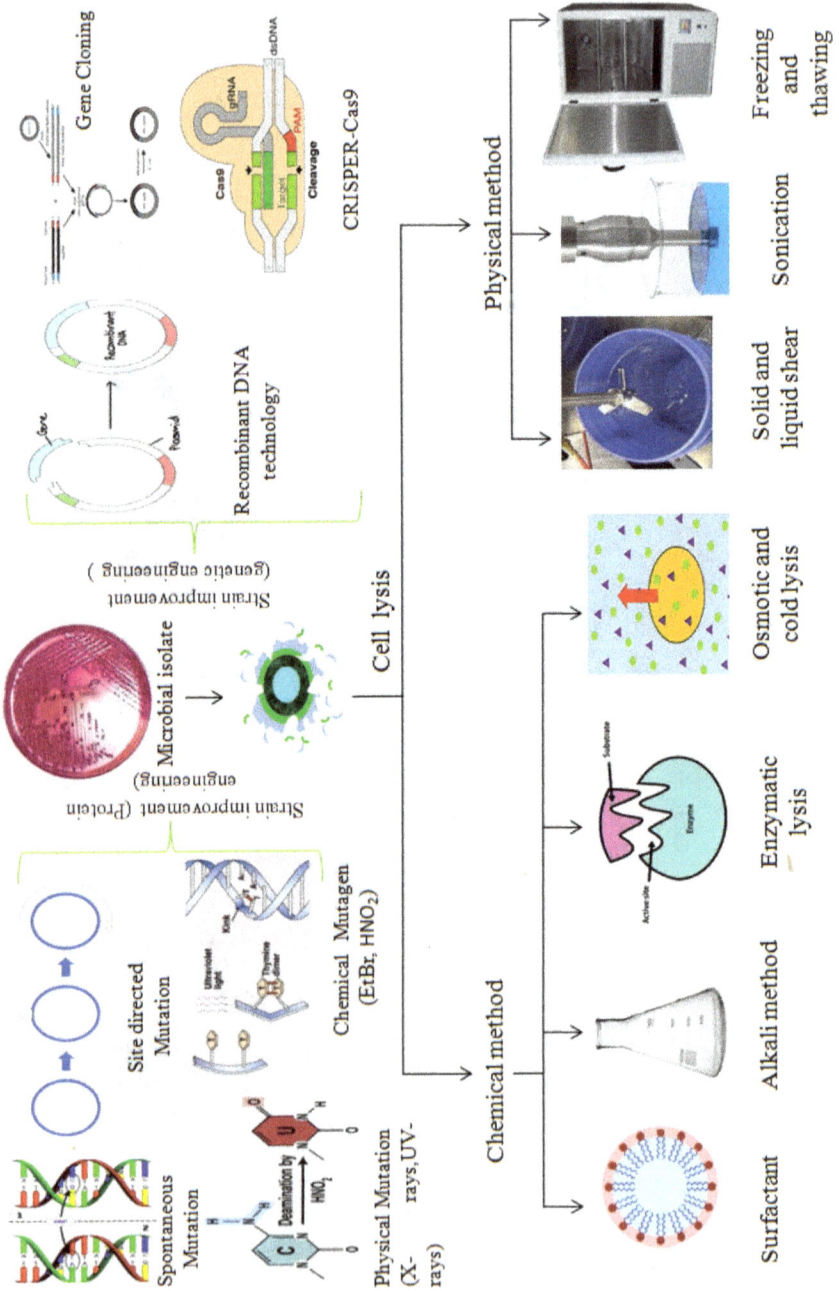

FIGURE 1.2 Microbial enzyme production methods and strain improvement strategies.

of the total enzyme production are US-based DuPont, Denmark-based Novozymes, and Switzerland-based Roche (Li et al., 2012). Engineered microbial enzymes are employed in the global market for the synthesis of a large variety of value-added products. Microbes can be modified and cultivated in vast quantities, utilizing recombinant DNA technology to achieve rising demands (Liu et al., 2013). Furthermore, through metagenomics and protein engineering, several novel enzymes have been developed (Figure 1.2). Metagenomics has discovered about 36 extremozymes from extreme environments, such as the arctic, deep seas, worm guts, rumen, and gold ores. To enhance the performance and quality of microbial enzymes for industrial application, several molecular techniques have been introduced (Nigam, 2013). Due to the wide range of applications of microbial enzymes, their production under varied physical and chemical conditions at the industrial scale is very important. Table 1.1 summarizes microbial enzymes and their applications in various industries.

1.2 PHARMACEUTICAL AND ANALYTICAL INDUSTRY

In the pharmaceutical industry, microbial enzymes play a substantial and critical role. Several researchers have reported medicinal applications of microbial enzymes, which includes cure of dead skin and burns by proteases; clot removal by nattokinase (Cho et al., 2010); and treatment of tooth decay, alimentary dyspepsia, and cyanide poisoning by dextranase, acid proteases, and rhodanase, respectively (Velappan and Thangaraj, 2014). Nowadays, therapeutic proteases available in the market are mainly serine proteases. Proteases contribute to maintaining physiological functioning and regulating the concentration and production of cytokines, growth factors, chemokines, and cellular receptors. Improper regulation may cause diseases such as inflammatory disorders, cancer, and others. Hence, proteases are of significant use in therapeutics for the treatment of such diseases (Craik et al., 2011). Serrapeptase isolated from *Serratia* sp. E 15 and *Serratia marcescens* is used in therapeutics for the removal of arterial plaques, blood clots, and other such dead protein debris. Moreover, it is used as an anti-inflammatory agent, therefore also becoming a treatment option for bronchitis, osteoarthritis, sinusitis, and rheumatoid arthritis. Proteases obtained from *B. polymyxa*, *A. oryzae*, and *Beauveria bassiana* aid in digestion in lytic enzyme deficiency syndromes (Mane and Tale, 2015). Chanalia and coworkers (2011) reported the synthesis of lysostaphin, a protease by *Staphylococcus simulans*, and its inhibitory activity against toxin-producing bacteria such as *S. carnosus*, *Staphylococcus aureus*, *S. saprophyticus*, and *S. epidermidis* (Chanalia et al., 2011). Lysostaphin has shown effectiveness against various toxins, diseases, and destruction of bacterial biofilms. Currently *Escherichia coli* and *Clostridium botulinum* are mostly used for the industrial production of therapeutic enzymes (Craik et al., 2011). L-glutaminase obtained from different yeast, bacteria, and fungi and L-asparaginase from *Erwinia chrysanthemi* and *E. coli* are used as anticancer agents. The recombinant form of L-glutaminase and L-asparaginase with increased efficiency has been patented and showed several therapeutic uses. Streptokinase produced by *Streptococcus* sp. is used to dissolve blood clots and is an important component of therapeutic formulation for myocardial infarction and coronary thrombosis (Chanalia et al., 2011). Collagenases from *Clostridium* sp. are

TABLE 1.1
Microbial Enzymes and Their Applications in Different Industries

Industry	Enzymes	Producer Microbes	Applications	References
Pharmaceutical	Penicillin, acylase, chitinase, peroxidase, tannase	Saccharomyces cerevisiae, Flavobacter sp., Sphinogomonas capsulate, Myxococcus Xanthus, Aspergillus niger, Vibrio proteolyticus, Bacillus proteolyticus, Staphylococcus simulans, Curvularia verruculosa, Pediooccus acidilactici, Erwinia chrysanthemi, Clostridium sporogenes, Mycoplasma hominis, Candida utilis, Arthrobacter protoformaia	Synthesis of 6APA and β-lactamantibiotics and antimicrobials; production of glucosamines and chitooligosaccharides; synthesis of fungicide and insecticide	(Dahiya et al., 2006)
Food, beverage, and dairy	Lipase, protease, lactase, invertases, phospholipase, glucose isomerase, amylase, laccase, glucose oxidase, inulinases, naringinase, pectinase	Bacillus sp., Pseudomonas sp., Clostridium sp., Penicillium sp., Aspergillus sp., Candida sp., Rhizopus sp., Mucor sp., Geobacillus sp.	Meat tenderization; processing of fruit juices; breakdown of proteins and starch into sugars; production of cheese, low-caloric beer, galactooligosaccharides, gluconic acid and lactose hydrolyzed milk products, high fructose syrup, and invert syrup; improving bread quality; debittering of citrus fruit juices	(Gurung et al., 2013; Singh et al., 2017; Dubey et al., 2017)
Detergents	Amylase, cellulase, lipase, protease, mannanase	Pseudomonas sp., Clostridium sp., Penicillium sp., Bacillus sp., Candida sp., Rhizopus sp., Pleurotus sp., Trichoderma sp., A. oryzae, A. flavus, Aspergillus niger, Fusarium solani	Removing stains of insoluble starch, protein, fats, and oils; increasing efficiency of detergents; used as cleaning agents	(Singh et al., 2019)
Textiles	Protease, amylase, keratinase, cellulase, pectinase, laccase catalase, peroxidase	Candida antarctica, Pseudomonas sp., Trametes versicolor, Penicillium sp., Phlebia radiata, Rhizopus sp., Bacillus subtilis, Bacillus licheniformis, Penicillium funiculosum, Clostridium histolyticum, Aspergillus niger	Wool treatment, dye degradation, cotton softening; degumming of raw silk, bio-polishing and bio-scouring, fabric finishing in denims	(Sharma et al., 2014; Singh et al., 2019; Liu et al., 2013)

	Enzymes	Microorganisms	Applications	References
Animal feed	Phytase, keratinases, pectinases, xylanase	*Thermoactinomyces thalophilus, Humicola insolens, Bacillus sp., Aspergillus niger, actinomycetes*	Enhance phosphorus content, feather meal processing for feed growth, increase digestibility, for production of animal feed	(Mitidieri et al., 2006; Singh et al., 2019; Sharma et al., 2013)
Paper and pulp	Cellulase, amylase, hemicellulase, lipase, ester ase, protease, laccase, ligninase, peroxidases, xylanase	*Steccherinum ochraceum, Polyporus versicolor, Aspergillus sp., Bacillus subtilis, Panus tigrinus, Trichoderma reesei, Candida antarctica, Thermoactinomyces thalophilus, Thermomyces lanuginosus, Humicola insolens, Penicillium sp., Aureobasidium pullulans*	Breakdown of starch to less viscosity; deinking, coating, and sizing of paper; improved drainage and promote ink removal; bio-pulping, degradation of wood components, and pre-bleaching and bio-bleaching of pulp; lipases reduce pitch; ligninase remove lignin to soften paper	(Polizeli et al., 2005; D'Annibale et al., 2006; Regalado et al., 2004)
Leather	Protease, keratinase, lipase	*Alcaligenes faecalis, Candida sp., Penicillium sp., Rhizopus, Mucor, Aspergillus flavus, Pseudomonas, Clostridium, Aspergillus niger, Bacillus subtilis, Aspergillus oryzae*	Unhearing, bating, depicking, leather manufacture	(Saha et al., 2009)

useful in wound debridement. In addition, polyphenol oxidases and lipases are useful in the treatment of Parkinson's disease and chitosanase in reducing high cholesterol and high blood pressure and controlling arthritis (Velappan and Thangaraj, 2014). Synthesis of single-cell proteins from chitooligosaccharides, shellfish waste, and glucosamines by chitinases is of immense use in the pharmaceutical industry (Karthik et al., 2017). Other enzymes having biotechnological applications include cholesterol oxidase and putrescine oxidase, which are used to detect cholesterol and biogenic amines, respectively (Le Roes-Hill and Prin, 2016). Tannases used for the synthesis of gallic acid have great importance in the medicinal area for antimicrobial drugs. Penicillin acylase, an essential therapeutic enzyme, has a key role in the production of β-lactam antibiotics. Some microbes for the industrial making of penicillin acylase are *P. chrysogenum*, *Achromobacter* sp., *B. megaterium*, *Acetobacter* sp., *E. coli*, *Streptomyces* sp., *Fusarium* sp., and *A. terreus* (Singh et al., 2019).

1.3   FOOD INDUSTRY

Microbial enzymes are widely employed in food industries for a variety of applications. Many enzymes such as lipase, phytases, xylanase, amylase, and transglutaminase are used to improve the quality, softness, texture, stability, shelf life, and conditioning of different products. Production of baked goods, fruit juices, and vegetable fermentation could be done with lipases. *Rhizomucor miehi*, *Thermomyces lanuginosus*, and *Fusarium oxysporum* are used to obtain recombinant lipases in food industries (Méndez and Salas, 2001). The industrial synthesis of transglutaminase enzymes by a *Stretoverticillium* sp. variant enhances texture, flavor, and shelf life of Chinese noodles, udon noodles, soba noodles, and pasta products. Phytases produced by fungal species of *Aspergillus*, *Mucor*, and *Cladosporium* provide high-quality bread with more volume and improved texture. Cellulases, amylases, and pectinases are required in breweries for high yield and product quality during fruit juice extraction (Garg et al., 2016). Amylolytic enzymes have a wide range of uses, including large-scale manufacturing of glucose syrups, maltose syrup, sugar syrup viscosity reduction, producing clarified juices with long-term storage, and starch solubilization in the brewing industry (Pandey et al., 2000). *B. amyloliquefaciens*, *Bacillus licheniformis*, *Aspergillus oryzae*, *Rhizopus* sp., and *Aspergillus niger* are efficient producers of thermostable amylase (Gupta et al., 2003). Tannase generated by *A. niger* aids in the reduction of sedimentation and bitterness in fruit juices, production of instant tea (Zhang et al., 2015), food preservation, and synthesis of propyl gallate (food antioxidant) in the food and beverage industry. Fungal glucose oxidases catalyze the synthesis of gluconic acid, which is used as a color stabilizer, chelator, and antioxidant in beverages and baked food items. Moreover, it contributes to the production of bread, beer, and wine in the baking and beverage industry. Microbial inulinases have tremendous application in the synthesis of single cell oil, fructose syrup, citric acid, 2,3-butanediol, lactic acid, ethanol, sorbitol, pullulan, single-cell proteins, and mannitol. Invertases are mostly used in the production of jams, candies, chocolate cherries, and artificial honey (Singh et al., 2019). Debittering enzymes like naringinase and limoninase have been utilized to dissolve bitter components and improve the quality of citrus juices, as well as have antioxidant potential. Some most efficient producers are *Cochliobolus* sp., *Phanopsis* sp., *Penicillium* sp., *Aspergillus* sp., and *Rhizopus* sp. Pectinases, a heterogeneous group of enzymes produced mostly by the fungus *Aspergillus niger* (Pedrolli et al., 2009), have been widely reported for their vital function in the wine-making and fruit-juice-extraction process. Cellulases and pectinases form a macerating enzyme complex that aids in the processing of vegetable and fruit juices with improved quality attributes. In 2013, Du and colleagues identified three new xylanases from *Humicola* sp. with greater catalytic efficiency and significance in the brewing industry. Xylanases are used in quality improvement of baked food items and clarifying fruit vegetable juices. Pullulanases are efficient producers of panose syrup, maltotriose syrup, low-calorie beer, and isopanose syrups and are used as an antistaling agent in bakery products (Singh et al., 2019). Aspartase-catalyzing synthesis of fumaric acid or aspartic acid is used for making artificial sweetener aspartame and fruit jellies in the food industry, respectively. Microbial galactosidase enzymes boost the nutritional content of legume-based foods and help sugar crystallize in beet sugar syrup (Singh et al., 2019).

1.4 DAIRY INDUSTRY

Dairy enzymes provide a high yield and improve color, flavor, and aroma. Several enzymes, like lipases, catalase, proteases, esterases, aminopeptidase, lactoperoxidase, lysozyme, transglutaminase, and lactase, are of significant use in the dairy market. They function as coagulants and bio-protective enzymes for the prolonged shelf life of dairy products. Production of cheese, yogurt, and other customized milk products is done with dairy enzymes, such as microbial rennet, proteases, and lipases (Qureshi et al., 2015). Rennet, made up of pepsin and chymosin, aids in milk coagulation into solid curds for the production of cheese and liquid whey. Currently, about 33% of cheese production at the global level is done using microbial rennet (Qureshi et al., 2015). Some microbial producers of rennet-like proteases are *Rhizomucor pusillus*, *Irpex lactis*, *A. oryzae*, *Endothia parasitica*, and *Rhizopus miehei* (Qureshi et al., 2015). Degradation of casein by plasmin results in proteolysis that can be either detrimental or beneficial depending on the type of product and level of hydrolysis. For example, proteolysis can provide the required texture and flavor in cheese-making and also can result in unwanted gelation in pasteurized milk and milk processed at ultra-high temperature. Enzymes produced by psychrotrophic microbes can cause this kind of proteolysis during the refrigeration of milk. Lactic acid bacteria are commonly used as a starter culture for curd production. These bacteria are a rich source of peptidases and proteases as they require them for fast growth in milk (dos Santos Mathias et al., 2017). Protease enzymes obtained from members of genera *Lactococcus*, *Streptococcus*, and *Lactobacillus* help in the breakdown of major milk allergens, like β-lactoglobulin and α-lactalbumin (Atanasova et al., 2014). In the dairy industry, lipases are employed for milk fat hydrolysis, quick cheese production, and flavor enhancement. Lipases derived from *Aspergillus oryzae*, *Rhizomucor miehei*, and *Aspergillus niger* play an important role in the cheese-making process. Transglutaminase catalyzes milk protein polymerization, which enhances the functional qualities of dairy products (Kieliszek and Misiewicz, 2014). Glucose oxidases help in cheese and curd formation (Pal et al., 2016). Lactase-producing microbes, such as *Neurospora* sp., *Streptococcus* sp., *Bacillus* sp., *Mucor* sp., *E. coli*, *Lactobacillus* sp., *Aspergillus* sp., *Kluyveromyces* sp., and *Candida* sp. (Nivetha et al., 2017) are used to make prebiotics, lactose-hydrolyzed milk products, and whey in the dairy industry (Singh et al., 2019). The β-galactosidase enzyme helps in the catalytic breakdown of lactose to galactose and glucose and therefore works as a digestive aid in lactose-intolerant patients (Qureshi et al., 2015). Aspartate-producing bacterium *Propionibacterium* sp. is also used for Swiss cheese production. Invertase enzyme obtained from *Saccharomyces carlsbergensis* and *Saccharomyces cerevisiae* are used to commercially synthesize invert sugar, utilized as a sweetener in dairy products.

1.5 FEED INDUSTRY

Animal feed is the most expensive item in poultry and livestock production. For low cost, supplementation of feed with enzyme preparations allows producers to produce meat in large quantities, cheaper and faster. Protein, non-starch polysaccharides, and phytic acid are found in animal feed, such as agricultural and grain milling leftovers, cereals, and agricultural waste residues. Feed enzymes are employed in the composition of animal diets to improve the digestibility and protein nutritional value of poultry feeds (Choct, 2006). Xylanases, proteases, polygalacturonases, galactosidases, phytases, glucanases, and amylases are some of the most significant feed enzymes for poultry (Adrio and Demain, 2014). Besides, they are of great importance in the reduction of feed cost and enhancement of nutritional quality and meat quality (Adrio and Demain, 2014). Xylanases help in easy digestibility by pretreatment of forage crops for ruminant feeding (Kumar et al., 2017). Microbial phytase liberates and makes bound phosphorus available in cereal-based feed for monogastrics. For enhanced utilization of protein from vegetables, microbial proteases can be added with feed for better feed utilization and digestion by animals (Lehmann et al., 2000). Few microbial enzymes used as feed enhancers include lactase from *Aspergillus* sp., amylase from *Bacillus* sp., and phytase from *Aspergillus ficuum*.

1.6 PAPER AND PULP INDUSTRY

Microbial enzymes were investigated by researchers to lessen reliance on harsh chemicals and make the process more efficient, rapid, and cost-effective in the paper and pulp industry (Verma et al., 2021; Bhandari et al., 2021). Enzymes such as xylanases and ligninases are used to increase pulp quality by removing hemicelluloses and lignin (Maijala et al., 2008); amylases for deinking, cleanliness of paper, better drainage, and starch coating; lipases for pitch control and deinking (Pankaj et al., 2015). Furthermore, mannases are utilized in paper industries to improve brightness by decomposing glucomannan (Clarke et al., 2000), and cellulase is employed in bioprocesses to recycle used printed papers (Patrick, 2004). *A. niger* and *Trametes suaveolens* are the most common sources of cellulases for such processes. Pulp treatment has been reported using cellulase-free xylanases isolated from *Streptomyces roseiscleroticus* and *Saccharomonospora viridis*. For bio-bleaching of pulps, xylanases from *B. subtilis*, *Aureobasidium pullulans*, *Streptomyces lividans*, *T. reesei*, and

Thermomyces lanuginosus are utilized. Lipases from *Rhizopus* sp., *Pseudomonas alcaligenes*, *Aspergillus niger*, and *Candida cylindrica* minimizes pitch-related difficulties in wood pulp (high levels of wood resin, resin acids, triglycerides, and waxes). Pitch-control agents, such as EnzOx® PC, Resinase® A, and EnzOx® SEL, are commonly employed in the pulp industry. White rot fungus produces ligninolytic enzymes, which are mostly employed to delignify hardwood fibers during the bleaching process. Wood degradation by cellulose and lignin hydrolysis into water and carbon dioxide is done by peroxidases in the paper industry (Regalado et al., 2004). The microbial xylanases in the pulp industry have contributed to pre-bleaching process of kraft pulps, further enhancing the bio-bleaching of wood pulp and the quality of recycled fibers.

1.7 LEATHER INDUSTRY

In the leather industry, biodegradable enzymes have proven to be an effective alternative. For millennia, microbial enzymes have aided in the improvement of leather quality at many stages of leather processing, which involves tanning, curing, plucking, degreasing, liming, soaking, dehairing, and bating (Adrio and Demain, 2014). Dehairing is the most important and time-consuming process in leather manufacturing, and it requires a variety of enzymes including lipases, amylases, and proteases to make the process eco-friendly. This lessens the need for harsh chemicals, such as lime, sulfide, and amines (De Souza and Gutterres, 2012). In the soaking, bating, and liming processes, enzymes such as proteases and lipases are utilized (Jridi et al., 2014). Few proteases (like Clarizyme®) are obtained from *A. flavus*, *Bacillus* sp., and *Streptomyces* sp. have been employed for enzymatic dehairing operations of skin and hides. Proteolytic enzymes from bacteria are employed in the bating process to produce soft, flexible leather, which is used to make handbags and gloves. In the tannery, a variety of microbial lipases have been utilized to remove grease for successful dyeing and finishing.

1.8 TEXTILE INDUSTRY

Microbial enzymes are utilized in fiber processing, product quality improvement, and several other purposes in textile industries (Choi et al., 2015). An array of 75 enzymes, broadly classified as oxidoreductases and hydrolases, is currently used to their full potential in textile industries. For pretreatment and finishing of cotton, enzymes from the hydrolase group (such as amylase, protease, cellulase, pectinase, lipase, and cutinase) and oxidoreductase group (such as ligninases, laccase, catalase, peroxidase, and ligninase) are utilized (Bhatt et al., 2019, 2020a, 2020b, 2021a). These enzymes are also used in procedures like fabric bio-polishing and bio-scouring, wool antifelting, bleach termination, cotton softening, wool finishing, denim finishing, synthetic fiber modification, textile desizing, dye decolorization, and so on (Chen et al., 2013). From *B. licheniformis*, *B. amyloliquefaciens*, *B. stearothermophilus*, and *Aspergillus* sp., desizing agents such as amylases can be produced. Proteases, lipases, cutinases, xylanases, and pectinases are enzymes that are utilized in the bio-scouring of cotton and biopreparation and the degumming of jute, ramie, hemp,

and flax. (Kumar et al., 2017). *Melanocarpus albomyces* produces a few novel cellulases that function well in biostoning and prevent back-staining. Cellulase treatment
gives cotton and natural and artificial cellulosic fibers a bio-polished, silky appearance and prevents fibrillation. Proteases and transglutaminases are favored for maintaining wool qualities, such as handling and shrink resistance, and increasing fiber
strength. The application of *Streptomyces mobaraense* TGase alone or in combination with protease in textile industries resulted in a 25% increase in tensile strength.
Alkaline proteases are used for degumming silk fiber, resulting in increased tensile
strength, smoothness, and sheen. Lipases have been employed to remove size lubricants, allowing fabrics to absorb more color and produce better dyeing results (Singh
et al., 2019).

1.9 DETERGENT INDUSTRY

Microbial enzymes in detergent are of great significance as they include fewer
bleaching chemicals and phosphates and thus have a positive impact on human and
environmental health (Bhatt et al., 2021a, 2021b, 2021c, 2021d, 2021e). The hydrolase groups of enzymes, of which amylase and proteases are the most often, are
useful in detergent (Bhatt et al., 2021f, 2021g, 2021h, 2021i). Other enzymes, including pectinases, cellulases, and lipases, are also employed for color brightness, soft
fabrics, removal of small fibers, and fabric maintenance (Li et al., 2012). Cutinase,
a hydrolytic enzyme, is utilized in dishwashing and laundry detergents as a lipolytic
enzyme (Pio and Macedo, 2009). Organic stains, like blood, human sweat, and egg,
can be removed with proteases. *Bacillus* spp. contribute a major share in worldwide
production of alkaline protease used in the detergent industry. Proteases from *B.
amyloliquefaciens* and *B. lichiniformis* are most commonly used to digest organic
stains. Subtilisin from *B. licheniformis* was the first microbial protease used in the
detergent industry. After which, enzymes were increasingly used in detergent formulations. Subtilisins, a type of serine protease obtained from *Bacillus licheniformis*,
Bacillus amyloliqufaciens, *Bacillus subtilis*, and *Bacillus clausii* are used in detergents and as automatic dishwashing solutions (Joo et al., 2003). Proteases are stable in
high-temperature and alkaline conditions. Proteases obtained from *Nocardiopsis* sp.
and *Vibrio fluvialis* have a long shelf life and compatibility with other components in
the commercial mixture of detergents. A detergent-stable protease enzyme encoded
by fungus *Aspergillus clavatus* ES1 has been cloned in *E. coli* for commercial production (Hajji et al., 2010). *B. stearothermophilus*, *B. licheniformis*, *B. subtilis*, *B.
amyloliquefaciens*, and actinomycetes and certain fungi are all good amylase makers
used in dishwashing and destarching detergents. Lipases and amylases are efficient
for stain removal of starchy and fatty food items (Keshwani et al., 2015). Lipases are
efficient in the removal and decomposition of residues and stains of oil, butter, and
sauces. First, commercial lipase was isolated from *Humicola* sp. by Novo Nordisk.
Microbial lipases aid in the manufacture of domestic detergents, dishwashers, and
industrial laundry. Lipase producers include *Bacillus prodigiosus*, *Thermomyces
lanuginosus*, *Bacillus pyocyaneus*, *Acinetobacter radioresistens*, *Bacillus fluorescens*, *Pseudomonas mendocina*, *Pseudomonas alcaligenes*, and *Pseudomonas
gluma* (Hasan et al., 2007). Cold-active lipases (CLPs) have recently gained

popularity in the detergent industry because they are biodegradable, have a high rate of catalysis, and have poor thermal stability (Joseph et al., 2007). Recently, the application of CLPs in detergents obtained from, *M. phyllosphaerae, Pseudoalteromonas* sp. NJ 70, *Pichia lynferdii*, and *Bacillus sphaericus*, are found effective in cold water, showing lesser wear and tear and low energy consumption (Kavitha, 2016). Lipases synthesized by *Streptomyces* sp., *Pseudomonas paucimobilis*, and *Bacillus* sp. B207 are added in detergent formulations (Salihu and Alam, 2012). Alkaline lipases from *Staphylococcus arlettae* JPBW-1 and *B. cepacia* RGP-10 exhibited better stability toward surfactants and oxidizing agents, showing effective removal of stains from fabrics (Prakasan et al., 2016). A combination of lipases with oxidoreductases needs less amount of surfactants, making them more eco-friendly (Agobo et al., 2017). Pullanases are used as additives in dishwashing and laundry detergents (Schallmey et al., 2004).

1.10 CONCLUSION

The microbial enzymes are a potential tool due to their great application. The industrial sector consisted of broad applications associated with the microorganism and their respective enzymes. Therefore, the catalytic mechanism of these enzymes has been explored in this chapter with reference to specific industrial studies. The recent advancement into the technologies explained the depth of information in each of the applications. In the present era, omics-based tools are developing very rapidly, so more research is needed to explore the enzymes from the cultured and uncultured microbes. It might be possible the more potential application of enzymes of the uncultured microbes. Metagenomic and metaproteomic tools might be helpful to explore such organisms.

REFERENCES

Adrio, J. L., & Demain, A. L. (2014). Microbial enzymes: Tools for biotechnological processes. *Biomolecules, 4*(1), 117–139.
Agobo, K. U., Arazu, V. A., Uzo, K., & Igwe, C. N. (2017). Microbial lipases: A prospect for biotechnological industrial catalysis for green products: A review. *Ferment Technol, 6*(144), 2.
Atanasova, J., Moncheva, P., & Ivanova, I. (2014). Proteolytic and antimicrobial activity of lactic acid bacteria grown in goat milk. *Biotechnology & Biotechnological Equipment, 28*(6), 1073–1078.
Bhatt, P., Bhatt, K., Huang, Y., Ziqiu, L., & Chen, S. (2020b). Esterase is a powerful tool for the biodegradation of pyrethroid insecticides. *Chemosphere, 244*, 125507. doi: 10.1016/j.chemosphere.2019.125507.
Bhatt, P., Bhatt, K., Sharma, A., Zhang, W., Mishra, S., & Chen, S. (2021g). Biotechnological basis of microbial consortia for the removal of pesticides from the environment. *Critical Reviews in Biotechnology, 41*(3), 317–338.
Bhatt, P., Gangola, S., Bhandari, G., Zhang, W., Maithani, D., Mishra, S., & Chen, S. (2021h). New insights into the degradation of synthetic pollutants in contaminated environments. *Chemosphere, 268*, 128827.
Bhatt, P., Huang, Y., Zhan, H., & Chen, S. (2019). Insight into microbial applications for the biodegradation of pyrethroid insecticides. *Frontiers in Microbiology, 10*, 1778.

Bhatt, P., Joshi, T., Bhatt, K., Zhang, W., Huang, Y., & Chen, S. (2021f). Binding interaction of glyphosate oxidoreductase and C-P lyase: Molecular docking and molecular dynamics simulation studies. *Journal of Hazardous Material*, *409*, 124927.

Bhatt, P., Rene, E. R., Kumar, A. J., Gangola, S., Kumar, G., Sharma, A., Zhang, W., & Chen, S. (2021a). Fipronil degradation kinetics and resource recovery potential of *Bacillus* sp. strain FA4 isolated from a contaminated agriculture fields of Uttarakhand, India. *Chemosphere*, *276*, 130156.

Bhatt, P., Rene, E. R., Kumar, A. J., Kumar, A. J., Zhang, W., & Chen, S. (2020a). Binding interaction of allethrin with esterase: Bioremediation potential and mechanism. *Bioresource Technology*, *315*, 13845. doi: 10.1016/j.biortech.2020.123845.

Bhatt, P., Sharma, A., Rene, E. R., Kumar, A. J., Zhang, W., & Chen, S. (2021i). Bioremediation mechanism, kinetics of fipronil degradation using *Bacillus* sp. FA3 and resource recovery potential from contaminated environments. *Journal of Water Process Engineering*, *39*, 101712.

Bhatt, P., Sethi, K., Gangola, S., Bhandari, G., Verma, A., Adnan, M., Singh, Y., & Chaube, S. (2020b). Modeling and simulation of atrazine biodegradation in bacteria and its effect in other living systems. *Journal of Biomolecular Structure and Dynamics*, *40* (7), 3285–3295. doi: 10.1080/07391102.2020.1846623.

Bhatt, P., Verma, A., Gangola, S., Bhandari, G., & Chen, S. (2021d). Microbial glycoconjugates in organic pollutants bioremediation: Recent advances and application. *Microbial Cell Factories*, *20*, 72.

Bhatt, P., Zhou, X., Huang, Y., Zhang, W., & Chen, S. (2021c). Characterization of the role of esterases in the biodegradation of organophosphate, carbamate and pyrethroid group pesticides. *Journal of Hazardous Material*, *411*, 125026.

Bhandari, G., Bagheri, A. R., Bhatt, P., & Bilal, M. (2021). Occurrence, potential ecological risks, and degradation of endocrine disruptor, nonyphenol from the aqueous environment. *Chemosphere*. 230,130013. doi: 10.1016/j.chemosphere.2021.130013.

Chanalia, P., Gandhi, D., Jodha, D., & Singh, J. (2011). Applications of microbial proteases in pharmaceutical industry: An overview. *Reviews in Medical Microbiology*, *22*(4), 96–101.

Chen, S., Su, L., Chen, J., & Wu, J. (2013). Cutinase: Characteristics, preparation, and application. *Biotechnology Advances*, *31*(8), 1754–1767.

Cho, Y. H., Song, J. Y., Kim, K. M., Kim, M. K., Lee, I. Y., Kim, S. B., & Kim, B. S. (2010). Production of nattokinase by batch and fed-batch culture of *Bacillus subtilis*. *New Biotechnology*, *27*(4), 341–346.

Choct, M. (2006). Enzymes for the feed industry: Past, present and future. *World's Poultry Science Journal*, *62*(1), 5–16.

Choi, J. M., Han, S. S., & Kim, H. S. (2015). Industrial applications of enzyme biocatalysis: Current status and future aspects. *Biotechnology Advances*, *33*(7), 1443–1454.

Clarke, J. H., Davidson, K., Rixon, J. E., Halstead, J. R., Fransen, M. P., Gilbert, H. J., & Hazlewood, G. P. (2000). A comparison of enzyme-aided bleaching of softwood paper pulp using combinations of xylanase, mannanase and α-galactosidase. *Applied Microbiology and Biotechnology*, *53*(6), 661–667.

Craik, C. S., Page, M. J., & Madison, E. L. (2011). Proteases as therapeutics. *Biochemical Journal*, *435*(1), 1–16.

D'Annibale, A., Quaratino, D., Federici, F., & Fenice, M. (2006). Effect of agitation and aeration on the reduction of pollutant load of olive mill wastewater by the white-rot fungus *Panus tigrinus*. *Biochemical Engineering Journal*, *29*(3), 243–249.

Dahiya, N., Tewari, R., & Hoondal, G. S. (2006). Biotechnological aspects of chitinolytic enzymes: A review. *Applied Microbiology and Biotechnology*, *71*(6), 773–782.

De Souza, F. R., & Gutterres, M. (2012). Application of enzymes in leather processing: A comparison between chemical and coenzymatic processes. *Brazilian Journal of Chemical Engineering*, *29*(3), 473–482.

dos Santos Mathias, T. R., de Aguiar, P. F., Silva, J. B. A., de Mello, P. P. M., & Servulo, E. F. C. (2017). Brewery waste reuse for protease production by lactic acid fermentation. *Food Technology and Biotechnology*, *55*(2), 218–224.

Du, Y., Shi, P., Huang, H., Zhang, X., Luo, H., Wang, Y., & Yao, B. (2013). Characterization of three novel thermophilic xylanases from *Humicola insolens* Y1 with application potentials in the brewing industry. *Bioresource Technology*, *130*, 161–167.

Dubey, M. K., Zehra, A., Aamir, M., Meena, M., Ahirwal, L., Singh, S., & Bajpai, V. K. (2017). Improvement strategies, cost effective production, and potential applications of fungal glucose oxidase (GOD): Current updates. *Frontiers in Microbiology*, *8*, 1032.

Fiedurek, J., & Gromada, A. (2000). Production of catalase and glucose oxidase by Aspergillus niger using unconventional oxygenation of culture. *Journal of Applied Microbiology*, *89*(1), 85–89.

Garg, G., Singh, A., Kaur, A., Singh, R., Kaur, J., & Mahajan, R. (2016). Microbial pectinases: An ecofriendly tool of nature for industries. *3 Biotech*, *6*(1), 1–13.

Gupta, R., Gigras, P., Mohapatra, H., Goswami, V. K., & Chauhan, B. (2003). Microbial α-amylases: A biotechnological perspective. *Process Biochemistry*, *38*(11), 1599–1616.

Gurung, N., Ray, S., Bose, S., & Rai, V. (2013). A broader view: Microbial enzymes and their relevance in industries, medicine, and beyond. *BioMed Research International*, *2013*.

Hajji, M., Jellouli, K., Hmidet, N., Balti, R., Sellami-Kamoun, A., & Nasri, M. (2010). A highly thermostable antimicrobial peptide from *Aspergillus clavatus* ES1: Biochemical and molecular characterization. *Journal of Industrial Microbiology and Biotechnology*, *37*(8), 805–813.

Hasan, F., Shah, A., & Hameed, A. (2007). Purification and characterization of a mesophilic lipase from *Bacillus subtilis* FH5 stable at high temperature and pH. *Acta Biologica Hungarica*, *58*(1), 115–132.

Joo, H. S., Kumar, C. G., Park, G. C., Paik, S. R., & Chang, C. S. (2003). Oxidant and SDS-stable alkaline protease from *Bacillus clausii* I-52: Production and some properties. *Journal of Applied Microbiology*, *95*(2), 267–272.

Joseph, B., Ramteke, P. W., Thomas, G., & Shrivastava, N. (2007). Cold-active microbial lipases: A versatile tool for industrial applications. *Biotechnology and Molecular Biology Reviews*, *2*(2), 39–48.

Jridi, M., Lassoued, I., Nasri, R., Ayadi, M. A., Nasri, M., & Souissi, N. (2014). Characterization and potential use of cuttlefish skin gelatin hydrolysates prepared by different microbial proteases. *BioMed Research International*, *2014*.

Karthik, N., Binod, P., & Pandey, A. (2017). Chitinases, Current developments in biotechnology and bioengineering. *Production, isolation purification of industrial products*. Chapter, 15, 335–368. doi: 10.1016/B978-0-444-63662-1.00015-4.

Kavitha, M. (2016). Cold active lipases—an update. *Frontiers in Life Science*, *9*(3), 226–238.

Keshwani, A., Malhotra, B., & Kharkwal, H. (2015). Natural polymer based detergents for stain removal. *World Journal of Pharmaceutical Sciences*, *4*(4), 490–508.

Kieliszek, M., & Misiewicz, A. (2014). Microbial transglutaminase and its application in the food industry. A review. *Folia Microbiologica*, *59*(3), 241–250.

Kumar, D., Kumar, S. S., Kumar, J., Kumar, O., Mishra, S. V., Kumar, R., & Malyan, S. K. (2017). Xylanases and their industrial applications: A review. *Biochemical and Cellular Archives*, *17*(1), 353–360.

Le Roes-Hill, M., & Prins, A. (2016). Biotechnological potential of oxidative enzymes from Actinobacteria. *Actinobacteria-Basics and Biotechnological Applications*, 200–226.

Lehmann, M., Kostrewa, D., Wyss, M., Brugger, R., D'Arcy, A., Pasamontes, L., & van Loon, A. P. (2000). From DNA sequence to improved functionality: Using protein sequence comparisons to rapidly design a thermostable consensus phytase. *Protein Engineering*, *13*(1), 49–57.

Li, S., Yang, X., Yang, S., Zhu, M., & Wang, X. (2012). Technology prospecting on enzymes: Application, marketing and engineering. *Computational and Structural Biotechnology Journal*, *2*(3), e201209017.

Liu, L., Yang, H., Shin, H. D., Chen, R. R., Li, J., Du, G., & Chen, J. (2013). How to achieve high-level expression of microbial enzymes: Strategies and perspectives. *Bioengineered*, *4*(4), 212–223.

Maijala, P., Kleen, M., Westin, C., Poppius-Levlin, K., Herranen, K., Lehto, J. H., & Hatakka, A. (2008). Biomechanical pulping of softwood with enzymes and white-rot fungus *Physisporinus rivulosus*. *Enzyme and Microbial Technology*, *43*(2), 169–177.

Mane, P., & Tale, V. (2015). Overview of microbial therapeutic enzymes. *International Journal of Current Microbiology and Applied Sciences*, *4*(4), 17–26.

Méndez, C., & Salas, J. A. (2001). Altering the glycosylation pattern of bioactive compounds. *Trends in Biotechnology*, *19*(11), 449–456.

Mitidieri, S., Martinelli, A. H. S., Schrank, A., & Vainstein, M. H. (2006). Enzymatic detergent formulation containing amylase from *Aspergillus niger*: A comparative study with commercial detergent formulations. *Bioresource Technology*, *97*(10), 1217–1224.

Nigam, P. S. (2013). Microbial enzymes with special characteristics for biotechnological applications. *Biomolecules*, *3*(3), 597–611.

Nivetha, A., & Mohanasrinivasan, V. (2017). Mini review on role of β-galactosidase in lactose intolerance. In *IOP Conference Series: Materials Science and Engineering* (Vol. 263, No. 2, p. 022046). IOP Publishing, United Kingdom.

Pal, P., Kumar, R., & Banerjee, S. (2016). Manufacture of gluconic acid: A review towards process intensification for green production. *Chemical Engineering and Processing: Process Intensification*, *104*, 160–171.

Pandey, A., Nigam, P., Soccol, C. R., Soccol, V. T., Singh, D., & Mohan, R. (2000). Advances in microbial amylases. *Biotechnology and Applied Biochemistry*, *31*(2), 135–152.

Pandey, A., Soccol, C. R., Nigam, P., Soccol, V. T., Vandenberghe, L. P., & Mohan, R. (2000). Biotechnological potential of agro-industrial residues. II: Cassava bagasse. *Bioresource Technology*, *74*(1), 81–87.

Pankaj, B., Bisht, T. S., Pathak, V. M., Barh, A., & Chandra, D. (2015). Optimization of amylase production from the fungal isolates of Himalayan region Uttarakhand. Ecology, Environment and Conservation, *21*(3), 1517–1521.

Patrick, K. (2004). Enzyme technology improves efficiency, cost, safety of stickies removal program. *Paper Age*, *120*, 22–25.

Pedrolli, D. B., Monteiro, A. C., Gomes, E., & Carmona, E. C. (2009). Pectin and pectinases: Production, characterization and industrial application of microbial pectinolytic enzymes. *Open Biotechnology Journal*, 9–18.

Pio, T. F., & Macedo, G. A. (2009). Cutinases: Properties and Industrial Applications. *Advances in Applied Microbiology*, *66*, 77–95.

Polizeli, M. L. T. M., Rizzatti, A. C. S., Monti, R., Terenzi, H. F., Jorge, J. A., & Amorim, D. S. (2005). Xylanases from fungi: Properties and industrial applications. *Applied Microbiology and Biotechnology*, *67*(5), 577–591.

Prakasan, P., Sreedharan, S., Faisal, P. A., & Benjamin, S. (2016). Microbial lipases-properties and applications. *Journal of Microbiology, Biotechnology and Food Sciences*, *6*(2), 799–807.

Qureshi, M. A., Khare, A. K., Pervez, A., & Uprit, S. (2015). Enzymes used in dairy industries. *International Journal of Applied Research*, *1*(10), 523–527.

Regalado, C., García-Almendárez, B. E., & Duarte-Vázquez, M. A. (2004). Biotechnological applications of peroxidases. *Phytochemistry Reviews*, *3*(1), 243–256.

Saha, B. C., Jordan, D. B., & Bothast, R. J. (2009). Enzymes, industrial (overview). In *Encyclopedia of Microbiology*. Academic Press, Oxford.

Salihu, A., & Alam, M. Z. (2012). Production and applications of microbial lipases: A review. *Scientific Research and Essays*, *7*(30), 2667–2677.

Saranraj, P., & Naidu, M. A. (2014). Microbial pectinases: A review. *Global Journal of Traditional Medicinal System*, *3*(1), 1–9.

Schallmey, M., Singh, A., & Ward, O. P. (2004). Developments in the use of *Bacillus species* for industrial production. *Canadian Journal of Microbiology*, *50*(1), 1–17.

Sharma, A., Thakur, V. V., Shrivastava, A., Jain, R. K., Mathur, R. M., Gupta, R., & Kuhad, R. C. (2014). Xylanase and laccase based enzymatic kraft pulp bleaching reduces adsorbable organic halogen (AOX) in bleach effluents: A pilot scale study. *Bioresource Technology*, *169*, 96–102.

Sharma, N., Rathore, M., & Sharma, M. (2013). Microbial pectinase: Sources, characterization and applications. *Reviews in Environmental Science and BioTechnology*, *12*(1), 45–60.

Singh, R. S., & Chauhan, K. (2018). Production, purification, characterization and applications of fungal Inulinase. *Current Biotechnology*, *7*(3), 242–260.

Singh, R. S., Chauhan, K., & Kennedy, J. F. (2017). A panorama of bacterial Inulinase: Production, purification, characterization and industrial applications. *International Journal of Biological Macromolecules*, *96*, 312–322.

Singh, R. S., Singh, T., & Pandey, A. (2019). Microbial enzymes—an overview. *Advances in Enzyme Technology*, 1–40.

Velappan, S., & Thangaraj, P. (2014). Phytochemical constituents and antiarthritic activity of *Ehretia laevis* Roxb. *Journal of Food Biochemistry*, *38*(4), 433–443. doi: 10.1111/jfbc.12071.

Verma, S., Bhatt, P., Verma, A., Mudila, H., Prasher, P., & Rene, E. R. (2021). Microbial technologies for heavy metal remediation: effect of process conditions and current practices. *Clean Technologies and Environmental Policy*. doi: 10.1007/s10098-021-02029-8.

Zhang, S., Gao, X., He, L., Qiu, Y., Zhu, H., & Cao, Y. (2015). Novel trends for use of microbial tannases. *Preparative Biochemistry and Biotechnology*, *45*(3), 221–232.

2 Production, Purification, and Application of the Microbial Enzymes

Anupam Pandey, Ankita H. Tripathi, and Priyanka H. Tripathi

ABSTRACT

Microbial enzymes are fascinating biocatalysts that have received a lot of consideration due to their assistances over chemical catalysts, such as higher selectivity and the ability to operate under mild reaction conditions. Furthermore, some industrial sectors demand enzymes with unique properties for use in the processing of raw materials and different substrates. Enzymes, because of their tremendous specificity, have the potential to change the whole industrial sector. All living organisms require enzymes as biocatalysts for both synthesis and degradation processes. Because of these biocatalysts, many chemically toxic reactions have been replaced by environmentally favorable biological activities. Enzyme technology is advancing rapidly. Novel enzymes with a wide range of uses and specificity have been discovered as technology has advanced, and new application areas are continually being researched. Microorganisms, such as yeast, bacteria, and fungi, as well as their enzymes, find applications in many arenas, such as chemicals, agriculture, fermentation, pharmaceuticals, and food production. Microbial enzymes have procured huge attention for their extensive uses in medicine and industries due to their catalytic activity, stability, and easier production and optimization/standardization than animal and plant enzymes. The commercial manufacturing of enzymes has been established with the use of advanced technologies. The demand for enzymes has lately increased due to an increase in the number of goods available in response to expanding markets. Due to considerable improvements in bioprocess technology, industrial enzyme manufacturing technology has seen tremendous success in recent years. With the advent of genetic engineering and protein engineering, enzymes with desired characteristics and better functionality might be produced. Due to the increasing role of microbial enzymes in regulating various industrially important chemical reactions, this chapter is an attempt to provide an overview of industrial microbial enzyme production, formulation, purification, and their commercial applications.

DOI: 10.1201/9781003202998-2

CONTENTS

2.1   Introduction.. 20
2.2   Use of Microbial Enzymes: Globally 21
2.3   Discovery of Novel Microbial Enzymes.. 22
      2.3.1   Genome Mining 22
      2.3.2   Genome Hunting.. 22
      2.3.3   Data Mining 22
2.4   Metagenomic Screening 23
2.5   Exploiting the Miscellany of Extremophiles 23
2.6   Improvements of Microorganisms.. 23
      2.6.1   Recombinant DNA Technology (rDNA) 23
      2.6.2   Mutation... 24
      2.6.3   Protein Engineering 24
2.7   Production of Microbial Enzymes: Methods.. 24
      2.7.1   Solid State Fermentation (SSF) 24
      2.7.2   Submerged Fermentation (SmF).. 25
2.8   Purification Methods of Microbial Enzymes.. 25
      2.8.1   Selective Precipitation 27
      2.8.2   Ion-Exchange Chromatography (IEC).. 27
      2.8.3   Gel Exclusion Chromatography 28
      2.8.4   Bio-Affinity Chromatography.. 29
      2.8.5   Hydrophobic Interaction 29
2.9   Application of Microbial Enzymes.. 29
      2.9.1   Industrial Applications of Enzyme Catalysis..................................... 29
      2.9.2   Microbial Enzymes Used in Food Sector 32
              2.9.2.1   Proteases 32
              2.9.2.2   α-Amylases 33
              2.9.2.3   Glucoamylases 33
              2.9.2.4   α-Acetolactate Decarboxylase 33
              2.9.2.5   Asparaginases 33
              2.9.2.6   Laccases 34
              2.9.2.7   Xylanases 34
2.10 Pharmaceutical and Therapeutic Applications of Enzymes............................ 34
2.11 Conclusion 35
References.. 36

2.1   INTRODUCTION

The International Union of Biochemistry has classified enzymes into six classes, namely ligases, lyases, oxidoreductases, isomerases, transferases, and hydrolases. Enzymes are biomolecules having a wide range of industrial applications (**Altan, 2004**). They reduce the utilization of various organic solvents, severe temperatures, and pH while also improving product purity, substrate specificity, enzyme activity, and termination and causing less impact on the environment. Microorganisms are the primary source of microbial enzymes and secrete a large number of extracellular enzymes/proteins that are easy to screen, require low-cost maintenance, and involve ease of genetic modifications. This is why microbes are gaining popularity and attention for their potential use in a variety of industrial applications. Since ancient times, microorganisms have been used to ferment a wide range of foods (**Soccol and Larroche, 2005**). The availability of microbiological sources has enabled the evolution of microbial enzymes to a large extent. Microbial enzymes are more stable than enzymes produced by plants or animals (**Asad et al., 2011**). These enzymes are used in a variety of industries, including textiles, food, paper, detergents, and many more. They are used in the preparation of various syrups (e.g., glucose, maltose, fructose, corn), detergents, and papers (**Pandey et al., 2000**). The majority of microorganisms utilized in industries for enzyme production belong to the genera *Bacillus* (*B. licheniformis* and *B. subtilis*) and *Streptomyces* and the fungi *Mucor, Aspergillus*, and *Rhizopus*. Established fermentation technologies can culture huge quantities of microorganisms in a comparatively short period. Due to low-cost growth media and short cycles of fermentation, microbial enzyme manufacturing is cost-effective on a wide scale.

Microbes produce a large number of enzymes, although only around 5% of them are exploited economically (**Binod et al., 2013**). The majority of microbial enzymes are used for making detergents, leathers, biofuels, paper, and textiles. Some are used in manufacturing chemicals, foods, animal feed, pharmaceuticals, and so on. Natural enzymes produced by microorganisms are often unsuitable for biocatalysis and must be modified before being used in commercial applications. The microbial strains have been genetically engineered to increase yield and improve enzyme properties (**Sanchez and Demain, 2017**). In comparison to un-modified microorganisms, the use of recombinant DNA technology improves production ten to hundredfold by cloning microbial enzyme coding genes. As a result, the enzyme industry quickly adopted the method and shifted the production of enzymes from non-industrial strains to industrial strains (**Galante and Formantici, 2003**). The use of technologies such as genomics, proteomics, rDNA technology, and genetic engineering aids in the discovery of novel enzymes produced by various microbes and their modification into superior versions. Metagenomics has been utilized to create a variety of new enzymes (**Ferrer et al., 2007**). Directed evolution of proteins include unidirectional and mutagenic reassembly, Y-ligation-based block shuffling, whole-genome shuffling, and DNA shuffling, heteroduplex, and non-homologous recombination (**Siehl et al., 2005; Reetz, 2009; Yuan et al., 2005; Bershstein and Tewfic, 2008**). These techniques improve the specificity, stability, activity, and solubility of microbial enzymes. Directed evolution enhanced the thermostability and activity of glyphosate-N-acetyltransferase by around 5-fold and 10,000-fold, respectively, using these approaches.

2.2   USE OF MICROBIAL ENZYMES: GLOBALLY

Enzymes have long been utilized as biocatalysts in a variety of industrial applications, including food processing, detergent manufacturing, beer fermentation, and pickling, as well as managing and speeding up catalytic reactions to yield a variety of critical end products quickly and precisely (**Singh et al., 2019**). Enzymes are active biocatalysts that have been investigated for large-scale catalysis processes as an industrial application for a number of reasons, including their ability to function under milder reaction conditions, astounding selectivity of the products, and reduced physiological and environmental toxicity (**Bommarius and Paye, 2013; Choi** *et al.,* **2015**). The benefits outlined above were proved to translate into decreased operating costs when they were utilized as biocatalysts in chemical processes (**DiCosimo** *et al.,* **2013**). As a result, they use less energy, produce less waste, and streamline their production routes (**Madhavan** *et al.,* **2017, Prasad and Roy, 2018**). They have been partially used in the pharmaceutical, food, and beverage industries (**Sun** *et al.,* **2018; Patel** *et al.,* **2017; Huisman and Collier, 2013**). However, further study is needed to demonstrate that biocatalysis is cost-effective in other applications, such as the conversion of natural gas and production of biofuel (**Pellis** *et al.,* **2018; Strong** *et al.,* **2016; Noraini** *et al.,* **2014**). According to Mordor Intelligence's study "Industrial Enzymes Market: Growth, Trends, and Forecast (2019–2024)," the global industrial enzymes market is predicted to increase at a compound yearly growth rate of 6.8% from 2019 to 2024 (**Kamma** *et al.,* **2001**). The health industry will see the most development in the proteases market due to the numerous benefits proteases provide, such as stimulating the immune system, preventing inflammatory bowel illnesses, and healing skin burns and stomach ulcers. Carbohydrases were worth $2.5 billion in 2016, and their market is expected to grow by more than a third by 2024 (**Mabrouk** *et al.,* **2011**).

2.3   DISCOVERY OF NOVEL MICROBIAL ENZYMES

The fact that only around 1% of the microorganisms in the biosphere can be cultured in the laboratory using traditional procedures makes screening of natural microorganisms for the enzymes tough. Three strategies, including genome mining, metagenomic sorting, and exploiting the miscellany of extremophiles, can be used to obtain new enzymes from nature (**Sanchez and Demain, 2017**).

2.3.1   GENOME MINING

Genome mining includes screening through genome sequence present in databases (NCBI) (**NCBI Microbial Genomes, 2013**) for genes that code for novel enzymes. More than 2,000 microbial genome sequences and microbial draft assemblies are available in these databases. For genome mapping, two approaches are utilized to discover new enzymes: genome hunting and data mining.

2.3.2   GENOME HUNTING

Open reading frames (ORF) are searched in the genome of a microorganism during genome hunting. Cloning and overexpression of the ORF coding for a valuable enzyme are followed by activity and functional screening (**Sanchez and Demain, 2017**).

2.3.3   DATA MINING

For microbial enzymes, data mining requires the use of a homology alignment tool such as BLAST. BLAST is used to find homologous protein sequences that could be evaluated as possible protein candidates by identifying conserved areas among sequences (**Sanchez and Demain, 2017**).

2.4   METAGENOMIC SCREENING

Metagenomic screening comprises creating a genomic DNA library from ambient DNA, then screening for open reading frames that potentially encode novel enzymes in a systematic manner (**Gilbert and Dupont, 2011; Uchiyama and Miyazaki, 2009**). Enzymes have been discovered through metagenomic screening of specific ecosystems (cow rumen, arctic tundra, marine environments, volcanic vents, and termite guts), which include amidase, epoxide hydrolase, lipase, amylase, oxidoreductase, beta-glucosidase, and decarboxylase. *Escherichia coli* is the most extensively used system for screening foreign genes, followed by improved expression systems like *Rhizobium leguminosarum, Streptomyces lividans*, and *Pseudomonas putida* (**Sanchez and Demain, 2017**).

2.5   EXPLOITING THE MISCELLANY OF EXTREMOPHILES

Extremophiles of the genera *Bacillus, Clostridium, Thermus*, and *Thermotoga* and hyperthermophiles (*Archea, Methanopyrus, Pyrococcus*, and *Methanopyrus*) can survive extreme environmental conditions including salinity, pressure, pH, and radiation. These extremophiles produce stable enzymes (**Sanchez and Demain, 2017**). For example, Taq polymerase is isolated from *Thermus aquaticus* (thermophile). Thermophilic amylases, cellulases, and proteases are already in use in industry. In the textile industry, enzymes isolated from psychrophiles are employed. Psychrophile-derived xylanases and cellulases are utilized in the paper and pulp industries to make biofuels (second generation). They are also utilized in the removal of oils and hydrocarbons from water. Lipases, xylanases, amylases, and proteases are among the enzymes isolated from halophiles (*Halobacterium, Halobacillus*, and *Halothermothrix*) (**Van Den Burg, 2003**). Enzymes isolated from halophiles are used as additives in detergents (**Shukla et al., 2009**).

2.6   IMPROVEMENTS OF MICROORGANISMS

It has been seen that the majority of naturally occurring microbes either do not generate enzymes in sufficient numbers for commercial use or do not have suitable features for the application. As a result, enormous efforts have been made to improve strains using traditional or molecular technologies in order to develop the essential traits and obtain overproducing strains (**Pandey et al., 2010**).

2.6.1   RECOMBINANT DNA TECHNOLOGY (rDNA)

Enzymes having a role in the various industrial application are usually isolated from several normal to extreme environments. Novel microbial enzymes have been discovered using high-throughput screening technologies; however, the problem is that these bacteria are difficult to cultivate in the lab and produce a low amount of enzyme. As a result, companies have used rDNA technology to boost the yield of specific enzymes by cloning the enzyme coding gene and using heterologous expression. Insulin, for example, has been produced using rDNA technology. The genetically altered *E. coli* produces 100 times more insulin than native microbial strains, making insulin more readily available, at a low cost, and in large quantities **(Chiang, 2004)**. As a result, industries are turning to genetically modified microbial strains to produce enzymes like lipases and amylases, which have a wide range of applications in the food industry **(Olempska-Beer *et al.*, 2006)**.

2.6.2   MUTATION

For the goal of scientific investigation, scientists may also purposefully introduce mutant sequences through DNA manipulation. In industry, the process of mutation, which includes the insertion of random mutations in the microbial genome followed by strain selection, is employed for a variety of purposes. For instance, mutant strain RUT C-30 (*Trichoderma reesei*) is considered the best and highest producer of cellulase enzyme **(Patel *et al.*, 2017)**.

2.6.3   PROTEIN ENGINEERING

Protein engineering is a technique for altering the sequence of a protein in order to obtain a specific result, such as enhanced stability to temperature, chemical solvents, and high and low pH. Protein engineering includes site-directed mutagenesis, which involves the substitution of particular amino acids, which requires a significant quantity of knowledge about the biocatalyst being modified, with modulation in its three-dimensional structure and mechanism involved in the chemical process. Engineered enzymes with improved performance have already been used in industrial operations, such as proteinases, cellulases, amylases, lipases, and glucoamylases **(Patel *et al.*, 2017)**.

2.7   PRODUCTION OF MICROBIAL ENZYMES: METHODS

Fermentation is a cost-effective, less-time-consuming, and space-saving method of producing microbial enzymes with high consistency in which the microbial enzymes can be optimized and modified easily **(Gurung *et al.*, 2013)**. The most commonly used methods for the production of microbial enzymes include solid-state and submerged fermentation.

2.7.1   SOLID STATE FERMENTATION (SSF)

Solid-state fermentation (SSF) is ideal for microorganisms that require low moisture content. The substrate used in SSF includes waste materials, such as bagasse, paper pulp, and bran, that are rich in nutrients. These substrates are constantly utilized by the microorganisms but at a very slow speed. SSF is a beneficial approach since it is less time-consuming and easier to use, has a high product recovery rate, and produces fewer effluents. As a result, SSF is a widely acknowledged approach for commercial

enzyme production (**Pant *et al.*, 2015**). Microbial enzyme synthesis in SSF procedure is influenced by various factors, such as particle size of the substrate, moisture
content, inoculum size, temperature regulation, cultivation time period, rate of CO_2
evolution, and O_2 consumption. The ease of SSF makes it ideal for the production of
largescale microbial enzymes. SSF method has been utilized to produce industrially
important enzymes, such as amylases, cellulases, xylanases, proteases, ligninases,
and glucoamylases. Currently, efforts are being made to produce enzymes via the
SSF technique, which includes major enzymes, such as phenolic acid esterases, phytases, microbial rennet, tannin acyl hydrolase, tannase, aryl alcohol oxidases, inulinase, oligosaccharide oxidases, and α-L-arabinofuranosidases (**Patel *et al.*, 2017**).

2.7.2 SUBMERGED FERMENTATION (SMF)

The submerged fermentation method (SmF) has been used for the production of
enzymes using microorganisms. Liquid substrates (free-flowing), such as broths and
molasses, are among the substrates utilized in SmF. The process requires the continuous addition of substrate as the substrate is utilized very rapidly. The technique
is suitable for the extraction of secondary metabolites produced by bacteria at high
moisture content conditions (**Patel *et al.*, 2017**). The procedure includes easy regulation of moisture, pH, oxygen transfer, temperature, and aeration. In SmF, fermenters with volumes ranging from thousands to hundreds of thousands of liters
are employed. These large-scale SmF fermenters are built in such a way that they
can control dissolved oxygen, pH, temperature, and foam in real time. In these fermenters, heat dissipation and mass transfer are also well-controlled. The microbes
in the SmF can be grown in four different ways. Batch culture, continuous culture,
perfusion-batch culture, and fed-batch culture are the four types of culture. Microbes
are inoculated into a fixed volume of culture medium in batch culture. In continuous
culture, during the exponential development phase of microorganisms, fresh culture
medium is continuously added to the batch culture, with the culture medium containing the desired end product being removed simultaneously. Nutrient medium is
gradually given to the culture in fed-batch culture. The addition of the culture and
extraction of an equivalent volume of consumed cell-free media is conducted in per-
fusion batch culture (**Patel *et al.*, 2017**).

2.8 PURIFICATION METHODS OF MICROBIAL ENZYMES

Enzymes are complex structures that catalyze all of the major chemical reactions
that are required for human survival and reproduction. Biocatalysis is the chemical process by which enzymes produce energy-efficient valuable chemical products
(**Jäckel and Hilvert, 2010**). Based upon the targeted reaction, whole-cell system or
purified enzymes are used as biocatalysts either in free or immobilized form. The
process of purifying an enzyme is always productive in terms of catalytic activity, high concentration, and responsiveness to receptor molecules that increase or
decrease the activity (**Kohls *et al.*, 2014**). Quantity, purity, and association of the
enzyme of interest with other biological molecules are important factors that must
be considered before embarking upon the purification of enzymes. After these as

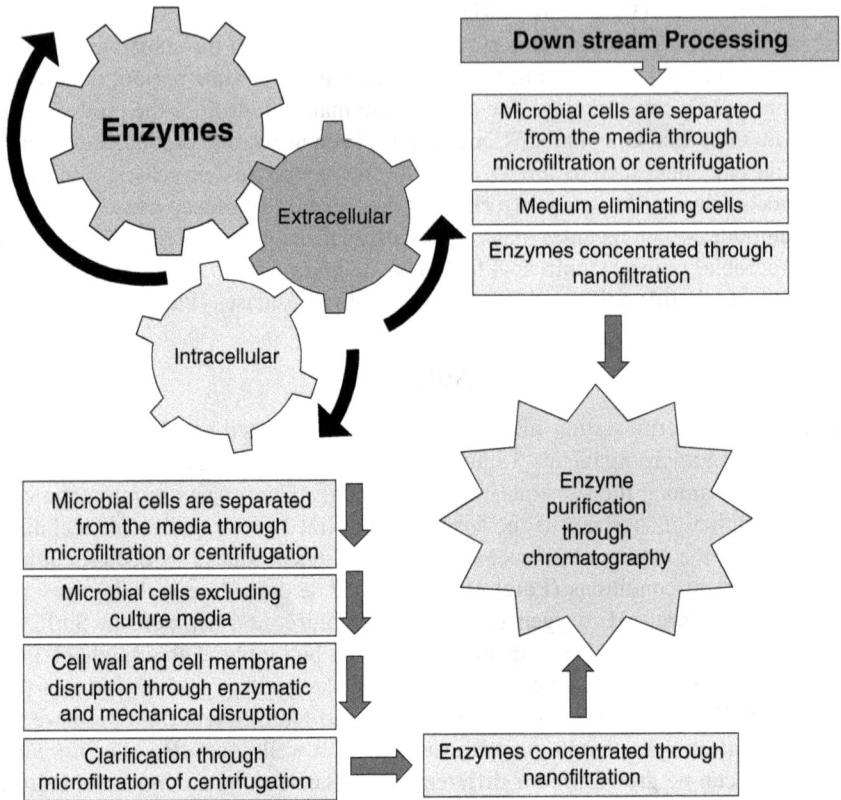

FIGURE 2.1 Purification of microbial enzymes.

benchmarks are considered, a suitable purifying process must be chosen along with the source of enzyme, extraction process of enzymes from sources, and storage conditions of enzymes (**Linn, 2009**). Microbial cells are the major source of industrially important enzymes, such as proteases, amylases, cellulases, lipases, lyases, and ureases (**Vijayaraghavan *et al.*, 2016**). Enzymes can be present intracellular or can be secreted in culture media during the growth of microorganisms (**Figure 2.1**).

The extraction of the intracellular enzyme is more tedious as they require breakage of the cell wall. Various mechanical (e.g., bead milling, homogenization, ultra-sonication, osmotic shock) and chemical (e.g., acid and alkali treatments) techniques are the most familiar processes for breaking the cell wall. The enzymatic hydrolysis of the cell wall uses lytic enzymes, such as lyase. Esterase is an effective technique that can be used alone or in combination with physical and chemical treatments. After breaking the cell wall, enzymes present in the supernatant can be easily separated using centrifugation from unbroken cell pellets and large insoluble molecules. Secreted proteins that act as enzymes can be separated directly from the cell pellet using centrifugation for lab scale and by filtration for pilot scale purification. A cocktail of various protease inhibitors, such as EDTA, leupeptin, pepstatin A, and PMSF, has been used to prevent enzyme inactivation and degradation during protein extraction. The protocol for the purification of enzymes is largely based upon the need of the enzyme. For example, the alkaline protease, which is used as an ingredient in the detergent industry, requires partial fractionation compared to streptokinase, which has application in therapeutics needs 99% purity (**Burgess, 2009**).

After obtaining a cell-free extract, different separation methods can be used in logical order to achieve the targeted goal (**Kornberg, 1990**). These purification methods are mainly divided into the following categories:

a) Selective precipitation
b) Separation based on charge
c) Separation based on adsorption properties
d) Separation based upon affinity to ligands
e) Separation based upon the molecular size of the enzymes

2.8.1 SELECTIVE PRECIPITATION

As partial purification is beneficial before proceeding to chromatographic procedures, selective precipitation using salt (ammonium sulfate precipitation), organic solvents (ethanol or acetone), and isoelectric precipitation (polyethylene glycol) is an early step in most of the enzyme purification process. The principle behind this process is that the individual proteins fractionate at different concentrations of salts and solvents. Ammonium sulfate, because of its high solubility, low cost, and less toxicity toward enzymes, behaves as an ideal salt for precipitation (**Vijayaraghavan, 2016; Burgess, 2009**). The fractionation range could be optimized using the hit and trial method. An experiment with a small sample volume up to 1 mg protein/ml in different aliquots can be performed with the addition of ammonium sulfate ranges between 0 and 90% saturation to check the initial and final precipitation concentration. Immediately after optimizing the precipitation range, the bulk of the cell extract is precipitated and is used for further purification. If required, elimination of ammonium sulfate can be achieved by dialysis, ultrafiltration, or gel filtration (e.g., with desalting columns). Acetone is generally used to precipitate alkaline proteases at various concentrations (**Kim *et al.*, 1996**). Enzyme precipitation can also be accomplished using water-soluble neutral polymers, such as polyethylene glycol (**Larcher *et al.*, 1996**).

2.8.2 ION-EXCHANGE CHROMATOGRAPHY (IEC)

In food science, for the purification of proteins, the widely used method is ion-exchange chromatography (IEC) which distinguishes the entities based on their charge. The stationary phase in IEC carries charged functional groups fixed by chemical bonds, which are associated with exchangeable counter ions. In the cation-exchange chromatography, the fixed groups are of positive charge, while the reverse is true for these groups in anion-exchange chromatography. This works on the principle that protein binds to exchanger at pH value either below or above their isoelectric point depending on the type of ion chromatography. The method involves the equilibration of resin with a low salt buffer followed by the sample application and its absorption, which is the complete desorption of bound protein. Desorption is carried out by increased salt concentration of elution buffer. The final steps involve the cleaning of a column to remove the strongly bounded substance. This method is widely employed because of its high binding capacity, which allows elution in concentrated form and separates protein with high resolution (**Rossomando, 1990**).

2.8.3 Gel Exclusion Chromatography

It is also known as a molecular sieve, gel filtration, and size exclusion chromatography. It separates proteins completely on the basis of molecular size. Purification is performed using a porous stationary phase to which the molecules have different degrees of access. Smaller molecules elute faster compared to larger molecules from the matrix. The liquid phase plays a major role which measures volumes like the void volume that is liquid between the beads and internal volume that is liquid within the pores of the bead. Molecules bigger than the bead size elute first, while the smaller ones elute later (**Stellwagen, 2009**).

TABLE 2.1
Methods of Enzyme Purification

S. No.	Purification Method	Enzyme Property	References
1.	Electrophoresis	Polarity/hydrophobicity/charge	(**Patel *et al.*, 2017**;
2.	Chromatofocusing		Rossomando, 1990)
3.	Isoelectrofocusing		
4.	Hydrophobic chromatography		
5.	Ion-exchange chromatography		
6.	Gel filtration	Mass/size	
7.	Ultrafiltration		
8.	Centrifugation		
9.	Dialysis		
10.	Column chromatography	Structural properties/specific binding	
11.	Affinity chromatography		
12.	Dye-ligand chromatography		
13.	Immobilized metal ion chromatography		
14.	Affinity elution		
15.	Immunoadsorption		
16.	Ionic strength changes	Solubility	
17.	Modulation in pH		
18.	Lowering of dielectric constant		

2.8.4 BIO-AFFINITY CHROMATOGRAPHY

It is a highly selective method as it utilizes the specific, reversible interactions between biomolecules. This method works on the association of the enzyme with the immobilized ligands, such as coenzymes, substrates, activators, and inhibitors. The enzymes are purified on the basis of their association. The non-bounded enzymes are washed off first, followed by bounded enzymes, by changing the elution condition. Bio-affinity chromatography provides the highest degree of purification because of its specific interactions (**Urh, 2009**).

2.8.5 HYDROPHOBIC INTERACTION

Protein surface carries hydrophobic groups or clusters of hydrophobic groups, which contributes to their surface hydrophobicity that allows the interactions with other proteins as well as with the column carrying hydrophobic groups. Interaction with non-polar compounds in a polar environment is enhanced and gains entropy by liberating water molecules, which drives the clustering of the hydrophobic groups. The hydrophobic interactions are affected by altering the structure of water by dissolved salts. For example, kosmotropic salts enhance the strength of the hydrophobic interactions while chemotropic salts weaken it. The resins commonly used for HIC are substituted with n-butyl, n-octyl, or phenyl groups. For an uncharacterized protein, phenyl-substituted resin is usually preferred over the highly hydrophobic octyl-substituted resins. The phenyl ligand is moderately hydrophobic in comparison to n-butyl and n-pentyl and will bind to aromatic amino acids through π–π interactions. To purify a protein from a crude extract with an ammonium sulfate precipitation at a concentration of ammonium sulfate that leaves desired protein in solution and its precipitates by centrifugation. The clarified solution is used for HIC elutes by decreasing the level of kosmotropic salts in the buffer. The proteins are eluted based on their hydrophobicity, while those tightly bounded ones which are difficult to elute with negative gradient are eluted with a positive gradient of organic solvent (**McCue, 2009**). The most common methods for analyzing the final purity of the enzymes are sodium dodecyl sulfate-polyacrylamide gel electrophoresis (SDS-PAGE), analytical gel filtration, and mass spectrometry (**Rhodes and Laue, 2009**).

2.9 APPLICATION OF MICROBIAL ENZYMES

Microbial enzymes have various advantages over plant- and animal-synthesized enzymes because the processing is easy and cost-effective and involves constant synthesis. In this chapter, we'll look at a few of the most important microbial enzymes that have a wide range of industrial applications.

2.9.1 INDUSTRIAL APPLICATIONS OF ENZYME CATALYSIS

Enzymes are highly effective biocatalysts that have been studied for industrial-scale catalysis due to a variety of benefits, including their ability to operate at softer reaction conditions, superior product selectivity, and decreased environmental and physiological toxicity (**Bommarius and Paye, 2013; Prasad and Roy, 2018**). When they were used as biocatalysts in chemical processes, the benefits described above were demonstrated to translate into lower operating costs. As a result, the pharmaceutical, food, and beverage industries have partially realized their decreased energy requirements, waste reduction, and streamlined manufacturing methods (**Sun et al., 2018; Patel, 2018**). The lack of enzyme stability at high temperatures or in turbulent flow regimes, as well as in potentially toxic solvents, limits the use of enzyme catalysis in chemical processes throughout the many industries where biocatalysis can be used (**Grigoras, 2017; Cao et al., 2016; Misson et al., 2015**). As a result, multidisciplinary approaches are concentrating on the development and manufacturing of robust, stable biocatalysts appropriate for use in a wider range of industrial contexts (**Chao et al., 2013**). Applications of industrially manufactured microbial enzymes—microbial enzymes are widely employed in a variety of industries, owing to a large number of sources available. Microbial enzymes can be genetically manipulated and are more cost-effective than plant and animal enzymes (**Vittaladevaram et al., 2017**). The production of microbial enzymes via fermentation processes necessitates microbial multiplication in order to obtain the desired output.

TABLE 2.2
Microbial Enzymes Used in Industries

S. No.	Microbial Enzyme	Isolated From	Application	References
1.	Proteases	*Aspergillus usamii*	Improvement of bread quality, milk coagulation, tenderization of meat	(**Sanchez and Demain, 2017**; **Patel et al., 2017**; **Raveendran et al., 2018**; Chan et al., 2013)
2.	α-amylase	*Bacillus licheniformis, Bacillus amyloliquefaciencs, Bacillus stearothermophilus.*	Starch liquefaction, improvement of bread quality, fruit juice clarification	
3.	Glucoamylases	*Rhizopus oryzae, Aspergillus awamori, Aspergillus niger*	Improvement of bread quality, light beer production, high-fructose and high-glucose syrups	
4.	Xylanases	*Streptomyces* sp., *Fusarium* sp., *Aspergillus* sp., *Penicillium* sp., *Pseudomonas* sp., *Bacillus* sp.	Improvement of bread quality, fruit juice clarification	

S. No.	Microbial Enzyme	Isolated From	Application	References
5.	Catalases	*Aspergillus niger*	Preservation of food and hydrogen peroxide removal from milk, bio-polishing of textiles	
6.	Laccases	*Funalia trogii, Bacillus licheniformis*	Synthesis of chemicals, paper pulp, bio-bleaching, textile finishing, bioremediation, wine stabilization, and biosensing	
7.	Asparaginases	*Bacillus licheniformis*	Degradation of asparagine	
8.	Naringinase	*Aspergillus niger, Eurotium, Fusarium, Circinella, Penicillium, Trichoderma, Rhizopus, Pseudomonas paucimobilis, Bacillus* sp., *Bacteriodes distasonis, Burkholderia cenocepacia, Thermomicrobium roseum*	Debittering enzyme for removal of bitterness from citrus fruits	
9.	α-acetolactate decarboxylase	*Brevibacillus brevis, Saccharomyces cerevisiae*	Speeding up beer maturation	
10.	Peroxidases	*Phanerochaete chrysosporium, Streptomyces viridosporus T7A*	Reduction of peroxides and oxidation of organic and inorganic compounds	
11.	Glucose oxidase	*Penicillium glaucum, Penicillium adametzii, Aspergillus niger*	Removal of oxygen from beer, foods, fruits, juices, dairy products, etc.	
12.	Pectinases	*Penicillium* sp., *Aspergillus niger*	Fermentation of coffee bean, clarification and filtration of fruit juices, wine, etc.	
13.	Cellulases	*Paenibacillus, Bacillus, Trichoderma, Aspergillus*	Used in biofuel production, recycling pulp and paper	
14.	Lipoxygenases	*Anabaena, Nostoc, Lasiodiplodia theobromae*	Used in the food and pharmaceutical industries	

(*Continued*)

TABLE 2.2 (*Continued*)

S. No.	Microbial Enzyme	Isolated From	Application	References
15.	Phospholipases	*Fusarium oxysporum*	Used in oil production, dairy, baking industry	
16.	Esterases	*Bacillus licheniformis*	Used for enhancing the flavors of fruit juices by modifying the fats and oils present in the beverages	
'17.	Lipases	*Penicillium camemberti, Aspergillus niger, A. oryzae, Candida cylindracea* Ay30, *Candida antarctica* (CALB), *Helvina lanuginosa, Geotrichum candidum, Pseudomonas* sp.	Used in various industries, including food, detergents, biofuel, and animal feed; also used in textile, leather, and paper processing	
18.	Lactases	*Kluyveromyces lactis*	Hydrolysis of lactose to produce galactooligosaccharides (GOS)	
19.	Hemicellulases	*T. reesei, Aspergillus niger, Penicillium* sp.	Used in food industries for the concentration of coffee and also used in the biofuel industry	
20.	Phytases	*P. funiculosum, Aspergillus* sp., *A. ficuum, Xanthomonas oryza, Bacillus* sp., *Pseudomonas*	Used in the animal feed industry	
21.	Invertases	*Saccharomyces*	Soft-center fondants and candies	
22.	Dextrinase	Fungal spp.	Making of corn syrup	

2.9.2 MICROBIAL ENZYMES USED IN FOOD SECTOR

2.9.2.1 Proteases

Plants, microorganisms (including fungi and bacteria), and animals have all been found to have proteases. The hydrolysis of peptide bonds in polypeptides and proteins is carried out by these enzymes. Protease accounts for over 60% of all industrial enzymes sold worldwide. Protease is used in the baking industry to make waffles, bread, crackers, and baked goods. They are used to improve the flavor and texture of bread by adjusting gluten strength and lowering mixing time, as well as to improve the homogeneity and consistency of dough. Acidic protease is produced by *Aspergillus usamii* and is used commercially to improve the function of wheat gluten. By breaking enough peptide bonds in proteins and releasing amino acids, acidic protease improves beer fermentation. They're also employed in the dairy industry. By managing the allergenic qualities of milk products, natural proteases improve the flavor of cheese and speed up the ripening process (**Raveendran *et al.*, 2018**).

2.9.2.2   α-Amylases

α-amylases are enzymes involved in the degradation of starch. The enzyme hydrolyses polysaccharide-1,4 glycosidic linkages to produce short-chain dextrins. Baking, starch liquefaction, brewing, and digestive assistance are just a few of the many uses for α-amylases in the food industry. They're utilized to improve bread quality by adding flavor and functioning as an anti-staling agent. α-amylases are added to bread baking to convert starch to dextrin, which improves color, flavor, and toasting. Gelatinization, liquefaction, and saccharification are the three phases in the enzymatic degradation of starch. Gelatinization is the process of dissolving starch granules to create a viscid suspension. The next step is liquefaction, which involves partial hydrolysis and decreases viscosity. Saccharification produces more maltose and glucose. The process of enzymatic conversion of starch requires heat-stable α-amylases, which are isolated from *Bacillus licheniformis*, *Bacillus amyloliquefaciencs*, and *Bacillus stearothermophilus*. α-amylases catalyze the conversion of starch to fermentable sugars, which are then fermented to alcohol by *Saccharomyces cerevisiae* (**Raveendran *et al.*, 2018**).

2.9.2.3   Glucoamylases

Glucoamylases are exo-enzymes, also known as saccharifying enzymes, that hydrolyze polysaccharides (starch) from the non-reducing end, removing glucose. The main microbial sources of glucoamylases are *Rhizopus oryzae*, *Aspergillus awamori*, and *Aspergillus niger*. The enzyme is utilized in the production of high-glucose/high-fructose syrups. They're also utilized to improve bread and flour's color, texture, flavor, and shelf life. The enzyme breaks down polysaccharide starch in the flour to produce fermentable sugars and maltose. *Saccharomyces cerevisiae* converts glucose to ethanol through fermentation. Light beers, soy sauce, and sake are all made with them (**Raveendran *et al.*, 2018**).

2.9.2.4   α-Acetolactate Decarboxylase

Brevibacillus brevis produces α-acetolactate decarboxylase, which may also be isolated from the recombinant *Saccharomyces cerevisiae*. The enzyme is used to accelerate the aging of beer. To make the final product, the enzyme first decarboxylates the substrate into enol form, then protonates it. The enzyme, therefore, eliminates α-aceto-hydroxy-butyrate and α-acetolactate, which are involved in the regulation of the rate-limiting stage in beer maturation, and thereby speeds up the process (**Zhao *et al.*, 2017**).

2.9.2.5   Asparaginases

Asparaginases are a type of microbially produced enzyme that has a wide range of applications in medicine, industry, and nutraceuticals. Asparaginases is an asparagine-depleting enzyme that catalyzes the breakdown of asparagine to produce ammonia and aspartic acid. Asparagine is a nonessential amino acid, yet it is required for malignant cells. As a result, asparagine scavenging may have an influence on the genesis and growth of malignant cells and is thus regarded as an anticancer agent. Various processes, such as deep frying in oil or baking, convert asparagine to acrylamide (a carcinogen). Depleting asparagine using enzymes could thereby prevent acrylamide production by 97% (**Krishnapura** *et al.*, **2016**).

2.9.2.6 Laccases

Laccases are blue oxidases, a subdivision of multicopper enzymes that consists of a group of oxidases. They are capable of oxidizing phenols, ascorbate, and aromatic amines. They are used as biocatalysts for the synthesis of chemicals, paper pulp, bio-bleaching, textile finishing, bioremediation, wine stabilization, and biosensing. Laccases are a secondary metabolite produced by a variety of fungus species, including Basidiomycetes, Deuteromycetes, and Ascomycetes. Laccase is produced by *Funalia trogi* through fermentation and by *Bacillus licheniformis* for industrial purposes. The enzyme has been used to avoid the production of haze by oxidizing polyphenolic substances. They're also used to remove polyphenols from beer and to keep wine stable (**Madhavi and Lele, 2009**).

2.9.2.7 Xylanases

Xylanases are enzymes synthesized by microorganisms for breaking xylans, a chief integral part of hemicellulose. Exoxylanases break β-1,4 bonds from non-reducing ends of hemicellulose yielding xylooligosaccharides, endoxylanases break β-1,4 bonds, and β-xylosidases break xylooligosaccharides and xylobiose to produce xylose. All three enzymes act together to break down xylans in a synergistic manner. Xylanases are synthesized by actinomycetes (*Streptomyces* sp.), fungi (*Fusarium* sp., *Aspergillus* sp., and *Penicillium* sp.), and bacteria (*Pseudomonas* sp., and *Bacillus* sp.). Xylanases are widely used for preparing bread by providing them appropriate volume, enhancing bread quality, and increasing water-binding property of water during dough preparation. They are also used in the clarification of fruit juices (**Raveendran** *et al.*, **2018**).

2.10   PHARMACEUTICAL AND THERAPEUTIC APPLICATIONS OF ENZYMES

Meghwanshi *et al.* (**2020**) have been found that enzymes are used in a variety of industrial applications, including the synthesis of pharmaceuticals, such as drugs (**Sun** *et al.*, **2018**); the processing of grain juices into lager and wine; the leavening of dough for bread production; the production of agrochemicals, artificial flavors, and biopolymers; and waste remediation (**Choi** *et al.*, **2015, Madhavan** *et al.*, **2017, Prasad and Roy, 2018**). Microorganisms provide the majority of industrial enzymes since they are the most convenient sources, allowing for faster production, easier scale-up, recovery and purification, strain manipulation for overexpression, enzyme activity, specificity modulations, and so on (**Brahmachari, 2016**). Approximately 200 varieties of microbial enzymes are currently employed commercially, out of a total of 4,000 known enzymes (**Liu and Kokare, 2017**). Several medications and

TABLE 2.3
Enzymes for Therapeutic Uses

Generic Name	Mechanism of Action	Medical Condition Treated	References
Pegademase bovine	Replace the enzyme in SCID Patient	For enzyme replacement therapy for ADA in patients with SCID	Huang and Manton, 2005
Dornase alpha	Cleave DNA release from neutrophils	To reduce mucous viscosity and enable the clearance of airway secretions in patients with CF	Grasemann et al., 2004
Imiglucerase	Catalyze the hydrolysis of glucocerebroside to glucose and ceramide	Replacement therapy in patients with types I, II, and III Gaucher's disease	Capablo et al., 2007
Sacrosidase	Replace sucrose in people lacking this enzyme	Treatment of congenital sucrase-isomaltase deficiency	Robayo-Torres et al., 2009
Rasburicase	Catalyze enzymatic oxidation of poorly soluble uric acid	Treatment of malignancy-associated or chemotherapy-induced hyperuricemia	Liu et al., 2005
Agalsidase beta (Fabrazyme)	Source of alpha-galactosidase A	Treatment of Fabry's disease	Germain et al., 2015
Nattokinase	Dissolve blood clots by cleaving plasmin and fibrin	Support healthy blood clotting, circulation, and platelet function	Hsu et al., 2009

pharmaceutical formulations contain active pharmaceutical ingredients that are manufactured employing enzymes as critical production components (**Mitchell, 2017**). The use of enzymes to treat enzyme deficits and other medical disorders in humans is known as enzyme therapy. Enzymes aid in food digestion, bodily purification, immune system strengthening, muscle contraction, and stress reduction on critical organs, such as the pancreas, in humans. Enzyme therapy can be used to treat a variety of medical conditions, including pancreatic insufficiency, cystic fibrosis, metabolic problems, and lactose intolerance, and to eliminate dead tissues, malignancies, tumors, and so on.

2.11 CONCLUSION

Despite the fact that enzyme technology is a well-established field of study, it is continually evolving. Several chemical-based technologies are being phased out of our culture in favor of technology that is more environmentally friendly. As a result, natural enzyme sources are being used for large-scale microbial enzyme production and purification. Simultaneously, researchers are searching for new enzymes based on an existing enzyme's potential utility. Enzymes have already proven their ability to guide us toward numerous biological reactions and roles as biocatalysts. The use of microbial enzymes for diverse chemical processes has provided several benefits for large-scale commercial applications, including specificity, environmental feasibility, and many more. Enzymes have a significant impact on every major industry, particularly the textile, food, and pharmaceutical industries. In terms of their potential and benefits, the discovery of new enzymes and the modification of existing enzymes will play a larger role in the future.

REFERENCES

Altan, A. (2004). Isolation and molecular characterization of extracellular lipase and pectinase producing bacteria from olive oil mills (Master's thesis, İzmir Institute of Technology).

Asad, W., Asif, M., & Rasool, S. A. (2011). Extracellular enzyme production by indigenous thermophilic bacteria: Partial purification and characterization of α-amylase by Bacillus sp. WA21. *Pakistan Journal of Botany*, *43*(2), 1045–1052.

Bershstein, S., & Tewfic, D. S. (2008). Advances in laboratory evolution of enzymes. *Current Opinion in Chemical Biology*, *12*, 151–158.

Binod, P., Palkhiwala, P., Gaikaiwari, R., Nampoothiri, K. M., Duggal, A., Dey, K., & Pandey, A. (2013). Industrial enzymes-present status and future perspectives for India. *Journal of Scientific & Industrial Research*, *72*.

Bommarius, A. S., & Paye, M. F. (2013). Stabilizing biocatalysts. *Chemical Society Reviews*, *42*(15), 6534–6565.

Brahmachari, G. (2016). *Biotechnology of microbial enzymes: Production, biocatalysis and industrial applications*. Academic Press, Oxford.

Burgess, R. R. (2009). Protein precipitation techniques. *Methods in Enzymology*, *463*, 331–342.

Cao, S., Xu, P., Ma, Y., Yao, X., Yao, Y., Zong, M., & Lou, W. (2016). Recent advances in immobilized enzymes on nanocarriers. *Chinese Journal of Catalysis*, *37*(11), 1814–1823.

Capablo, J. L., Franco, R., De Cabezón, A. S., Alfonso, P., Pocovi, M., & Giraldo, P. (2007). Neurologic improvement in a type 3 Gaucher disease patient treated with imiglucerase/miglustat combination. *Epilepsia*, *48*(7), 1406–1408.

Chan, F. K. M., Moriwaki, K., & De Rosa, M. J. (2013). Detection of necrosis by release of lactate dehydrogenase activity. In *Immune homeostasis* (pp. 65–70). Humana Press, Totowa, NJ.

Chao, F. A., Morelli, A., Haugner III, J. C., Churchfield, L., Hagmann, L. N., Shi, L., & Seelig, B. (2013). Structure and dynamics of a primordial catalytic fold generated by in vitro evolution. *Nature Chemical Biology*, *9*(2), 81–83.

Chiang, S. J. (2004). Strain improvement for fermentation and biocatalysis processes by genetic engineering technology. *Journal of Industrial Microbiology and Biotechnology*, *31*(3), 99–108.

Choi, J. M., Han, S. S., & Kim, H. S. (2015). Industrial applications of enzyme biocatalysis: Current status and future aspects. *Biotechnology Advances*, *33*(7), 1443–1454.

DiCosimo, R., McAuliffe, J., Poulose, A. J., & Bohlmann, G. (2013). Industrial use of immobilized enzymes. *Chemical Society Reviews*, *42*(15), 6437–6474.

Ferrer, M., Beloqui, A., Golyshina, O. V., Plou, F. J., Neef, A., Chernikova, T. N., . . . Golyshin, P. N. (2007). Biochemical and structural features of a novel cyclodextrinase from cow rumen metagenome. *Biotechnology Journal: Healthcare Nutrition Technology*, *2*(2), 207–213.

Galante, Y. M., & Formantici, C. (2003). Enzyme applications in detergency and in manufacturing industries. *Current Organic Chemistry*, *7*(13), 1399–1422.

Germain, D. P., Charrow, J., Desnick, R. J., Guffon, N., Kempf, J., Lachmann, R. H., & Wilcox, W. R. (2015). Ten-year outcome of enzyme replacement therapy with agalsidase beta in patients with Fabry disease. *Journal of Medical Genetics*, *52*(5), 353–358.

Gilbert, J. A., & Dupont, C. L. (2011). Microbial metagenomics: Beyond the genome. *Annual Review of Marine Science*, *3*, 347–371.

Grasemann, H., Lax, H., Treseler, J. W., & Colin, A. A. (2004). Dornase alpha and exhaled NO in cystic fibrosis. *Pediatric Pulmonology*, *38*(5), 379–385.

Grigoras, A. G. (2017). Catalase immobilization—A review. *Biochemical Engineering Journal*, *117*, 1–20.

Gurung, N., Ray, S., Bose, S., & Rai, V. (2013). A broader view: Microbial enzymes and their relevance in industries, medicine, and beyond. *BioMed Research International*, *2013*.

Hsu, R. L., Lee, K. T., Wang, J. H., Lee, L. Y. L., & Chen, R. P. Y. (2009). Amyloid-degrading ability of nattokinase from Bacillus subtilis natto. *Journal of Agricultural and Food Chemistry*, *57*(2), 503–508.

Huang, H., & Manton, K. G. (2005). Newborn screening for severe combined immunodeficiency (SCID): A review. *Frontiers in Bioscience*, *10*(1–3), 1024.

Huisman, G. W., & Collier, S. J. (2013). On the development of new biocatalytic processes for practical pharmaceutical synthesis. *Current Opinion in Chemical Biology*, *17*(2), 284–292.

Jäckel, C., & Hilvert, D. (2010). Biocatalysts by evolution. *Current Opinion in Biotechnology*, *21*(6), 753–759.

Kamma, J. J., Nakou, M., & Persson, R. G. (2001). Association of early onset periodontitis microbiota with aspartate aminotransferase activity in gingival crevicular fluid. *Journal of Clinical Periodontology*, *28*(12), 1096–1105.

Kim, W., Choi, K., Kim, Y., Park, H., Choi, J., Lee, Y., . . . Lee, S. (1996). Purification and characterization of a fibrinolytic enzyme produced from Bacillus sp. Strain CK 11–4 screened from Chungkook-Jang. *Applied and Environmental Microbiology*, *62*(7), 2482–2488.

Kohls, H., Steffen-Munsberg, F., & Höhne, M. (2014). Recent achievements in developing the biocatalytic toolbox for chiral amine synthesis. *Current Opinion in Chemical Biology*, *19*, 180–192.

Kornberg, A. (1990). Why purify enzymes? *Methods in Enzymology*, *182*, 1–5.

Krishnapura, P. R., Belur, P. D., & Subramanya, S. (2016). A critical review on properties and applications of microbial l-asparaginases. *Critical Reviews in Microbiology*, *42*(5), 720–737.

Larcher, G., Cimon, B., Symoens, F. O., Tronchin, G., Chabasse, D., & Bouchara, J. P. (1996). A 33 kDa serine proteinase from Scedosporium apiospermum. *Biochemical Journal*, *315*(1), 119–126.

Linn, S. (2009). Strategies and considerations for protein purifications. *Methods in Enzymology*, *463*, 9–19.

Liu, C. Y., Sims-McCallum, R. P., & Schiffer, C. A. (2005). A single dose of rasburicase is sufficient for the treatment of hyperuricemia in patients receiving chemotherapy. *Leukemia Research*, *29*(4), 463–465.

Liu, X., & Kokare, C. (2017). Microbial enzymes of use in industry. In *Biotechnology of microbial enzymes* (pp. 267–298). Academic Press, Oxford.

Mabrouk, H., Mechria, H., Mechri, A., Rahali, H., Douki, W., Gaha, L., & Najjar, F. (2011, November). Butyrylcholinesterase activity in schizophrenic patients. *Annales de biologie Clinique*, *69*(6), 647–652.

Madhavan, A., Sindhu, R., Binod, P., Sukumaran, R. K., & Pandey, A. (2017). Strategies for design of improved biocatalysts for industrial applications. *Bioresource Technology*, *245*, 1304–1313.

Madhavi, V., & Lele, S. S. (2009). Laccase: Properties and applications. *BioResources*, *4*(4), 1694–1717.

McCue, J. T. (2009). Theory and use of hydrophobic interaction chromatography in protein purification applications. *Methods in Enzymology*, *463*, 405–414.

Meghwanshi, G. K., Kaur, N., Verma, S., Dabi, N. K., Vashishtha, A., Charan, P. D., & Kumar, R. (2020). Enzymes for pharmaceutical and therapeutic applications. *Biotechnology and Applied Biochemistry*, *67*(4), 586–601.

Misson, M., Zhang, H., & Jin, B. (2015). Nanobiocatalyst advancements and bioprocessing applications. *Journal of the Royal Society Interface*, *12*(102), 20140891.

Mitchell, J. B. (2017). Enzyme function and its evolution. *Current Opinion in Structural Biology*, *47*, 151–156.

NCBI Microbial Genomes, 2013. Available online: <www.ncbi.nlm.nhi.gov/genomes/microbial> (accessed 10.07.13).

Noraini, M. Y., Ong, H. C., Badrul, M. J., & Chong, W. T. (2014). A review on potential enzymatic reaction for biofuel production from algae. *Renewable and Sustainable Energy Reviews*, *39*, 24–34.

Olempska-Beer, Z. S., Merker, R. I., Ditto, M. D., & DiNovi, M. J. (2006). Food-processing enzymes from recombinant microorganisms—a review. *Regulatory Toxicology and Pharmacology*, *45*(2), 144–158.

Pandey, A., Binod, P., Ushasree, M. V., & Vidya, J. (2010). Advanced strategies for improving industrial enzymes. *Chemical Industry Digest*, *23*, 74–84.

Pandey, A., Nigam, P., Soccol, C. R., Soccol, V. T., Singh, D., & Mohan, R. (2000). Review-advances in microbial amylases. *Biotechnology and Applied Biochemistry*, *31*, 135–152.

Pant, G., Prakash, A., Pavani, J. V. P., Bera, S., Deviram, G. V. N. S., Kumar, A., . . . Prasuna, R. G. (2015). Production, optimization and partial purification of protease from Bacillus subtilis. *Journal of Taibah University for Science*, *9*(1), 50–55.

Patel, A. K., Singhania, R. R., & Pandey, A. (2017). Production, purification, and application of microbial enzymes. In *Biotechnology of microbial enzymes* (pp. 13–41). Academic Press, Oxford.

Patel, R. N. (2018). Biocatalysis for synthesis of pharmaceuticals. *Bioorganic & Medicinal Chemistry*, *26*(7), 1252–1274.

Pellis, A., Cantone, S., Ebert, C., & Gardossi, L. (2018). Evolving biocatalysis to meet bioeconomy challenges and opportunities. *New Biotechnology*, *40*, 154–169.

Prasad, S., & Roy, I. (2018). Converting enzymes into tools of industrial importance. *Recent Patents on Biotechnology*, *12*(1), 33–56.

Raveendran, S., Parameswaran, B., Beevi Ummalyma, S., Abraham, A., Kuruvilla Mathew, A., Madhavan, A., . . . Pandey, A. (2018). Applications of microbial enzymes in food industry. *Food Technology and Biotechnology*, *56*(1), 16–30.

Reetz, M. T. (2009). Directed evolution of enantioselective enzymes: An unconventional approach to asymmetric catalysis in organic chemistry. *Journal of Organic Chemistry*, *74*, 5767–5778.

Rhodes, D. G., & Laue, T. M. (2009). Determination of protein purity. *Methods in Enzymology*, *463*, 677–689.

Robayo-Torres, C. C., Opekun, A. R., Quezada-Calvillo, R., Xavier, V., Smith, E. B., Navarrete, M., & Nichols, B. L. (2009). 13C-breath tests for sucrose digestion in congenital sucrase isomaltase deficient and sacrosidase supplemented patients. *Journal of Pediatric Gastroenterology and Nutrition*, *48*(4), 412.

Rossomando, E. F. (1990). [24] Ion-exchange chromatography. *Methods in enzymology*, *182*, 309–317.

Sanchez, S., & Demain, A. L. (2017). Useful microbial enzymes—an introduction. In *Biotechnology of microbial enzymes* (pp. 1–11). Academic Press.

Shukla, A., Rana, A., Kumar, L., Singh, B., & Ghosh, D. (2009). Assessment of detergent activity of Streptococcus sp. AS02 protease isolated from soil of Sahastradhara, Doon Valley, Uttarakhand. *Asian Journal of Microbiology, Biotechnology & Environmental Sciences*, *11*, 587–591.

Siehl, D. L., Castle, L. A., Gorton, R., Chen, Y. H., Bertain, S., Cho, H. J., . . . Lassner, M. W. (2005). Evolution of a microbial acetyltransferase for modification of glyphosate: A novel tolerance strategy. *Pest Management Science: Formerly Pesticide Science*, *61*(3), 235–240.

Singh, R. S., Singh, T., & Singh, A. K. (2019). Enzymes as diagnostic tools. In *Advances in enzyme technology* (pp. 225–271). Elsevier, Oxford.

Soccol, C. R., & Larroche, C. (2005). Enzyme technology. *Asiatech, India*, 297318, Kaur P, Singh B, Ber E, Straube N, Piontek M, Satyanarayana T, Kunze.

Stellwagen, E. (2009). Gel filtration. In *Methods in enzymology* (Vol. 463, pp. 373–385). Academic Press, Oxford.

Strong, P. J., Kalyuzhnaya, M., Silverman, J., & Clarke, W. P. (2016). A methanotroph-based biorefinery: Potential scenarios for generating multiple products from a single fermentation. *Bioresource Technology*, *215*, 314–323.

Sun, H., Zhang, H., Ang, E. L., & Zhao, H. (2018). Biocatalysis for the synthesis of pharmaceuticals and pharmaceutical intermediates. *Bioorganic & Medicinal Chemistry*, *26*(7), 1275–1284.

Uchiyama, T., & Miyazaki, K. (2009). Functional metagenomics for enzyme discovery: Challenges to efficient screening. *Current Opinion in Biotechnology*, *20*(6), 616–622.

Urh, M., Simpson, D., & Zhao, K. (2009). Affinity chromatography: General methods. *Methods in Enzymology*, *463*, 417–438.

Van Den Burg, B. (2003). Extremophiles as a source for novel enzymes. *Current Opinion in Microbiology*, *6*(3), 213–218.

Vijayaraghavan, P., Raj, S. R. F., & Vincent, S. G. P. (2016). Industrial enzymes: Recovery and purification challenges. In *Agro-industrial wastes as feedstock for enzyme production* (pp. 95–110). Academic Press, Amsterdam.

Vittaladevaram, V. (2017). Fermentative production of microbial enzymes and their applications: Present status and future prospects. *Journal of Applied Biology and Biotechnology*, *5*(4), 090–094.

Yuan, L., Kurek, I., English, J., & Keenan, R. (2005). Laboratory-directed protein evolution. *Microbiology and Molecular Biology Reviews*, *69*, 373–392.

Zhao, F., Wang, Q., Dong, J., Xian, M., Yu, J., Yin, H., . . . Wang, J. (2017). Enzyme-inorganic nanoflowers/alginate microbeads: An enzyme immobilization system and its potential application. *Process Biochemistry*, *57*, 87–94.

3 Recent Advancements in Microbial Enzymes and Their Application in Bioremediation of Xenobiotic Compounds

*Saurabh Gangola, Pankaj Bhatt, Samiksha Joshi,
Saurabh Kumar, Narendra Singh Bhandari, Samarth
Terwari, Om Prakash, and Amit Kumar Mittal*

ABSTRACT

Due to anthropogenic activity, our surrounding environment is contaminated day by day, and it is causing severe health effects to humans. Different physicochemical methods are implicated in removing the xenobiotic compounds, but these are not efficient and expensive too. The application of these methods ends up with the production of secondary pollutants. Severe symptoms have been seen due to these xenobiotic compounds, such as allergy, itching, respiratory problems, digestive problems, paralysis, and cancer. For the last few years, microorganisms have been used to remove hazardous chemicals from our surrounding environment. Microorganisms work at the optimum environment and metabolize xenobiotic compounds as their source of carbon and energy. Microbes secrete extracellular enzymes, such as oxygenase, esterase, laccase, and dehydrogenase, and degrade toxic compounds to nontoxic and environmentally safe forms. Using microbial enzymes as a new biological tool for biodegradation is the current interest of the researcher. This chapter summarizes all the microbial enzymes (fungal and bacterial) and their applications in biodegradation and bioremediation processes.

CONTENTS

3.1 Introduction 42
3.2 Microbial Enzymes in Bioremediation 45
    3.2.1 Cytochrome P450 45
    3.2.2 Laccase 46
    3.2.3 Dehalogenase 47
    3.2.4 Dehydrogenase 48
    3.2.5 Hydrolase 48
    3.2.6 Protease 49
    3.2.7 Lipase 50
References 51

DOI: 10.1201/9781003202998-3

3.1 INTRODUCTION

Anthropogenic activities, such as contemporary farming techniques, industrialization, overcrowding, and unhealthy rivalry for supremacy, are wreaking havoc on the planet. These anthropogenic activities contribute to generating unprecedented amounts of pollutants, such as pesticides, polyaromatic hydrocarbons (PAHs), azo dyes, polychlorinated chemicals, phenols, heavy metals, and other pollutants. These compounds are resistant to biodegradation and can persist in the environment for a long time, causing acute and chronic harm to the ecosystem's biotic components (Gangola et al., 2018, 2021). These contaminants are already reported for their teratogenic, carcinogenic, mutagenic, and toxic effects in humans and their surrounding ecological environment (Liu et al., 2019). As a result, issues about the disposal of organic waste from the environment are crucial. Several conventional methods are being used to reduce the toxic effects of xenobiotic compounds, such as disposal of garbage in a pit and pouring it in, incineration, chemical breakdown, and ultraviolet (UV) disposal. However, due to a shortage of space, expensive costs, complicated procedures, stringent regulatory restrictions imposed on decontamination by various countries, and widespread public opposition, these physical and chemical techniques are losing ground. These procedures come with a number of drawbacks, including the production of vast amounts of sludge, which necessitates safe disposal, and the development of harmful secondary pollutants (Karigar and Rao, 2011). As a result, an eco-friendly biological method for the removal of xenobiotic compounds is required, which could replace the conventional physicochemical techniques successfully. Bioremediation is the method to remove hazardous pollutants from the environment by utilizing the biological system (bacteria and fungi) through the process of mineralization and detoxification. The use of microbial enzymes to remediate persistent organic pollutants is regarded as environmentally friendly, cost-effective, innovative, and promising (Bhatt et al., 2021a). The bioremediation procedure, on the other hand, has significant drawbacks. It's a long process, and only a few bacteria species capable of manufacturing certain enzymes have demonstrated their ability to breakdown contaminants thus far. As a result, we choose genetically engineered microbes for bioremediation since they produce a huge quantity of needed enzymes under ideal conditions (Bhatt et al., 2021b). Selecting the specific characteristics of microorganisms producing xenobiotic degrading enzymes to metabolize xenobiotic compounds as a source of their carbon and energy (Gangola et al., 2019) (**Figure 3.1**). This process converts harmful substances into harmless substances and, in some cases, unique products (Phale et al., 2019).

The metabolic potential of bacteria makes them versatile for survival throughout the biosphere. They can grow in a variety of environments and create enzymes. Enzymes synthesized by different aerobic bacteria, like *Alcaligenes*, *Bacillus*, *Pseudomonas*, *Sphingomonas*, *Mycobacterium*, and *Rhodococcus*, aid in pesticide degradation (Gangola et al., 2021). In addition, enzymes obtained from the anaerobic bacterial population helps in the breakdown of trichloroethylene, chloroform, and polychlorinated biphenyls (PCBs) (Sharma, 2012). Some common enzymes used for bioremediation include esterases, laccases, cytochrome P450, hydrolases, dehydrogenases, dehalogenases, lipases, and proteases (Table 3.1).

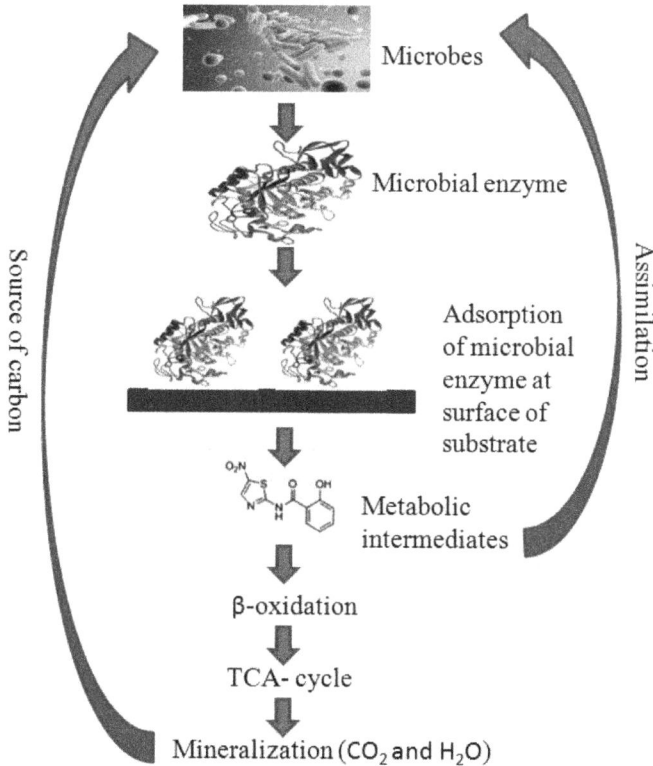

FIGURE 3.1 Complete degradation mechanism of microbial enzymes for the removal of xenobiotic compounds from the contaminated site.

TABLE 3.1
Application of Microbial Enzymes in Biodegradation of Toxic Contaminants

Microbial Enzymes	Organisms	Substrates	Applications	References
Esterase	Bacillus subtilis 1D	Cypermethrin, fipronil, imidaloprid	Metabolizing toxic pesticides into nontoxic products	Gangola et al., 2018
Laccase	Bacillus subtilis 1D	Cypermethrin, fipronil, imidaloprid, sulfosulfurone	Metabolizing toxic pesticides into nontoxic products	Gangola et al., 2018
Laccase	Pseudomonas putida F6	Dye	Degradation of synthetic dyes	(McMahon et al., 2007)
Laccase	Streptomyces cyaneus	Bisphenol A, diclofenac, mefenamic acid	Oxidation of micropollutants such as bisphenol A, diclofenac, and mefenamic acid	Margot et al., 2013

(Continued)

TABLE 3.1 (*Continued*)

Microbial Enzymes	Organisms	Substrates	Applications	References
Laccase	*Geobacillus thermocatenulatus*	Dyes (Congo red and bromophenol)	Decolorization of textile dyes, especially Congo red and bromophenol blue	Verma and Shirkot (2014)
Laccase	*Bacillus safensis*	Wastewater treatment	Bioremediation of wastewater	(Yanmış et al., 2016)
P450	*Rhodococcus rhodochrous*	hexahydro-1,3,5-trinitro-1,3,5-triazine (RDX)	Degradation of RDX	Seth-Smith et al. (2002)
P450	*Bacillus megaterium*	Polychlorinated di-benzo-p-dioxins (PCCDs)	Hydroxylation of PCCDs	Sulistyaningdyah et al. (2004)
Amylase	*Bacillus* sp.; *Geobacillus*	Starch	Starch liquefaction	Nigam (2013)
Lipase	*Bacillus subtilis*	Wastewater	Bioremediation of waste water	Mazhar et al. (2017)
Lipase	*Bacillus pumilus*	Palm oil in wastewater	Industrial wastewater treatment contaminated with palm oil	(Saranya et al., 2019)
Lipase	*Chromobacter iumviscosum*	Polybutylene succinate-co-adipate (PBSA), Poly(ε-caprolactone) (PCL), Polybutylene succinate (PBS)	Degradation of PBSA, PCL, PBS	Hoshino and Isono (2002)
Lipase	*Bacillus subtilis*	Oil or grease	Removal of trough oil or grease stains from detergent	Saraswat et al. (2017)
Lipase	*Sphingobacterium* sp. strain S2	Polylactic acid (PLA)	Degradation of PLA	Satti et al. (2019)
Dehydrogenase	*Pseudomonas putida*	Xylenol	Catabolism of 2,4-xylenol	Chen et al. (2014)
Aldehyde dehydrogenase	*Bacillus cereus* 2D	Cypermethrin, fipronil, imidaloprid	Metabolizing toxic pesticides into nontoxic products	Gangola et al. (2021)
Dehydrogenase	*S. rhizophila*	Polyvinyl alcohol	Polyvinyl alcohol degradation	Wei et al. (2018)
Protease	*Bacillus subtilis*	Casein and feather	Degradation of casein and feather	Suh and Lee (2001)
Protease	*Bacillus pumilus*	Chicken feathers	Degradation of feathers	Riffel et al. (2003)
Protease	*Streptomyces thermoviolaceus*	Fibrin, muscle, collagen, nail, and hair	Hydrolysis of fibrin, muscle, collagen, nail, and hair	Chitte et al. (1999)

Microbial Enzymes	Organisms	Substrates	Applications	References
Protease	*Termoanaerobacter*, *Keratinophilus*	Keratin fibers	Degradation of keratin fibers	Riessen and Antranikian (2001)
Dehalogenase	*Ancylobacter aquaticus*	Halogen acid ester	Degradation of halogen acid ester	Kumar et al. (2016)
Dehalogenase	*Bacillus* sp.	Tributyl phosphate	Degradation of tributyl phosphate (TBP)	Zu et al. (2012)
Dehalogenase	*Ochrobactrum*	Tetrabromobisphe-nol A	Degradation oftetrabromobisphe-nol A (TBBPA)	Liang et al. (2019)
Dehalogenase	*Pseudomonas* sp. TL	Halogen acid	Degradation of halogen acid	Liu et al. (1994)
Esterase	*Pseudomonas nitroreducens* CW7	Allethrin	Ester bond degradation	Bhatt et al. (2020)
Esterase	Indigenous bacterial cultures	Pyrethroid, carbamate, and organophosphates	Ester bond catalysis	Bhatt et al. (2021c)

3.2 MICROBIAL ENZYMES IN BIOREMEDIATION

3.2.1 CYTOCHROME P450

Cytochrome P450 is a superfamily of heme enzymes obtained from bacteria, archaea, and eukaryota (Bak et al., 2011). These enzymes perform a variety of functions like drug metabolism, conversion of toxic compounds into nontoxic ones and synthesis of complex natural products (Li et al., 2020). Xenobiotic degradation by cytochrome P450s has been reported (Anzenbacher and Anzenbacherova, 2001) via dealkyla-tions, aliphatic hydroxylations, dehalogenation, and epoxidations reactions.

To bioremediate organic contaminants and hydrocarbons, many protein engi-neering and nonengineering research on microbial P450s have been conducted. The oxidation potential of PAHs (fluoranthene, pyrene, and phenanthrene into quinones and phenols) by P450 (P450BM3) isolated from *B. megaterium* CYP102A1 have been analyzed by protein engineering studies among the known microbial P450s. Some microbial P450s have been shown to have the ability to bioremediate poly-halogenated aromatics (Carmichael and Wong, 2001). Lamb and coworkers investi-gated genetically altered *Acinetobacter calcoaceticus* strain BD413 expressing P450 enzyme (CYP105D1) isolated from *Streptomyces griseus* ATCC 13273, making microorganisms capable of surviving on herbicides, recalcitrant compounds, pes-ticides, and agrochemicals (Lamb et al., 2000). Kumar and coworkers discovered modified CYP102A1, which displayed improved activity toward PAHs, polychlori-nated biphenyls (PCBs), and linear alkanes, which are commonly utilized in toxic chemical bioremediation, gaseous alkane detoxification, and terpenes (Kumar et al., 2012). Mutants of heme monooxygenase CYP101A1 (P450cam) from *P. putida* are F87W, F98W, Y96F, and V247L. These mutants showed strong degradation activity against recalcitrant halogenated pentachlorobenzenes that are resistant to dioxygenases. Mutants are more potent to oxidize and convert polychlorinated benzenes into chlorophenol, which are then destroyed by diverse bacteria. As a result, the CYP101A1 mutations could be used to develop new polychlorinated benzene bioremediation systems (Jones et al., 2001).

Similarly, Chakraborty and Das found catabolic genes, plasmids, and genomes expressing P450s in *Rhodococcus*, *Gordonia*, *Mycobacterium*, and *Pseudomonas* for remediation of persistent organic pollutants (POPs) from the environment (Chakraborty and Das, 2016). Immobilization using Pt/TiO_2-Cu nanoparticles P450 BM3 (CYP102A1) obtained from *Bacillus* sp., engineered from *E. coli* BL21, is found to effectively degrade several organic gaseous pollutants. The YC-JY1bisdB strain was created in *E. coli* to investigate the involvement of the P450 enzyme (Jia et al., 2020). Kan et al. also discovered remediation of PAHs by CYP108J1 isolated from *Rhodococcus* sp. P14.

3.2.2 LACCASE

Laccases (benzenediol oxygen oxidoreductases) is a copper-containing extracellular enzymes which found in plants, bacteria, and fungi (Shekher et al., 2011; Gangola et al., 2022). They are made up of monomeric, dimeric, and tetrameric glycoproteins. Microbial laccase from various microorganisms is largely recognized, defined, and investigated, especially *Streptomyces* laccase from actinomycetes. *S. cyaneus*, *S. coelicolor*, *S. bikiniensis*, and *S. ipomoea* are some of these species, with *S. coelicolor* being the most well-studied (Guan et al., 2018). Laccase synthesis was boosted by lignin and phenolic substances found in agricultural wastes like rice bran, banana peel, and sawdust (Muthukumarasamy et al., 2015). Laccase catalyzes the oxidation of nonphenolic compounds (less soluble and more stable) (Gianfreda et al., 1999), phenolic compounds (ortho- and paradiphenol), aromatic amines, and their substituted forms (Chandra and Chowdhary, 2015).

Laccase is usually a thermostable enzyme with a long half-life, as observed in CotA from *B. subtilis* at 75°C (Chandra and Chowdhary, 2015). Laccase can degrade xenobiotics and create polymeric compounds that can be employed in bioremediation. Polycyclic aromatic hydrocarbons (PAHs), the most well-known contaminant, are dispersed equally in nature and consist of a linearly structured benzene ring (Li et al., 2010). Because of their carcinogenicity, persistence, toxicity, and mutagenicity in nature, such pollutants, as well as their intermediates, are of major environmental concern. They are created as a result of the incomplete combustion of industrial wastes and fossil fuels. These aromatic hydrocarbons are xenobiotic due to their low water solubility and slow breakdown rate (Ihssen et al., 2015). Laccase transforms PAHs to their quinone form, which it then degrades further to carbon dioxide. However, when used with the most powerful laccase mediator, 1-hydroxybenzene triazole (HBT), it changes 1,8-naphthalic acid and acenaphthylene to 1,2-acenapthalenedione (Madhavi and Lele, 2009). Efficient degradation and detoxification of chlorolignin, distillery effluents, postmethanated distillery effluents, and dyes have been reported for the laccase enzyme (Sondhi et al., 2015). Decolorization of textile effluents has been reported using recombinant laccase enzyme (CotA) from *E. coli*.

Purified and crude recombinant CotA laccase decolorized seven structurally distinct dyes with equal efficiency. Purified and crude CotA laccase decolorization rates were greater when simulated textile effluents (STE) was buffered at neutral pH (Wang and Zhao, 2017).

Similarly, Lu et al. showed that alkaline laccase generated from a recombinant strain of *Bacillus licheniformis* degraded synthetic colors like carmine and reactive black completely in one hour. The pure recombinant laccase showed effective decolorization of 93% of colours in four hours at pH 9 (Lu et al., 2013). Genetically modified laccase obtained by *Bacillus vallismortis* strain fmb103 has shown bioremediation of aquaculture wastewater (Sun et al., 2017). Laccase CopA obtained from soil bacterium *Pseudomonas putida* F6 showed effective degradation of Remazol Brilliant Blue, bromocresol purple, amido black 10B, reactive black 5, and Evans blue (Mandic et al., 2019). Further, more laccase from *Bacillus* sp. converted bisphenol A to 4-ethyl-2-methoxy phenol (Rajeswari and Bhuvaneswari, 2016). Similarly, decolorization of reactive brilliant orange K-7R, reactive deep blue, and reactive azo dyes was displayed within one hour by laccase Lac15 obtained from a marine microbial metagenome (Guan et al., 2014). As a result, laccase has a lot of promise for treating wastewater, including phenolic and nonphenolic chemicals, PHAs, synthetic dyes, pesticides, and other developing contaminants.

3.2.3 DEHALOGENASE

Because of its important role in the bioremediation of halogenated organic chemicals, microbial dehalogenase is of immense use. The breakdown of C-X bonds by dehalogenase enzyme [89, 90] can occur via reductive, hydrolytic, and oxygenolytic reactions (Allpress and Gowland, 1998).

Zu et al. identified a pure *Bacillus* sp. GZT strain from waste-recycling site sludge that has a high capacity to mineralize and debrominate TBP at the same time. Debromination can take place in two methods, with reductive bromination being the more common and methyl bromination being the less common (Zu et al., 2012). Liang et al. proved that dehalogenase activity showing debromination activity is an intracellular enzyme. Similarly, the unique bacterial strain *Ochrobactrum* sp. developed a recombinant strain *E. coli* BL21 when cloned with tbbpaA (DE3). TBBPA degrading ability of purified dehalogenase enzyme from recombinant strain was found to be high (78%) [31]. Bioremediation of haloalkane was reported by haloalkane dehydrogenase (Cairns et al., 1996) of *Rhizobium* sp. and halohydrin dehydrogenase of *Pseudomonas umsongensis* (Xue et al., 2018).

Ancylobacter aquaticus strain UV5 (Kumar et al., 2016), *Rhizobium* sp., *Pseudomonas* sp. (Liu et al., 1994), and *Pseudomonas umsongensis* YCIT1612 (Xue et al., 2018) produced enzymes degrading halogenated compounds. Fricker et al. employed *Dehalococcoides mccartyi* to dechlorinate the toxic pentachlorophenol into 3,5-dichlorophenol (DCP) (Fricker et al., 2014). Nelson et al. showed three *Dehalobacter* sp. strains (14DCB1, 13DCB1 and 12DCB1) to dehalogenate tetrachloroethene, dichlorotoluenes, and chlorobenzenes. All strains dehalogenate 3,4-DCT to mono chlorotoluene, with 14DCB1 being the only one to dehalogenate parasubstituted chlorines. *Dehalobacter* spp. are flexible dehalogenators, according to these findings (Nelson et al., 2014). According to Zhang et al., dehalogenases 1 and 2 were discovered in the marine bacteria *Pseudomonas stutzeri* DEH130, both enzymes were tested for dehalogenase activity, and dehalogenase 2 was more active toward the substrate L-2-CPA. Compared to dehalogenase 1, dehalogenase 2 was more stereospecific for halogenated substrate (Zhang et al., 2013). Boyer et al. identified *Desulfitobacterium frappieri* PCP-1 2,4,6-trichlorophenol reductive dehalogenase, which dechlorinated pentachlorophenol (PCP) into 3-chlorophenol at the ortho, meta, and para locations. Reductive dehalogenation transforms these environmentally harmful halogenated chemicals into less poisonous and easily biodegradable forms (Boyer et al., 2003).

3.2.4 DEHYDROGENASE

Dehydrogenases are oxidoreductases found in a variety of organisms. Microbial alcohol dehydrogenases convert alcohol to aldehyde or ketone in the presence of NAD^+/$NAD(P)^+$, heme, pyrroloquinoline quinine, or cofactor F420, whereas aldehyde dehydrogenase converts aldehyde to a carboxylic acid in the presence of $NAD(P)^+$ (Sophos and Vasiliou, 2003).

Polyethylene glycol dehydrogenase isolated from bacterial cell-free extract was detected to degrade polyethylene glycol of different molecular weights (Kawai and Yamanaka, 1989). Several reports showed the activity of dehydrogenases obtained from *Sphingomonas* sp. for the breakdown of polyethylene glycol (Sugimoto et al., 2001). Polypropylene glycol dehydrogenase found in the membrane or periplasm of *Stenotrophomonas maltophilia* is an efficient degrader of high molecular weight compounds, whereas when present in the cytoplasm, it metabolizes low-molecular-weight PPG (Tachibana et al., 2002). In addition, water-soluble polyvinyl alcohol is degraded by genetically altered polyvinyl alcohol dehydrogenases. Glycols like 2,4-pentanediol, 1,3-butane/cyclohexanediol, and polypropylene glycols are oxidized by this enzyme, but not primary or secondary alcohols (Hirota-Mamoto et al., 2006).

In the metabolism of aromatic chemicals, aldehyde dehydrogenase is discovered to be active. *Rhodococcus* sp. NCIMB12038 uses naphthalene as its main carbon source; heterologous production of the NCIMB12038 cis-naphthalene dihydrodiol dehydrogenase showed 39% similarity with *Pseudomonas putida* G7 cis-naphthalene dihydrodiol dehydrogenase (Kulakov et al., 2000). This enzyme displayed strong activity with variety of substrates like cis-2,3-dihydro-2,3-dihydroxybiphenyl (2,3-DDB) and cis-1,2-dihydro-1,2-dihydroxy-naphthalene (1,2-DDN). In the investigation of the metabolic intermediate (Ji et al., 2020), it was discovered that the aldehyde dehydrogenase (NidD) catalyzes the breakdown of 1-hydroxy-2-naphthaldehyde to 1-hydroxy-2-naphthoic.

3.2.5 HYDROLASE

Hydrolases use water to break chemical bonds and transform larger molecules into smaller ones, reducing the toxicity of contaminants. They aid in the cleavage of C-O, C-C, C-N, S-N, S-P, S-S, C-P, and other bonds by water, as well as catalyze a variety of related reactions, such as condensations and alcoholics. The key advantages of this enzyme class are their widespread availability, low cost, low environmental impact, absence of cofactor stereoselectivity, and tolerance for the addition of water-miscible solvents (Karigar and Rao, 2011).

Numerous hydrolytic enzymes of microbial origin, such as esterases, lipases, amylases, nitrilases, cellulases, peroxidases, proteases, and cutinases, are used to remediate oil-contaminated sites, plastics, food waste, pesticides, and biofilm deposits. Hydrolytic enzymes have a wide range of possible applications in feed additives, biological sciences, and the chemical industry (Kumar and Sharma, 2019). Esterases, amidases, and proteases may break down ester, amide, and peptide bonds, resulting in compounds with low or no toxicity. Carbamate or parathion hydrolase from *Achromobacter, Pseudomonas, Flavobacterium, Nocardia*, and *Bacillus cereus* has been effectively employed in the hydrolysis of pollutants such as carbofuran, carbaryl or parathion, diazinon, cypermethrin, fipronil, sulfosulfuron, imidacloprid, and coumaphos (Sutherland et al., 2002; Gangola et al., 2021).

Organophosphate (OP) chemicals are neurotoxins with high lethality. They are commonly used as pesticides in agriculture, and they pose a hazard to the biotic environment. The hydrolysis of phosphodiester links can detoxify organophosphate pesticides, whereas the hydrolysis of carboxyl ester bonds can detoxify pyrethroids and malathion. Organophosphate pesticides often contain phosphorus in the form of a phosphate ester or a phosphonate. The crucial step in detoxification is hydrolysis of the P-O-aryl and P-O-alkyl bonds (Singh, 2014). Drinking water, groundwater, fruits, and grains are contaminated with organophosphates, causing toxicity (Kapoor and Rajagopal, 2011). Microorganisms that degrade malathion include *Bacillus cereus, Alicyclobacillus tengchogenesis, Bacillus licheniformis*, and *Brevibacillus* sp. (Littlechild et al., 2015). *B. licheniformis* synthesizes hydrolytic enzymes for using malathion as a carbon source and further helps in its remediation (Xie et al., 2013). Enzymes like organophosphate acid anhydrolases, methyl parathion hydrolases (MPH), and organophosphate hydrolases are reported as efficient degraders of organophosphate contaminated sites (Schenk et al., 2016). The organophosphate hydrolase gene was isolated from *Pseudomonas diminuta* and *E. coli* and demonstrated its detoxifying capabilities, which degrade methyl parathion 10–80% and 3.6–45%, respectively (Kapoor and Rajagopal, 2011). Su et al. found that the organophosphate hydrolase enzyme associated with outer membrane vesicles of gram-negative bacteria showed a high-rate breakdown of parathion compounds (Su et al., 2017).

Cutinase is a biocatalyst for the breakdown of plastics and polycaprolactone that was isolated from the bacterium *Fusarium solani f. pisi* (Singh et al., 2016). Dang et al. reported a high-rate degradation of oxo-biodegradable and biodegradable plastics by hydrolase enzymes secreted by *Bacillus* sp. BCBT21 isolated from composting agricultural residuals (Dang et al., 2018). Three strains of *Pseudomonas* sp. (i.e., PKDM2, PKDE1, and PKDE2) were found as potential degraders of di-(2-ethylhexyl phthalate) (DEHP) into phthalic acid.

3.2.6 PROTEASE

Proteases belong to the hydrolase family and are commonly produced by *Amycolatopsis* sp., *Aspergillus* sp., and *Bacillus* sp. Microbial proteases are important due to their high production, activity, and reduced cost. The total production of microbial proteases is around two-thirds of commercial proteases. Proteases are employed in the degradation of different polymers by breaking lipase γ-ω-bonds, α-ester linkages and poly-hydroxybutyrate (PHB) depolymerase β-ester bonds (Haider et al., 2019).

Animal molting and shedding, horns, nails, and poultry feces are resistant to breakdown due to the presence of insoluble keratin protein. Along with their unpleasant odor, they are responsible for environmental degradation. Keratinase is a protease enzyme that degrades keratin proteins and can be used in the bioremediation of poultry wastes. Keratinase enzyme developed from *Stenotrophomonas maltophilia* KB13 and *Bacillus* sp. FPF-1 has revealed a considerable breakdown activity of chicken feathers (Bhange et al., 2016), indicating its capacity to degrade recalcitrant keratinous waste biomass from the agriculture sector (Nnolim et al., 2020). Disulfide reductase and serine protease from *Stenotrophomonas* sp. worked together to successfully degrade keratin, resulting in a 50-fold increase in keratinolytic activity over protease alone. The breaking of disulfide bonds in the keratin protein is catalyzed by the disulfide reductase (Yamamura et al., 2002). Keratinase enzymes synthesized by *Pseudomonas* sp. and *Bacillus* sp. hydrolyzes keratinous wastes obtained from different origins (Mazotto et al., 2011). Keratinase is employed in leather industries for the most crucial dehairing step to replace the chemical treatment of Na_2S and CaO (Akhter et al., 2020).

In addition, protease enzymes are also employed in the bioremediation of marine crustacean wastes. *Bacillus licheniformis* MP1 produces an alkaline protease that deproteinizes shrimp waste to a 75% level (Jellouli et al., 2011). Various reports showed 84% deproteinization of crab shell proteins by proteases of *S. marcescens* FS-3, 72% of shrimp and crab shell powder, 78% of natural shrimp shells, and 45% of acid-treated shrimp and crab shell powder by proteases of *P. aeruginosa* K-187 (Oh et al., 2000). *Bacillus subtilis* producing chitinase enzymes (BsChi) effectively degraded chitin of crab shell wastes to N-acetyl-D-glucosamine (Wang et al., 2018).

Proteases produced by *Pseudomonas chlororaphis* and *Pseudomonas fluorescens* displayed potent degradation of Impranil substrate (water-dispersible polyurethane) (Howard and Blake, 1998). Proteases help to clean up the environment by decomposing and turning keratinous wastes and waste from marine crustaceans into beneficial compounds. Similarly, the enzyme's action has replaced the usage of harmful chemicals, lowering the amount of waste released into the environment. In the bioremediation of a polluted environment, protease plays a critical function.

3.2.7 LIPASE

Lipases are biocatalysts known for the conversion of triglycerides into glycerol and fatty acids (Casas-Godoy et al., 2012). They belong to the serine hydrolases family. Lipase destroys lipids generated from microbes, animals, and plants, decreasing hydrocarbon levels in contaminated soil (Karigar and Rao, 2011). Microbial lipase has numerous applications, such as bioremediation of petroleum compounds, greasy effluents, therapeutics, oil residues, and contaminants from pulp and paper and cosmetic industries.

Lipases from *Acinetobacter* sp., *Pseudomonas* sp., *Pseudomonas aeruginosa*, *Rhodococcus* sp., and *Mycobacterium* sp. have been used in the remediation of oil spills, soil contaminated with industrial waste oil, crude-oil-contaminated waste-water, and castor oil (Verma et al., 2012). One of the primary environmental challenges is that mineral oil hydrocarbons generated from petroleum products pollute the soil. Lipase, synthesized by bacterial isolates from soil contaminated by automotive engine oil, may breakdown the hydrocarbons that are the primary soil pollutants (Mahmood et al., 2017). Lipases are also used in domestic laundry to remove difficult oil and grease stains and reduce pollution. In detergent formulations, +e crude lipase from *Bacillus subtilis* strain is used to reduce phosphate-based chemicals (Saraswat et al., 2017).

Lipase produced by *Pseudomonas* carries out transesterification reaction and breakdown of para hydroxybenzoic acid esters in activated sludge when ethanol or methanol is present as the source of carbon. These findings suggested that transesterification could be a key mechanism for paraben pollution reduction in aquatic environments (Wang et al., 2018). Biodegradable polymers have been used as one of the techniques to solve the issue of environmental contamination, which is growing in popularity. Lipase PL from *Alcaligenes* sp. catalyzes the hydrolysis of PLA polymers into oligomers, then monomers (Hoshino and Isono, 2002). On the other hand, *Pseudomonas* lipase was able to degrade the copolymers (polycaprolactone/polylactic acid) into a variety of soluble degradation products. Similarly, the remediation of polycaprolactone (PCL) by lipases synthesized from *Lactobacillus plantarum*, *Lactobacillus brevis*, and their coculture, which showed that the degradation ability of lipases from *L. plantarum*, was highest (Wang et al., 2019).

Conclusion: Increasing contamination in our surrounding environment is a serious concern. The use of a microbial system is a recent and advanced technique to reduce the pollution level. The microbial system secreted stress-responsive extracellular enzymes (esterases, laccases, dehydrogenases, dehalogenases, peroxidases, lipases, proteases) that help to degrade the pollutant from the environment. Microorganisms utilize pollutants as a source of carbon and energy. These enzymes mineralize the pollutants into a nontoxic form or an environmentally safe form. The use of degradative enzymes from the microbial system is a cheap and eco-friendly approach for the remediation of contaminated sites.

REFERENCES

Akhter, M., Wal Marzan, L., Akter, Y., & Shimizu, K. (2020). Microbial bioremediation of feather waste for keratinase production: An outstanding solution for leather dehairing in tanneries. *Microbiology Insights*, *13*, 1178636120913280.

Allpress, J. D., & Gowland, P. C. (1998). Dehalogenases: Environmental defence mechanism and model of enzyme evolution. *Biochemical Education*, *26*(4), 267–276.

Anzenbacher, P., & Anzenbacherova, E. (2001). Cytochromes P450 and metabolism of xenobiotics. *Cellular and Molecular Life Sciences CMLS*, *58*(5), 737–747.

Bak, S., Beisson, F., Bishop, G., Hamberger, B., Höfer, R., Paquette, S., & Werck-Reichhart, D. (2011). Cytochromes P450. *The Arabidopsis Book/American Society of Plant Biologists*, *9*.

Bhange, K., Chaturvedi, V., & Bhatt, R. (2016). Feather degradation potential of *Stenotropho-monas maltophilia* KB13 and feather protein hydrolysate (FPH) mediated reduction of hexavalent chromium. *3 Biotech, 6*(1), 42.

Bhatt, P., Gangola, S., Bhandari, G., Zhang, W., Maithani, D., Mishra, S., & Chen, S. (2021b). New insights into the degradation of synthetic pollutants in contaminated environments. *Chemosphere, 268*, 128827.

Bhatt, P., Rene, E. R., Kumar, A. J., Gangola, S., Kumar, G., Sharma, A., . . . Chen, S. (2021a). Fipronil degradation kinetics and resource recovery potential of *Bacillus* sp. strain FA4 isolated from a contaminated agricultural field in Uttarakhand, India. *Chemosphere, 276*, 130156.

Bhatt, P., Rene, E. R., Kumar, A. J., Zhang, W., & Chen, S. (2020) Binding interaction of allethrin with esterase: Bioremediation potential and mechanism. *Bioresource Technology, 315*, 123845.

Bhatt, P., Zhou, X., Huang, Y., Zhang, W., & Chen, S. (2021c). Characterization of the role of esterases in the biodegradation of organophosphate, carbamate, and pyrethroid pesticides. *Journal of Hazardous Materials, 411*, 125026.

Boyer, A., Pagé-BéLanger, R., Saucier, M., Villemur, R., Lépine, F., Juteau, P., & Beaudet, R. (2003). Purification, cloning and sequencing of an enzyme mediating the reductive dechlorination of 2, 4, 6-trichlorophenol from *Desulfitobacterium frappieri* PCP-1. *Biochemical Journal, 373*(1), 297–303.

Cairns, S. S., Cornish, A., & Cooper, R. A. (1996). Cloning, sequencing and expression in *Escherichia coli* of two *Rhizobium* sp. genes encoding haloalkanoate dehalogenases of opposite stereospecificity. *European Journal of Biochemistry, 235*(3), 744–749.

Carmichael, A. B., & Wong, L. L. (2001). Protein engineering of *Bacillus megaterium* CYP102: The oxidation of polycyclic aromatic hydrocarbons. *European Journal of Biochemistry, 268*(10), 3117–3125.

Casas-Godoy, L., Duquesne, S., Bordes, F., Sandoval, G., & Marty, A. (2012). Lipases: An overview. *Lipases and Phospholipases*, 3–30.

Chakraborty, J., & Das, S. (2016). Molecular perspectives and recent advances in microbial remediation of persistent organic pollutants. *Environmental Science and Pollution Research, 23*(17), 16883–16903.

Chandra, R., & Chowdhary, P. (2015). Properties of bacterial laccases and their application in bioremediation of industrial wastes. *Environmental Science: Processes & Impacts, 17*(2), 326–342.

Chen, Y. F., Chao, H., & Zhou, N. Y. (2014). The catabolism of 2, 4-xylenol and p-cresol share the enzymes for the oxidation of para-methyl group in *Pseudomonas putida* NCIMB 9866. *Applied Microbiology and Biotechnology, 98*(3), 1349–1356.

Chitte, R. R., Nalawade, V. K., & Dey, S. (1999). Keratinolytic activity from the broth of a feather-degrading thermophilic *Streptomyces thermoviolaceus* strain SD8. *Letters in Applied Microbiology, 28*(2), 131–136.

Dang, T. C. H., Nguyen, D. T., Thai, H., Nguyen, T. C., Tran, T. T. H., Le, V. H., . . . Nguyen, Q. T. (2018). Plastic degradation by thermophilic *Bacillus* sp. BCBT21 isolated from composting agricultural residual in Vietnam. *Advances in Natural Sciences: Nanoscience and Nanotechnology, 9*(1), 015014.

Fricker, A. D., LaRoe, S. L., Shea, M. E., & Bedard, D. L. (2014). Dehalococcoides mccartyi strain JNA dechlorinates multiple chlorinated phenols including pentachlorophenol and harbors at least 19 reductive dehalogenase homologous genes. *Environmental Science & Technology, 48*(24), 14300–14308.

Gangola, S., Joshi, S., Kumar, S., & Pandey, S. C. (2019). Comparative analysis of fungal and bacterial enzymes in biodegradation of xenobiotic compounds. In *Smart Bioremediation Technologies* (pp. 169–189). Academic Press, Cambridge.

Gangola, S., Joshi, S., Kumar, S., Sharma, B., & Sharma, A. (2021). Differential proteomic analysis under pesticides stress and normal conditions in *Bacillus cereus* 2D. *Plos One, 16*(8), e0253106.

Gangola, S., Sharma, A., Bhatt, P., Khati, P., & Chaudhary, P. (2018). Presence of esterase and laccase in *Bacillus subtilis* facilitates biodegradation and detoxification of cypermethrin. *Scientific Reports*, *8*(1), 1–11.

Gangola, S., Sharma, A., Joshi, S., Bhandari, G., Prakash, O., Govarthanan, M., . . . Bhatt, P. (2022). Novel mechanism and degradation kinetics of pesticides mixture using Bacillus sp. strain 3C in contaminated sites. *Pesticide Biochemistry and Physiology*, *181*, 104996.

Gianfreda, L., Xu, F., & Bollag, J. M. (1999). Laccases: A useful group of oxidoreductive enzymes. *Bioremediation Journal*, *3*(1), 1–26.

Guan, Z. B., Luo, Q., Wang, H. R., Chen, Y., & Liao, X. R. (2018). Bacterial laccases: Promising biological green tools for industrial applications. *Cellular and Molecular Life Sciences*, *75*(19), 3569–3592.

Guan, Z. B., Zhang, N., Song, C. M., Zhou, W., Zhou, L. X., Zhao, H., . . . Liao, X. R. (2014). Molecular cloning, characterization, and dye-decolorizing ability of a temperature-and pH-stable laccase from *Bacillus subtilis* X1. *Applied Biochemistry and Biotechnology*, *172*(3), 1147–1157.

Haider, T. P., Völker, C., Kramm, J., Landfester, K., & Wurm, F. R. (2019). Plastics of the future? The impact of biodegradable polymers on the environment and on society. *Angewandte Chemie International Edition*, *58*(1), 50–62.

Hirota-Mamoto, R., Nagai, R., Tachibana, S., Yasuda, M., Tani, A., Kimbara, K., & Kawai, F. (2006). Cloning and expression of the gene for periplasmic poly (vinyl alcohol) dehydrogenase from *Sphingomonas* sp. strain 113P3, a novel-type quinohaemoprotein alcohol dehydrogenase. *Microbiology*, *152*(7), 1941–1949.

Hoshino, A., & Isono, Y. (2002). Degradation of aliphatic polyester films by commercially available lipases with special reference to rapid and complete degradation of poly (L-lactide) film by lipase PL derived from *Alcaligenes* sp. *Biodegradation*, *13*(2), 141–147.

Howard, G. T., & Blake, R. C. (1998). Growth of pseudomonas fluorescens on a polyester—polyurethane and the purification and characterization of a polyurethanase—protease enzyme. *International Biodeterioration &Biodegradation*, *42*(4), 213–220.

Ihssen, J., Reiss, R., Luchsinger, R., Thöny-Meyer, L., & Richter, M. (2015). Biochemical properties and yields of diverse bacterial laccase-like multicopper oxidases expressed in Escherichia coli. *Scientific Reports*, *5*(1), 1–13.

Jellouli, K., Ghorbel-Bellaaj, O., Ayed, H. B., Manni, L., Agrebi, R., & Nasri, M. (2011). Alkaline-protease from *Bacillus licheniformis* MP1: Purification, characterization and potential application as a detergent additive and for shrimp waste deproteinization. *Process Biochemistry*, *46*(6), 1248–1256.

Ji, D., Mao, Z., He, J., Peng, S., & Wen, H. (2020). Characterization and genomic function analysis of phenanthrene-degrading bacterium *Pseudomonas* sp. Lphe-2. *Journal of Environmental Science and Health, Part A*, *55*(5), 549–562.

Jia, Y., Eltoukhy, A., Wang, J., Li, X., Hlaing, T. S., Aung, M. M., . . . Yan, Y. (2020). Biodegradation of bisphenol A by *Sphingobium* sp. YC-JY1 and the essential role of cytochrome P450 monooxygenase. *International Journal of Molecular Sciences*, *21*(10), 3588.

Jones, J. P., O'Hare, E. J., & Wong, L. L. (2001). Oxidation of polychlorinated benzenes by genetically engineered CYP101 (cytochrome P450cam). *European Journal of Biochemistry*, *268*(5), 1460–1467.

Kapoor, M., & Rajagopal, R. (2011). Enzymatic bioremediation of organophosphorus insecticides by recombinant organophosphorous hydrolase. *International Biodeterioration & Biodegradation*, *65*(6), 896–901.

Karigar, C. S., & Rao, S. S. (2011). Role of microbial enzymes in the bioremediation of pollutants: A review. *Enzyme Research*, *2011*.

Kawai, F., & Yamanaka, H. (1989). Inducible or constitutive polyethylene glycol dehydrogenase involved in the aerobic metabolism of polyethylene glycol. *Journal of Fermentation and Bioengineering*, *67*(4), 300–302.

Kulakov, L. A., Allen, C. C., Lipscomb, D. A., & Larkin, M. J. (2000). Cloning and characterization of a novel cis-naphthalene dihydrodiol dehydrogenase gene (narB) from *Rhodococcus* sp. NCIMB12038. *FEMS Microbiology Letters*, *182*(2), 327–331.

Kumar, A., Pillay, B., & Olaniran, A. O. (2016). L-2-haloacid dehalogenase from *Ancylobacter
aquaticus* UV5: Sequence determination and structure prediction. *International Journal
of Biological Macromolecules*, *83*, 216–225.

Kumar, A., & Sharma, S. (2019). *Microbes and Enzymes in Soil Health and Bioremediation* (pp. 353–366). Springer.

Kumar, S., Jin, M., & Weemhoff, J. L. (2012). Cytochrome P450-mediated phytoremediation
using transgenic plants: A need for engineered cytochrome P450 enzymes. *Journal of
Petroleum & Environmental Biotechnology*, *3*(5).

Lamb, D. C., Kelly, D. E., Masaphy, S., Jones, G. L., & Kelly, S. L. (2000). Engineering of
heterologous cytochrome P450 in *Acinetobacter* sp.: Application for pollutant degradation. *Biochemical and Biophysical Research Communications*, *276*(2), 797–802.

Li, X., Lin, X., Zhang, J., Wu, Y., Yin, R., Feng, Y., & Wang, Y. (2010). Degradation of polycyclic aromatic hydrocarbons by crude extracts from spent mushroom substrate and its
possible mechanisms. *Current Microbiology*, *60*(5), 336–342.

Li, Z., Jiang, Y., Guengerich, F. P., Ma, L., Li, S., & Zhang, W. (2020). Engineering cytochrome
P450 enzyme systems for biomedical and biotechnological applications. *Journal of Biological Chemistry*, *295*(3), 833–849.

Liang, Z., Li, G., Mai, B., Ma, H., & An, T. (2019). Application of a novel gene encoding bromophenol dehalogenase from *Ochrobactrum* sp. T in TBBPA degradation. *Chemosphere*, *217*, 507–515.

Littlechild, J. A. (2015). Archaeal enzymes and applications in industrial biocatalysts.
Archaea, *2015*.

Liu, J. Q., Kurihara, T., Hasan, A. K., Nardi-Dei, V., Koshikawa, H., Esaki, N., & Soda,
K. (1994). Purification and characterization of thermostable and nonthermostable
2-haloacid dehalogenases with different stereospecificities from *Pseudomonas* sp. strain
YL. *Applied and Environmental Microbiology*, *60*(7), 2389–2393.

Liu, L., Bilal, M., Duan, X., & Iqbal, H. M. (2019). Mitigation of environmental pollution by
genetically engineered bacteria—Current challenges and future perspectives. *Science of
the Total Environment*, *667*, 444–454.

Lu, L., Wang, T. N., Xu, T. F., Wang, J. Y., Wang, C. L., & Zhao, M. (2013). Cloning and
expression of thermo-alkali-stable laccase of *Bacillus licheniformis* in Pichia pastoris
and its characterization. *Bioresource Technology*, *134*, 81–86.

Madhavi, V., & Lele, S. S. (2009). Laccase: Properties and applications. *BioResources*, *4*(4),
1694–1717.

Mahmood, M. H., Yang, Z., Thanoon, R. D., Makky, E. A., & Rahim, M. H. A. (2017). Lipase
production and optimization from bioremediation of disposed engine oil. *Journal of
Chemical and Pharmaceutical Research*, *9*(6), 26–36.

Mandic, M., Djokic, L., Nikolaivits, E., Prodanovic, R., O'Connor, K., Jeremic, S., . . . Nikodinovic-Runic, J. (2019). Identification and characterization of new laccase biocatalysts
from *Pseudomonas* species suitable for degradation of synthetic textile dyes. *Catalysts*, *9*(7), 629.

Margot, J., Bennati-Granier, C., Maillard, J., Blánquez, P., Barry, D. A., & Holliger, C.
(2013). Bacterial versus fungal laccase: Potential for micropollutant degradation. *AMB
Express*, *3*(1), 1–14.

Mazhar, H., Abbas, N., Ali, S., Sohail, A., Hussain, Z., & Ali, S. S. (2017). Optimized production of lipase from *Bacillus subtilis* PCSIRNL-39. *African Journal of Biotechnology*, *16*(19), 1106–1115.

Mazotto, A. M., de Melo, A. C. N., Macrae, A., Rosado, A. S., Peixoto, R., Cedrola, S. M., . . .
Vermelho, A. B. (2011). Biodegradation of feather waste by extracellular keratinases and

gelatinases from *Bacillus* spp. *World Journal of Microbiology and Biotechnology, 27*(6), 1355–1365.

McMahon, A. M., Doyle, E. M., Brooks, S., & O'Connor, K. E. (2007). Biochemical characterisation of the coexisting tyrosinase and laccase in the soil bacterium *Pseudomonas putida* F6. *Enzyme and Microbial Technology, 40*(5), 1435–1441.

Muthukumarasamy, N. P., Jackson, B., Joseph Raj, A., & Sevanan, M. (2015). Production of extracellular laccase from *Bacillus subtilis* MTCC 2414 using agroresidues as a potential substrate. *Biochemistry Research International, 2015*.

Nelson, J. L., Jiang, J., & Zinder, S. H. (2014). Dehalogenation of chlorobenzenes, dichlorotoluenes, and tetrachloroethene by three *Dehalobacter* spp. *Environmental Science & Technology, 48*(7), 3776–3782.

Nigam, P. S. (2013). Microbial enzymes with special characteristics for biotechnological applications. *Biomolecules, 3*(3), 597–611.

Nnolim, N. E., Okoh, A. I., & Nwodo, U. U. (2020). *Bacillus* sp. FPF-1 produced keratinase with high potential for chicken feather degradation. *Molecules, 25*(7), 1505.

Oh, Y. S., Shih, L., Tzeng, Y. M., & Wang, S. L. (2000). Protease produced by *Pseudomonas aeruginosa* K-187 and its application in the deproteinization of shrimp and crab shell wastes. *Enzyme and Microbial Technology, 27*(1–2), 3–10.

Phale, P. S., Sharma, A., & Gautam, K. (2019). Microbial degradation of xenobiotics like aromatic pollutants from the terrestrial environments. In *Pharmaceuticals and Personal Care Products: Waste Management and Treatment Technology* (pp. 259–278). Butterworth-Heinemann, United Kingdom, Elsevier.

Rajeswari, M., & Bhuvaneswari, V. (2016). Production of extracellular laccase from the newly isolated *Bacillus* sp. PK4. *African Journal of Biotechnology, 15*(34), 1813–1826.

Riessen, S., & Antranikian, G. (2001). Isolation of *Thermoanaerobacter keratinophilus* sp. nov., a novel thermophilic, anaerobic bacterium with keratinolytic activity. *Extremophiles, 5*(6), 399–408.

Riffel, A., Lucas, F., Heeb, P., & Brandelli, A. (2003). Characterization of a new keratinolytic bacterium that completely degrades native feather keratin. *Archives of Microbiology, 179*(4), 258–265.

Saranya, P., Selvi, P. K., & Sekaran, G. (2019). Integrated thermophilic enzyme-immobilized reactor and high-rate biological reactors for treatment of palm oil-containing wastewater without sludge production. *Bioprocess and Biosystems Engineering, 42*(6), 1053–1064.

Saraswat, R., Verma, V., Sistla, S., & Bhushan, I. (2017). Evaluation of alkali and thermotolerant lipase from an indigenous isolated *Bacillus* strain for detergent formulation. *Electronic Journal of Biotechnology, 30*, 33–38.

Satti, S. M., Abbasi, A. M., Marsh, T. L., Auras, R., Hasan, F., Badshah, M., . . . Shah, A. A. (2019). Statistical optimization of lipase production from *Sphingobacterium* sp. strain S2 and evaluation of enzymatic depolymerization of poly (lactic acid) at mesophilic temperature. *Polymer Degradation and Stability, 160*, 1–13.

Schenk, G., Mateen, I., Ng, T. K., Pedroso, M. M., Mitić, N., Jafelicci Jr, M., . . . Ollis, D. L. (2016). Organophosphate-degrading metallohydrolases: Structure and function of potent catalysts for applications in bioremediation. *Coordination Chemistry Reviews, 317*, 122–131.

Seth-Smith, H. M., Rosser, S. J., Basran, A., Travis, E. R., Dabbs, E. R., Nicklin, S., & Bruce, N. C. (2002). Cloning, sequencing, and characterization of the hexahydro-1, 3, 5-trinitro-1, 3, 5-triazine degradation gene cluster from *Rhodococcus rhodochrous*. *Applied and Environmental Microbiology, 68*(10), 4764–4771.

Sharma, S. (2012). Bioremediation: Features, strategies and applications. *Asian Journal of Pharmacy and Life Science*. ISSN, 2231, 4423.

Shekher, R., Sehgal, S., Kamthania, M., & Kumar, A. (2011). Laccase: Microbial sources, production, purification, and potential biotechnological applications. *Enzyme Research, 2011*.

Singh, B. (2014). Review on microbial carboxylesterase: General properties and role in organophosphate pesticides degradation. *Biochemistry and Molecular Biology, 2*, 1–6.

Singh, R., Kumar, M., Mittal, A., & Mehta, P. K. (2016). Microbial enzymes: Industrial progress in 21st century. *3 Biotech, 6*(2), 1–15.

Sondhi, S., Sharma, P., George, N., Chauhan, P. S., Puri, N., & Gupta, N. (2015). An extracellular thermo-alkali-stable laccase from *Bacillus tequilensis* SN4, with a potential to biobleach softwood pulp. *3 Biotech, 5*(2), 175–185.

Sophos, N. A., & Vasiliou, V. (2003). Aldehyde dehydrogenase gene superfamily: The 2002 update. *Chemico-Biological Interactions, 143*, 5–22.

Su, F. H., Tabañag, I. D. F., Wu, C. Y., & Tsai, S. L. (2017). Decorating outer membrane vesicles with organophosphorus hydrolase and cellulose binding domain for organophosphate pesticide degradation. *Chemical Engineering Journal, 308*, 1–7.

Sugimoto, M., Tanabe, M., Hataya, M., Enokibara, S., Duine, J. A., & Kawai, F. (2001). The first step in polyethylene glycol degradation by sphingomonads proceeds via a flavoprotein alcohol dehydrogenase containing flavin adenine dinucleotide. *Journal of Bacteriology, 183*(22), 6694–6698.

Suh, H. J., & Lee, H. K. (2001). Characterization of a keratinolytic serine protease from *Bacillus subtilis* KS-1. *Journal of Protein Chemistry, 20*(2), 165–169.

Sulistyaningdyah, W. T., Ogawa, J., Li, Q. S., Shinkyo, R., Sakaki, T., Inouye, K., . . . Shimizu, S. (2004). Metabolism of polychlorinated dibenzo-p-dioxins by cytochrome P450 BM-3 and its mutant. *Biotechnology Letters, 26*(24), 1857–1860.

Sun, J., Zheng, M., Lu, Z., Lu, F., & Zhang, C. (2017). Heterologous production of a temperature and pH-stable laccase from *Bacillus vallismortis* fmb-103 in *Escherichia coli* and its application. *Process Biochemistry, 55*, 77–84.

Sutherland, T., Russell, R., & Selleck, M. (2002). Using enzymes to clean up pesticide residues. *Pesticide Outlook, 13*(4), 149–151.

Tachibana, S., Kawai, F., & Yasuda, M. (2002). Heterogeneity of dehydrogenases of *Stenotrophomonas maltophilia* showing dye-linked activity with polypropylene glycols. *Bioscience, Biotechnology, and Biochemistry, 66*(4), 737–742.

Verma, A., & Shirkot, P. (2014). Purification and characterization of thermostable laccase from thermophilic *Geobacillus thermocatenulatus* MS5 and its applications in removal of textile dyes. *Scholars Academic Journal of Biosciences, 2*(8), 479–485.

Verma, S., Saxena, J., Prasanna, R., Sharma, V., & Nain, L. (2012). Medium optimization for a novel crude-oil degrading lipase from *Pseudomonas aeruginosa*SL-72 using statistical approaches for bioremediation of crude-oil. *Biocatalysis and Agricultural Biotechnology, 1*(4), 321–329.

Wang, D., Li, A., Han, H., Liu, T., & Yang, Q. (2018). A potent chitinase from *Bacillus subtilis* for the efficient bioconversion of chitin-containing wastes. *International Journal of Biological Macromolecules, 116*, 863–868.

Wang, L., Liu, T., Sun, H., & Zhou, Q. (2018). Transesterification of para-hydroxybenzoic acid esters (parabens) in the activated sludge. *Journal of Hazardous Materials, 354*, 145–152.

Wang, T. N., & Zhao, M. (2017). A simple strategy for extracellular production of CotA laccase in *Escherichia coli* and decolorization of simulated textile effluent by recombinant laccase. *Applied Microbiology and Biotechnology, 101*(2), 685–696.

Wang, X., Wang, W., Zhang, Y., Sun, Z., Zhang, J., Chen, G., & Li, J. (2019). Simultaneous nitrification and denitrification by a novel isolated Pseudomonas sp. JQ-H3 using polycaprolactone as carbon source. *Bioresource Technology, 288*, 121506.

Wei, Y., Fu, J., Wu, J., Jia, X., Zhou, Y., Li, C., . . . Chen, F. (2018). Bioinformatics analysis and characterization of highly efficient polyvinyl alcohol (PVA)-degrading enzymes from the novel PVA degrader *Stenotrophomonas rhizophila* QL-P4. *Applied and Environmental Microbiology, 84*(1), e01898-17.

Xie, Z., Xu, B., Ding, J., Liu, L., Zhang, X., Li, J., & Huang, Z. (2013). Heterologous expression and characterization of a malathion-hydrolyzing carboxylesterase from a thermophilic bacterium, *Alicyclobacillus tengchongensis*. *Biotechnology Letters, 35*(8), 1283–1289.

Xue, F., Ya, X., Tong, Q., Xiu, Y., & Huang, H. (2018). Heterologous overexpression of *Pseudomonas umsongensis* halohydrin dehalogenase in *Escherichia coli* and its application in epoxide asymmetric ring opening reactions. *Process Biochemistry, 75*, 139–145.

Yamamura, S., Morita, Y., Hasan, Q., Yokoyama, K., & Tamiya, E. (2002). Keratin degradation: A cooperative action of two enzymes from *Stenotrophomonas* sp. *Biochemical and Biophysical Research Communications, 294*(5), 1138–1143.

Yanmış, D., Adıgüzel, A., Nadaroğlu, H., Güllüce, M., & Demir, N. (2016). Purification and characterization of laccase from thermophilic *Anoxybacillus gonensis* P39 and its application of removal textile dyes. *Romanian Biotechnological Letter, 21*(3), 11485–11496.

Zhang, J., Cao, X., Xin, Y., Xue, S., & Zhang, W. (2013). Purification and characterization of a dehalogenase from *Pseudomonas stutzeri* DEH130 isolated from the marine sponge Hymeniacidon perlevis. *World Journal of Microbiology and Biotechnology, 29*(10), 1791–1799.

Zu, L., Li, G., An, T., & Wong, P. K. (2012). Biodegradation kinetics and mechanism of 2, 4, 6-tribromophenol by Bacillus sp. GZT: A phenomenon of xenobiotic methylation during debromination. *Bioresource Technology, 110*, 153–159.

4 Industrial Applications of Bacterial Enzymes

Md. Shahbaz Anwar

ABSTRACT

Bacteria have been employed for centuries for their biocatalytic potential to produce different goods without considering the nature of their ingredients on the basis of biochemical characteristics. Nowadays, interest in bacterial enzymes has risen because of their extensive application in various fields due to their effective catalytic action, efficient stability, easy production, and optimization processes, compared to animal and plant enzymes. Enzyme applications in a range of industries (such as the pharmaceutical industry, food industry, juice and baking industry, dairy industry, brewing and wine industry, pulp and paper industry, textile industry, detergent industry, leather industry, cosmetics industry, and waste management) increase day by day due to accounting various parameters like requirements of low energy, minimum processing time, cost-effectiveness, nontoxicity, and eco-friendliness, which show attractive features and good opportunities and also form a good impact in the various fields of technical applications. Here, in this chapter, enzymes obtained from bacteria and their prospects for the benefit of human beings are the center of attention. This information will assist in drawing attention toward effective potential, which is fast gaining popularity and supporting healthcare.

Keywords: bacterial enzymes, bioconversion, application, industry

CONTENTS

4.1 Introduction... 60
4.2 Application of Bacterial Enzymes 61
    4.2.1 Pharmaceutical and Therapeutic Potential... 61
        4.2.1.1 Cancer Treatment... 62
        4.2.1.2 Repair of Damaged Tissue 63
        4.2.1.3 Infectious Diseases Treatment 63
        4.2.1.4 Other Miscellaneous Treatments... 63
    4.2.2 Food Industry.. 65
        4.2.2.1 Dairy 66
        4.2.2.2 Baking Industry... 67
        4.2.2.3 Juice Industry 68
        4.2.2.4 Brewing Industry... 68
        4.2.2.5 Feed Industry 69

DOI: 10.1201/9781003202998-4

4.2.3   Paper and Pulp Industry.. 70
4.2.4   Textile Industry 70
4.2.5   Leather Industry... 72
4.2.6   Cosmetics Industry 73
4.2.7   Detergents Industry.. 75
4.2.8   Waste Treatment 77
References.. 78

4.1   INTRODUCTION

Enzymes are molecules that have biological origins and catalyze different chemical reactions. They are very specific in their nature; the rates of a specific reaction are accelerated by minimizing activation energy devoid of any fixed change in enzyme molecule, so they are considered essential molecules for living beings [1, 2, 3]. In the reactions of enzymes, substrates molecules are used at the process are beginning, that are transformed into diverse molecules by the action of enzymes, called products (Figure 4.1).

All reactions occurring in cells of living things require enzymes for the normal functioning of life. The selective nature of enzymes toward a substrate will accelerate simply a little reaction, as of various promises. Enzymes prepared in a cell decide the favorable passage of the metabolite that happens in that cell [4]. The rates of reactions of most enzymatic molecules are millions of times quicker compared to that unanalyzed response. Enzymes produced from bacteria are used from the earliest times in food fermentation and are still used in the preparation of many food items [5]. In food industries, they play a major role because of their more stable nature in comparison to animal and plant enzymes. They were produced in a money-spinning way through fewer requirements of time and space due to their high uniformity, easy optimization, and process modification; they can be completed in a quite easy manner [6]. In various industrial sectors, numerous applications have been reported by using these enzymes (e.g., the applications of amylolytic enzymes in various items like food, paper, detergent, and textile industries) [7]. For various product formations, like glucose syrups, crystalline glucose, maltose syrups, and high-fructose corn syrups,

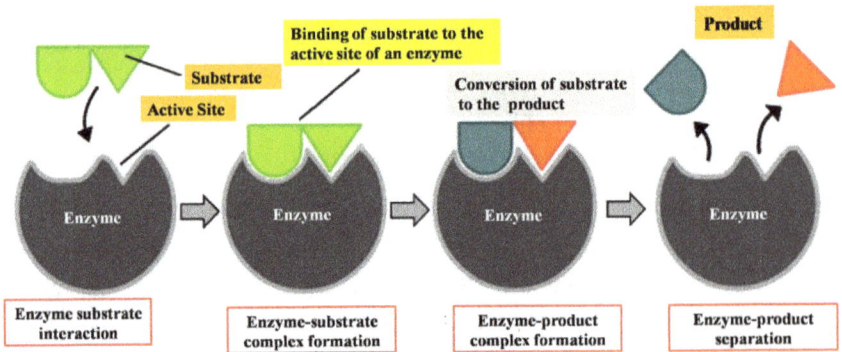

FIGURE 4.1 Catalytic mechanism of an enzyme.

bacterial enzymes are also used. In the detergent industry, bacterial enzymes are used as additives to take away stains related to starch. The paper industry utilizes these enzymes to reduce the starch-based viscosity for coating the paper appropriately. For the warp sizing of textile fibers in the textile industry, amylases are used [8]. In the dairy industry, organoleptic features, such as flavor, color, aroma, and the efficient yield of vast milk products are improved by using bacterial enzymes. Some enzymes with dairy applications are lactoperoxidase, lipase, protease, lactase, esterase, transglutaminase, catalase, aminopeptidase, and lysozyme. The enzymes also work as coagulants, help to improve good health as in a safety manner, and increase dairy products' life span. Among the prominent dairy products in which different bacterial enzymes are used are butter and yogurt [9]. In different industries, bacterial enzymes are used in various ways, such as pharmaceutical production, juice processing from grains and converting them into wine products, leavening processes for bread preparation, production of agrochemicals, and waste management and remediation [10]. The majority of enzymes of industrial interest are obtained from various microbes, considered as the best efficient resources that are helpful for quicker production, simple scale-up, recovery, and easy purification. Overexpression of microbial strains is easily manipulated that can enhance the enzyme activity and specificity modulations [11]. Presently, nearly about 200 different kinds of enzymes of microbial origin from nearly 4,000 identified enzymes molecules are utilized commercially [12]. Out of total enzymes, 75% are nearly produced through the three topmost enzyme companies, like the US-based DuPont, Denmark-based Novozymes, and Switzerland-based Roche [13]. Widespread applications of microorganisms in various bioprocesses convey a wide range of industrial products. The industrial production of enzymes of microbial origin is schematically represented in Figure 4.2 [14].

According to the report "Market of industrial enzymes: their growth, trends, and forecast [2019–2024]" from Mordor Intelligence, the annually expected growth rate of enzymes related to industrial markets globally is at a rate of 6.8% throughout 2019–2024 of the forecast period [15]. Some international players account in the market of enzymes related to industrial purposes are El du Pont de Nemours and Company, Associated British Foods (ABF), Koninklijke DSMNV, Novus International Inc., and AB Enzymes GmbH Industrial enzymes market to attain revenue of $12.8 billion by 2025 [16]. Here, attention is given to present distinct roles of different enzymes isolated from bacteria that are involved in various ways of different technical fields.

4.2 APPLICATION OF BACTERIAL ENZYMES

Enzymes are used in different processes industrially, like in pharmaceuticals, detergents, baking products, leather, textiles, brewing, foods and juices, dairy, cosmetics, pulp and paper, and waste management. Here is a range of methods presenting how enzymes obtained from bacteria are used in different broad categories.

4.2.1 Pharmaceutical and Therapeutic Potential

Bacterial enzymes have numerous imperative roles in the pharmaceutical and diagnostic industries. Therapeutic drugs are one of the most important ones that are

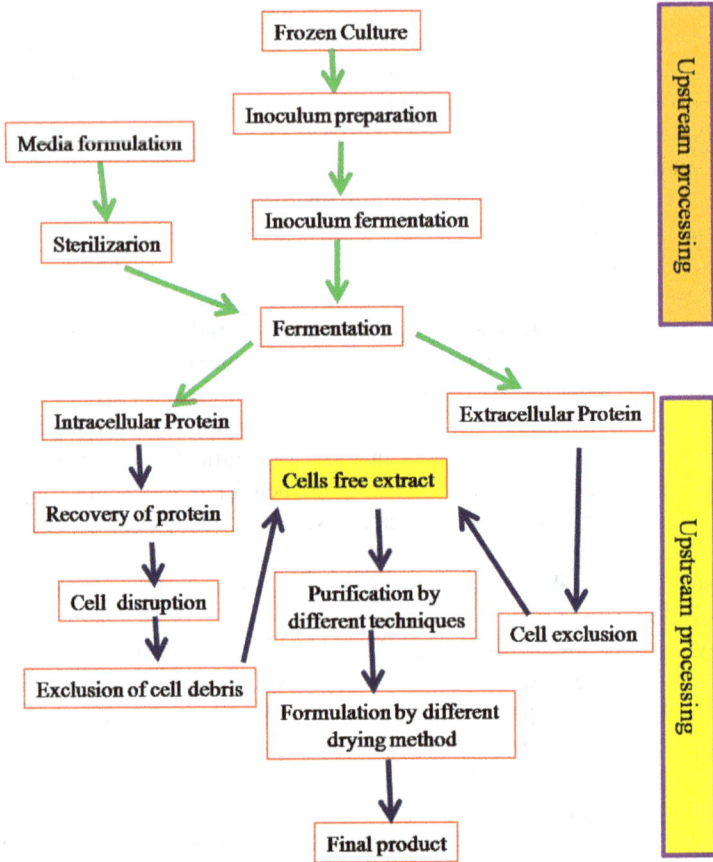

FIGURE 4.2 Industrial production of microbial enzymes.

linked with enzymatic deficiency, disorders related to the digestive system, different events like testing kits and diagnostic tools for ELISA, and diabetes [17]. Enzymes are utilized in the pharmaceutical industry extensively because they are growing and developing quickly. Here we are describing some applications of these enzymes in a few specific disciplines.

4.2.1.1 Cancer Treatment

Research on cancer has some good illustrations of the utilization of enzymes therapeutically. Numerous studies on cancer have confirmed that human melanoma and hepatocellular carcinomas can be inhibited by using the arginine-degrading enzyme [18]. Currently, Oncaspar (pegaspargase), a PEGylate enzyme is reported for the treatment of acute lymphoblastic leukaemia (ALL) as a component of multi-agent chemotherapy in paediatric and adult patients.

The asparagine is synthesized by normal cells, while the cancerous cells are unable to synthesize it. As a result, with the existence of the asparagine-degrading enzyme, the cancerous cell has died. So both the enzymes asparaginase and PEG asparaginase are considered the most useful adjuncts for the process of chemotherapy. The proliferation is an additional significant feature of oncogenesis. Thus, the utilization of chondroitinase AC and chondroitinase B for the removal of chondroitin sulfate proteoglycans has been reported to inhibit tumor growth, metastatic process, and neo-vascularization process [19].

4.2.1.2 Repair of Damaged Tissue

Numerous enzymes exhibiting proteolytic activities are obtained from different bacteria that had been reported to eliminate burned and dead skins. A wide range of superior classes and pure forms of enzymes are now in trials for clinical use. VibrilaseTM, an enzyme having proteolytic activity, is established to be successful in opposition to denatured proteins of damaged or burned skin. Chondroitinase, another enzyme, has been demonstrated to regenerate the wounded spinal cord by eradicating the glial scar, and thus, chondroitin sulfates accumulate and finally stop the growth of axons. [20]. Similar hydrolytic activity by using hyaluronidase has also been reported on chondroitin sulfate, which maintains the regeneration of damaged nerve tissue [21].

4.2.1.3 Infectious Diseases Treatment

A natural antibacterial agent, lysozyme is produced by different bacterial species, like *Staphylococcus* [22], that act against HIV, as RNase-A and urinary RNase-U specifically corrupt viral RNA [23].

4.2.1.4 Other Miscellaneous Treatments

Some important enzymes, like dextranase, rhodanese, and acid protease, are used for the treatment of tooth decay, cyanide poisoning, and alimentary dyspepsia, respectively [24].

Lipases [25] and polyphenol oxidases [26] produced from *Bacillus* and *Azospirillum*, respectively, synthesize an intermediate for diltiazem [2R,3S]-3-[4-methoxyphenyl] methyl glycidate and 3,4-dihydroxyphenyl alanine, for the treatment of Parkinson's disease, respectively [27]. Tyrosinase enzymes are engaged in the process of melanogenesis and in L-DOPA production, which is a pioneer for dopamine, a Parkinson's disease treatment, and to control the neurogenic injury of the myocardium, dopamine acts as a potent drug [28, 29]. Chitosanase [30], obtained from *Bacillus* sp. strain KCTC 0377BP, produces biologically active chitosan oligosaccharides (COSs) from the chitosan. COSs have multiple properties. For example, they are antimicrobial, working against different microbes. They also lower blood cholesterol and ultimately relieve high blood pressure. They are also antioxidants, control arthritis, and have antitumor properties [31, 32, 33, 34].

Pharmaceutical applications of bacterial enzymes to overcome different problems associated with health are illustrated in Table 4.1. (Similar tabulations are also reported [35, 36, 37, 38, 17].) Moreover, for the determination of diabetes and other health disorders, enzymes are generally applicable in clinical diagnostics, like urease from *E. coli* [39] for urea; lipase, carboxylesterase, and glycerol kinase obtained from *E. coli* are involved in the synthesis of triglycerides; urate oxidase [40] is involved

TABLE 4.1

Pharmaceutical Applications of Bacterial Enzymes to Overcome Different Health Problems

Enzyme	Application	Source	References
Nattokinase	Cardiovascular disease treatment	*Bacillus subtillis*	[60, 61]
Super oxide dismutase	Anti-inflamatory action	*Mycobacterium* and *Nocardia* sp.	[37]
L-methionase Arginine deiminase Asparaginase Glutaminase Phenylalanine ammonia lyase	Cancer treatment	*Clostridium sporogenes* *Mycpolasma* *Bacillus subtillus* *Bacillus subtillus* *Rhodotorula glutinis* and *Rhodosporidium toruloides*	[62, 63, 37, 64, 65, 66, 67]
Staphylokinase Urokinase	Anticoagulation actions	*Staphylococcus aureus* *Bacillus subtillus*	[68, 37, 69, 70]
α-amylase	Diabetes, pancreatitis, and cancer research	*Bacillus licheniformis* and *B. stearothermophilus*	[6]
Glutaminase	Leukemia	*E. coli*SFL-1 and *Acinetobacter*	[65]
Streptokinase	Blood clots	*Streptococci* sp.	[70]
β-lactamase	Antibiotic resistance	*Klebsiella, Citrobacter freundii,* and *S. marcescens pneumonia*	[71]
Asparaginase	Leukaemia	*E. coli*	[64]
Collagenase	Skin ulcers	*C. perfringens*	[72]
Chondroitinase	Tumor treatment	*Flavobacterium heparinum*	[19]
VibrilaseTM	Treatment of damaged tissue	*Vibrio proteolyticus*	[73]
Hyaluronidase	Regeneration of damaged nerve tissue	*Treponema pallidum* and *Treponema pertenue*	[73]
Superoxide dismutase (Serrapeptase)	Anti-inflammatory	*Lactobacillus plantarum* and *Nocardia* sp.	[14]
Streptokinase, urokinase	Anticoagulants	*Streptococci* sp. and *Bacillus subtilis*	[14]
Glutathione peroxidases, superoxide dismutases, catalase	Antioxidants	*Corynebacterium glutamicum* and *Lactobacillus plantarum*	[14]
Laccase, rhodanese	Detoxification	*Pseudomonas aeruginosa*	[14]
Rhodanase	Cyanide poisoning	*Sulfobacillus sibiricus*	[14]
Uricase	Hyperuricemia and gout	*Bacillus subtilis* and *Pseudomonas putida*	[75]
Streptodornase	Infections of the skin and oropharynx	Hemolytic streptococci	[76]

in the synthesis of uric acid; creatinase and sarcosine oxidases are involved in the synthesis of creatinine [41, 42].

Cholesterol oxidase has been reported to find and exchange cholesterol, a helpful biotechnological application. The majority of amylases obtained from different gram-positive bacteria, like *Bacillus amyloliquefaciens*, *Bacillus licheniformis*, and *Bacillus stereothermophilus*, which are essential for several medical circumstances, such as acute pancreatic inflammation, severe ulcer in peptic parts, strangulation in the ileus, twisting of ovarian cysts, macroamylasemia, and mumps [6].

Lipase obtained from genera including *Pseudomonas, Bacillus*, and *Burkholderia* is utilized in various processes, like dietary triglycerides, cell signaling, and inflammation [43, 44]. Enzymes for the production of 6-amino penicillanic acid antimicrobials by penicillin acylase are obtained from *Bacillus megaterium, Alcaligenes*, and *E. coli* [45, 46, 47, 48, 49, 50]. Adenosine deaminase (ADA) obtained from *Bacillus* sp. J-89 is an enzyme that helps in the metabolism of purine, making inosine and free ammonia by the hydrolytic breakdown of adenosine. ADA regulation is targeted as a probable therapeutic agent to treat various infections related to viruses and lymphoproliferative disorders [51]. Treatment of several diseases, like severe combined immunodeficiency disease (SCID), is reported by the help of Adenosine deaminase (ADA) enzyme [52]. This enzyme plays an important role in breakdown of excess adenosine present in the patients that leads to diminish their toxicity of the raised adenosine levels to the patient.

Serratiopeptidase (Serratia E15 protease) is an enzyme that is a promising effective anti-inflammatory agent [54, 55]. This is a type of protease linked with the trypsin family and has about 52 kDa molecular mass [56].

Nattokinase, a serine protease of the subtilisin family, despite its name, is actually not a kind of kinase; it is used for the inactivation of plasminogen activator inhibitor-I [PAI-I] by displaying a strong fibrinolytic action when exposed to human blood clots [57]. Dissimilar to different proteins, they are not inactivated or digested if ingested orally and passed through the gut [58]. Nattokinase also exhibits maximum stability and is economically feasible and effective, and it can be obtained from *B. subtilis natto* [59].

4.2.2 FOOD INDUSTRY

The current world population of 7.3 billion is expected to reach 8.5 billion by 2030, 9.7 billion in 2050 and 11.2 billion in 2100, according to a new UNDESA (United Nations Department of Economic and Social Affairs) report, hence, the food requirement is expected to rise (https://www.un.org/en/development/desa/news/population/2015-report.html). A supply of good-quality food can be achieved by the utilization of important enzymes in the food industry. These are helpful in the production of food with good aroma, color, flavor, appearance, texture, and nutritive value [77]. Enzymes' role in the food industry has enhanced the quality and safety of food and ingredients. Moreover, enzymes are now used in new areas like sweetener technology and modification of fat [13]. Enzymes are used in different areas of the food industry, like the dairy industry, baking industry, juice production, and brewing industry.

4.2.2.1 Dairy

Enzymes used in dairy, an important section of the food industry, facilitate the progression and development of different characteristics, like color, flavor, aroma, and yield. The application of enzymes like esterase, lactase, protease, lipase, aminopeptidase, lactoperoxidase, lysozyme, catalase, and transglutaminase in the dairy market is well documented. Their uses vary; for example, they are used as coagulants and bioprotectives to augment the safety of dairy products. These enzymes are generally utilized for the manufacture of yogurt and other dairy products [78]. Nowadays, lipases are involved in various parameters, like improved flavor, preparation of cheese in a faster manner, customized production of milk goods, and milk fat [79, 80]. *Bacillus* spp., *Arthrobacter* spp., *Achromobacter* spp., *Pseudomonas* spp., and *Alcaligenes* spp. are used for large-scale lipases production [81, 82].

Transglutaminase is an enzyme that catalyzes the polymerization process of milk proteins to get better usefulness of dairy products [83, 84]. Globally, approximately 70% of adults lack lactase in the intestine; this enzyme is essential for the digestion of lactose. Humans with lactose intolerance are unable to metabolize lactose due to the deficit of this enzyme. Lactase performs the breakdown of lactose to glucose and galactose and, consequently, proves digestive aid to milk products by enhancing the solubility and sweetness in the milk [85, 78]. Certain bacteria, like the lactic acid bacteria, *Bifidobacterium* and *Lactobacillus*, acquire β-galactosidase enzyme (i.e., bacterial lactase) that facilitate them to digest and utilize lactose [86]. So lactose availability becomes easier due to various bacterial sources [87, 88, 89]. Thus, in the absence of lactase, it is necessary to remove or minimize lactose content in the milk products for the people who do not have lactase to prevent severe tissue dehydration and diarrhea, and occasionally, consequences may be fatal [90, 91, 72]. Another benefit of milk treated with lactase is that it increases the sweetness of milk, thus eliminating the need for adding sugars in the production of flavored milk drinks. Ice cream, frozen dessert, and yogurt manufacturers generally use lactase to enhance scoop and creaminess, digestibility, and sweetness, to reduce sandiness owing to crystallization of lactose in intense preparations. Cheese prepared from milk, which is already hydrolyzed, ripens more rapidly compared to normal milk [93].

In lactic acid bacteria, the proteolytic system is very important in developing flavor in fermented milk products. The proteolytic system is based on proteinases, which cleave the protein present in the milk to peptides; peptidases convert the peptides to small peptides and amino acids; the transport system is fully responsible for the cellular uptake of these products. So the lactic acid bacteria convert the milk casein into the free amino acids and peptides that are necessary for their growth and development. These proteinases comprise all serine proteases, like aminopeptidases, extracellular proteinases, tripeptidases, endopeptidases, and peptidases specific to proline. In spite of lactic streptococcal proteinases, some other proteinases obtained from non-lactic streptococci have also been reported. There are some serine-type of proteinases (e.g., proteinases obtained from *Lactobacillus acidophillus*, *L. delbrueckii bulgaricus*, *L. lactis*, *L. plantarum*, and *L. helveticus*). For the development of flavor in the products obtained from fermented milk, aminopeptidases are essential and are able to produce single-amino-acid residues from oligopeptides due to the action of extracellular proteinase activity [93].

In dairy processing, some minor enzymes comprise some degree of applications, like catalase, aminopeptidase, superoxide dismutase, neutral proteinase, and lysozymes. Instead of pasteurization, catalase is often used for cheese processing in various points, like in making certain types of cheeses, like Swiss cheese, to preserve milk enzymes that are significant to the end product and development of flavor. Superoxide dismutase, an enzyme, acts as an antioxidant and generates H_2O_2, and it is more effective in the presence of catalase. For a faster cheese-ripening process, neutral proteinase and aminopeptidase, obtained from *Bacillus subtilis* and *Lactobacillus* sp., respectively, are generally used [14].

Lysozyme acts as an antimicrobial, lowering bacterial quantity in milk but not harming the *L. bifidus* activity, so it acts as a preservative. Some minor enzymes of milk and milk products have recently been reported for their scope of application [93].

4.2.2.2 Baking Industry

Enzymes are used in baking to provide flour development, improve texture, stabilize the dough, improve the dough's volume and color, uniform the crumb structure, prolong crumb softness, and lengthen the freshness of bread. Enzymes are used to fulfill the increasing demand for quality baking products and baked goods. Enzymes for the baking industry are expected to rise to $695.1 million in value by 2019, rising at a CAGR of 8.2% from 2013 to 2019 [94].

The bread-making process is considered one of the most familiar food-making processes worldwide. Enzyme applications in the bread-development process show their value in organizing the quality and efficiency of the products. Amylase, in pure form or when mixed with other enzymes, is incorporated into the flour to hold moisture more efficiently to improve the softness, freshness, and shelf life of bread. Amylase obtained from *Bacillus stereothermophilus* is generally supplemented to the dough to break down the starch and convert it to smaller dextrins that are further used by the yeast. So finally, Alpha Amylases augments the fermentation rate and lowers the viscosity, which results in the expansion in the texture and volume of the product [6].

Transglutaminase isolated from *Streptoverticillium* sp. and *Streptomyces* sp. Is used to improve laminated dough's strength through high molecular weight glutenins cross-linking [14]. Also, it enhances the flour quality, the texture and amount of bread, and the texture of cooked pasta [95, 96, 84]. The flavor quality of baked products is generally caused by releasing short-chain fatty acids during the esterification process, and the shelf life of the baked products is also improved by the use of lipases [97, 98, 99, 100, 101, 102, 13, 103]. Protease represents about 60% of enzymes related to industrial purposes in the market [14].

The demand for proteases in the market is growing globally, at a rate of about 5.3% from 2014 to 2019. They are extensively used in the banking industry for the manufacture of different baked foods, like bread, crackers, and waffles. These enzymes lower the mixing time, promote dough consistency and uniformity, regulate the strength of gluten in bread, and enhance the surface and flavor [104, 105]. They help improve different parameters, like nutritional value, solubility, and digestibility of proteins associated with them, and modify their functional properties, like emulsification and coagulation [106].

4.2.2.3 Juice Industry

Enzymes utilized in the juice industry help in processing to improve the operation efficiency, like shedding the outer layer [107]. Different enzymes, like pectinases, amylases, and cellulases, are used during the processing of fruits for several steps, like maceration (where organized tissue is distorted into a deferment of intact cells, resulting in pulpy products), liquefaction (where a liquid is generated from a solid), and clarification (an important step in fruit juice processing, mainly in order to eliminate pectin and other carbohydrates associated with juice), and improve yield and cost-effectiveness [108, 109].

The stability and quality of juices manufactured are improved by adding enzymes that digest different substances, like pectin, starch, cellulose, and proteins, of fruits and vegetables and provide better yields, and they also reduce time to process the fruits and vegetables [110]. In the juice industry, cellulases in pure form or mixed with other enzymes are applied to increase the process presentation, improve the yield, and improve the extraction, clarification, and stabilization of juices [111]. They lower the viscosity of nectar and puree obtained from different fruits, like apricot, papaya, mango, plum, pear, and peach. Also, they are applied in the extraction of different flavonoids from flowers and seeds. The method of extraction by using cellulase is more preferred compared to the methods used conventionally because of the attributes like better yield, less loss due to heat damage, and low processing time. Cellulases are also used for taking out the phenolic compounds from grape pomace [112]. β-glucosidases mixed with pectinase are used to modify the structure, flavor, and aroma of fruits and vegetables [113]. They are also reported to lower the citrus bitterness and improve the aroma and taste [114]. Amylases are utilized to take advantage of clear or cloudy juice production [115, 116].

Some other enzymes, like naringinase, limoninase, and debittering enzymes, generally hydrolyze bitter components present in the citrus juice and improve the quality [117, 13]. Structural polysaccharide pectin, present almost in all fruits, is generally required to maintain and regulate juices' cloudiness by using polygalacturonase, pectin lyase pectin esterases, and various arabanases [118, 119].

4.2.2.4 Brewing Industry

The beverage industry is divided into two groups and eight small subgroups. The first is the nonalcoholic group, which includes soft drinks, syrups, fruit juices, packaged water, tea, and coffee. The second is the alcoholic group, which comprises distilled spirits, beer, and wine [14].

Enzymes used in breweries generally help in processing and producing consistent and good-quality products. In the brewing industry, different microbial enzymes are applied to digest cell walls during the extraction of plant material to enhance yield, aroma, and color and clarify the products [120].

Amylases obtained from bacterial are used in distilled alcoholic beverages to metabolize starch into sugars earlier in fermentation and to get rid of starch turbidities. Enzymes break down barley in unmalted form and other starchy adjuncts and facilitate the lowering of the cost of brewing beer. Adding proteases controls the development of chill-hazes in the beer [24].

In the production of wine, cellulases mixed with other enzymes are used to improve yield and quality [121]. The benefits of using these enzymes are better color development, enhanced maceration, and must clarification and, finally, improvement in wine quality and stability [122]. Studies by [123] Oksanen et al. demonstrate that cellulases significantly reduce the viscosity of wort obtained from grains. The aroma of wine is improved by β-glucosidases in the course of the glycosylated precursor's conversions. Naringinase, combined with β-glucosidase and arabinosidase, is used to improve the aroma and quality of wine [124].

In beer, production haze formation is an important problem, as it affects the quality of the end products. Beer contains various ingredients, such as carbohydrates, polyphenols, proteins, fatty acids, amino acids, and nucleic acids. These ingredients can precipitate during haze formation [125]. So to overcome this problem, glucanase and protease obtained from different bacteria, like *Bacillus subtillis*, are used to restrict haze formation [14].

4.2.2.5 Feed Industry

To meet the demand for feed worldwide, development of feed enzymes happen continually. Enzyme application in the diets of animals began in the 1980s, and in the 1990s, it exploded. These enzymes are expanding in significance as they augment the digestive capability of nutrients that are highly consumed by animals [126]. In addition, the dietary value of protein for nourishment accessible to poultry is improved by the utilization of these feed enzymes [127]. The main feed enzymes that are available on the market are xylanases, phytases, and β-glucanases. Other enzymes, like α-galactosidases, mannanases, proteases, pectinases, and amylases, are also available to use.

The two main sectors, poultry and swine, are where feed enzymes are generally consumed; however, the use of enzymes in aquaculture and for feeding ruminants and pets is also planned in the near future [128, 103, 129, 130, 131]. Phytase enzyme, the principal section in the feed industry, that hydrolyzes phytic acid associated natural phosphorus in cereal-based seeds. Phytic acid is well known to possess anti-nutritional behavior due to strong affinity to chelate divalent ions. Therefore, it is extremely poor as a dietary source of phosphorous. To improve bio-availability of micronutrients, phytase enzyme is used [132, 133, 129].

Some animals like monogastric animals are incapable to assimilate plant based supply comprising elevated sum of cellulose and their feeds, therefore xylanases and β-glucanases enzymes fully degrade as well as digest elevated quantity of cellulose and hemicelluloses [129, 101].

Proteases are utilized in animal feed to promote nutrition by breaking down different proteins into their respective amino acids. Proteases are utilized in animal feed to promote nutrition by breaking down different proteins into their respective amino acids. Consequently, making the dietary value of the food better as well as improving the quality by tenderness of meat [134, 103].

Feed enzymes augment nutritive value, thus raising the efficiency of digestion. Animal feed enzymes also facilitate breaking down the indiscriminating factors (e.g., fiber, phytate) that are present naturally in a variety of feed ingredients. There may be consequences, such as reducing the production of meat or eggs, lowering the feed efficiency, and disturbing the digestive system as well. Adding exogenous enzymes generally augments the accessibility of nutrients from feed ingredients. Feed enzymes improve meat and egg production by enhancing nutrient utilization and reducing animal excreta [135].

4.2.3 PAPER AND PULP INDUSTRY

With increasing consciousness of sustainability issues, the application of enzymes obtained from bacteria for the paper and pulp industry has developed steadily to diminish the unfavorable consequence on the ecosystem. The exploitation of enzymes lowers the time of processing, expenditure of energy, and the number of chemicals used in processing. Enzymes are utilized for enhancing drinking water, bleaching paper, and treating various wastes by escalating biological oxygen demand (BOD) and chemical oxygen demand (COD) [136].

Xylanases are utilized in the paper and pulp industry to enlarge the value of the pulp by degrading lignin and hemicelluloses [137] and boosting bleach [14]. For wood processing in the paper industry, xylanase preparations are utilized, which must be devoid of activities of cellulose that provide paper brightness due to their special solubilization of xylans in plant materials and the removal of hemicelluloses selectively from the kraft pulp [138].

The paper and pulp industry generally requires the separation and degradation of lignin from various plant materials, where pre-action of wood pulp by using ligni-nolytic enzymes is essential for a cleaner strategy of lignin removal compared to chemical bleaching. Bleach enhancement of mixed wood pulp has been attained by means of mixed culture strategies during the joint activity of laccase and xylanase [139]. The application of laccase is generally used as an alternative to the constraint of a large amount of chlorine in the process of chemical pulping; afterward, it diminishes the quantity of wastes that cause ozone depletion and acidification [140]. In the industry, amylase is used to control deinking, improving paper cleanliness, and improving drainage [141].

Amylases are also used by the paper industry for reducing the viscosity of starch to achieve suitable covering of paper [142]. The treatment of coating of the paper promotes smoothness and sturdiness to improve the quality of writing. The viscosity of natural enzymes is very lofty for paper sizing so it can be modified by partially degrading polymer by application of alpha amylases [6]. Lipases are used for deinking regularly to enhance pitch control [143]. Before 1990s, some selected approaches related to enzymatic field have been used on a big scale for the process of paper making [144]. Cellulases are used for deinking of paper, which improves softness and drainage; they are also utilized for the growth of the bioprocess to use in recycling of printed papers [145].

In addition, mannases are used to metabolize glucomannan to improve the brightness of paper [146], and protease is used for the removal of biofilm.

4.2.4 TEXTILE INDUSTRY

The textile industry is responsible for the vast production of waste. Bleaching chemicals and dyes are two of the largest sources of environmental pollution [147]. To overcome these environmental pollution in such industries, enzymes are applied to allow the development of environmentally friendly technologies in fiber processing and strategies to improve the final product quality [148].

Principal enzymes in the pretreatment and finishing processes of cotton are hydrolase and oxidoreductase. The hydrolase group includes cellulase, amylase, cutinase, protease, pectinase, and lipase/esterase, which are used in bio-polishing and scouring of fabric, antifelting of wool, denim finishing, cotton softening, desizing, modification of synthetic fibers, wool finishing, and so on [149, 150].

Oxidoreductase, another group of enzymes, comprises laccase, catalase, peroxidase, and ligninase, which play an important role in various steps like bleach termination, bio-bleaching, dye decolorization fabric, and wool finishing [151]. A brief detail of applications of enzymes in textiles industries is shown in Table 4.2. Amylase is utilized in the textile industry in warp-sizing textile fibers [116].

Sometimes, sizing agents like starch are included in fabric production to make the weaving process quick and safe. Starch is preferred because it's easily available, very cheap, and easily washed out. The desizing process is generally applied to remove the starch from the fabric with the aim of strengthening the fabric to avoid the breaking of the warp yarn during the weaving progression [6]. The alpha amylases selectively take away the size without affecting the fibers [152].

Amylase obtained from *Bacillus* strains is used in the textile industry. Washing denim and high-quality cotton is done so as to give a well-used appearance. Potassium permanganate or sodium hypochlorite is used in stone-washing denim pants, giving the denim a faded look, and this impairs the texture and machine. Cellulases produce

TABLE 4.2
Detail of Applications of Enzymes in Textiles Industries

Enzyme	Application	Source
Amylase	Desizing, uniform wet processing in the textile industry	*B. licheniformis, Bacillus* sp.
Cellulase	Bio-polishing, cotton softening, denim finishing	*Pseudomonas fluorescens, Serratia marcescens, Bacillus subtilis, E. coli*
Catalase	Bleach termination	*Corynebacterium diphtheriae, Burkholderia cepacia*
Laccase	Improvement of the look of the denim clothes	*Bacillus subtilis*
Pectate lyase	Bio-scouring	*Pseudomonas* sp., *Bacillus* sp.
Protease	Removal of wool fiber scales, degumming of silk	*B. subtilis*
Lipase	Removal of size lubricants, denim finishing	*Bacillus, Pseudomonas, Burkholderia*
Lignin peroxidase	Degrade various types of textile dyes	*Acinetobacter calcoaceticus*
Collagenase	Wool finishing	*Clostridium histolyticum*

commercially from bacterial sources, which increases the effectiveness without disturbing the piece of clothing or the machine.

Cellulases remove the uncovered surface of the colored texture, leaving the insides unblemished [153].

Bio-polishing is a concluding process that improves fiber quality by reducing the pilling property of fiber made from cellulose. The objective of this process is to eliminate the cotton microfibrils by applying the cellulase enzyme. This treatment makes the fabric surface cleaner, softer, cooler, and more lustrous [153].

Catalase is generally used to eliminate hydrogen peroxide residue that remains after completing the bleaching of cotton fabrics. Bleaching with hydrogen peroxide is a basic stage that leads to the dyeing of cotton fabrics. Hydrogen peroxide acts as an oxidizing agent that removes reactive dyes [154, 155]. The remaining peroxide deposits on the fabric have a negative impact [156].

Thus, it is very important to remove the remaining bleaching baths from the fabric and machinery prior to supplying the dye. Successful washes of the fabric with water or chemicals are essential to weaken the remaining bleaching agents. Nevertheless, this requires a large quantity of water, a longer time to process, or the use of chemical products, which collectively harm the ecosystem, resulting in additional costs and the generation of more effluents. An alternative to overcome these is the use of enzyme catalases, which hydrolyze the decomposition of hydrogen peroxide [157].

Proteases break down the hydrolysis reaction of certain peptide bonds in protein molecules. That is why they are used to take away the rigid and dreary gum coating of sericin from the raw silk to improve luster and softness. Protease treatment can adjust the surface of wool and silk fibers to give a new look [158].

The finishing of denim has a significant value in the denim industry. The removal of indigo is a very important step to give an abrasion effect on the fabric. Laccase has the capacity to break down indigo; that's why laccase is more efficient in the final treatment of denim [159]. It also enhances the retention of the dye on wool. This method has the economic advantage of dropping the quantity of dye used when deeper colors are required [160]. It is also able to degrade and decolorize the industrially important wool azo dye Diamond Black PV 200 without the addition of redox mediators [161]. Enzymes are also used for roving treatment to improve the yarn consistency [162]. The benefit associated with this process is that it performs under mild reaction conditions that making it ecologically friendly [163].

Scouring is the process of removal of non-cellulosic fabric from outside of cotton. Usually, pectinase and cellulase are used in combination for bio-scouring. Pectinase destroys the structure of the cuticle of the cotton by removing the layer in between the cuticle and body of cotton fiber, while cellulase destroys the structure of the cuticle by removing the primary cellulosic wall just below the cuticle of cotton [153].

4.2.5 LEATHER INDUSTRY

The leather industry is more customary, that discharges various waste disposes at different stages of leather processing units, which create severe health issues and hazards environmental issues [164]. The degradation by different enzymes is a proficient alternative to improve leather quality and help to reduce waste [103]. Enzymes (e.g., proteases, amylases, lipases) are utilized in the initial step of skin treatment and also work as soaking auxiliaries [165]. The reward of using enzymes in the soaking process is quicker rehydration and elimination of inter-fibrillary resources [166]. Enzyme action is one of the best methods used to accelerate the soaking process, reducing the use of harmful unhairing chemicals [167, 168, 169].

A large number of bacterial species are used for various enzyme production and are efficiently utilized in the industry of leather, like lipases obtained from *Pseudomonas aeruginosa* strain BUP2 [170]. Lipases are commonly used in detergents. Microbes that produce lipases are *Bacillus flexus*XJU-1SG2, *Bacillus subtilis*JPBW-9, *Pseudomonas aeruginosa sanai*, *Bacillus licheniformis*, *Bacillus licheniformis* VS G1, *Bacillus pumilus Geobacillus* sp., *Serratia marcescens* DEPTK2 [171, 172], and *Geobacillus thermoleovorans* DA2 [173]. Proteases can be isolated from *Bacillus licheniformis* [174] and *Bacillus subtilis* [175]. Amylase was isolated from a wide range of microorganisms, like *Bacillus subtilis* [14]. Keratinase obtained from Antarctic *Actinomycete* [176] and from *Bacillus licheniformis* in recombinant *Bacillus subtilis* [177].

The dehairing process is the initial step for the application of enzymes in the leather industry. The principal process in leather preparation requires a bulk amount of different enzymes, like proteases, lipases, and amylases [178, 179, 180]. The application of enzymes in the dehairing process is very striking because it can conserve the hair and lower the organic load released as waste matter. Also, these methods reduce the reliance on destructive chemicals, like lime, sulfides, and amines [181, 182, 183].

Enzymes are required to make the procedure easy and improve the quality of leather during the different leather-processing stages, like curing, soaking, dehairing, liming, bathing, degreasing, picking, and tanning [151]. The enzymes used in leather processing are neutral proteases, alkaline proteases, and lipases. Alkaline proteases are used to eliminate nonfibrillar proteins throughout the soaking process to make the leather soft, supple, and pliable. Neutral and alkaline proteases work together in the dehairing process to lower water wastage [184]. In spite of these, lipases are also used at some stage in degreasing to remove fats [164, 79]. The advantages of using enzymes compared to chemicals are low BOD and COD in effluents, reduced odor, and enhanced hair recovery [185].

4.2.6 COSMETICS INDUSTRY

Biotechnology had an effect on cosmetics in quite a lot of ways. Cosmetics companies exploit biotechnology to determine, develop, and produce different important components of cosmetic formulations and to assess the actions of these components on different body parts. Thus, biotechnology signifies a good option for the development of active ingredients that are capable of slowing down aging [186, 187].

European Commission Regulation No. 1223/2009 defines a cosmetic product as any product or substance made to be used on the outer surface of the human body, like hair, nails, lips, external genital organs, epidermis, teeth, and mucous membranes of the oral cavity, with the purpose of cleaning, changing the appearance, perfuming, protecting, keeping in good condition, and promoting acceptable body odors [188]. Some chemical compounds used in cosmetics have potential effects on the body surfaces of different users. Due to these effects, plus the environmental impact, herbal cosmetics have more attracted the consumers' interest, making them very popular in the cosmetics industry since the early 1990s. Nowadays, certain compounds obtained using biotechnological processes have appealing skincare properties that are considered very helpful ingredients of cosmetics. Microorganisms obtained from various environmental sources are used in manufacturing processes. By applying recombinant DNA technology, it is possible to clone the genes of interest that possess different beneficial properties in the enzymes and are used in pharmaceuticals and cosmetics [189].

Applications of these enzymes allow a precise biochemical pathway that is more advantageous on the surface of a body part. Superoxide dismutase (SOD) is one such enzyme. It stops injuries caused by free radicals and other harmful pollutants [189]. It also manages a level of a diversity of reactive oxygen species (ROS) and nitrogen species that are formed by exposure to ultraviolet and other radiation, plus the usual products of cellular metabolism, reducing the latent toxicity of these molecules and controlling cellular characteristics that are synchronized by their signaling functions [190]. ROS also worsens the skin, and as a result, SOD is considered a very important anti-aging enzyme that eliminates ROSs from the human body. SOD also preserves the integral arrangement of keratin, encourages elasticity of the skin, and promotes a smooth feeling to the skin [191].

Seki et al. reported that subtilisin (serine protease), obtained from *Bacillus licheniformis*, is an efficient skin exfoliator [192]. Commercial proteases used in the cosmetics industry are basically obtained by using recombinant DNA technology [189]. Another significant enzyme that has a role in the DNA repair mechanism is photolyase. When DNA repair is deficient and melanin cannot protect the skin from the damage caused by solar radiation, the risk of accumulating mutations that create cancer may increase. DNA photolyases overturn these lesions by abolishing thymine dimers, which play a significant role in the DNA repair mechanism [193, 194].

The chromosomes of the cyanobacterium *Synechocystis* sp. PCC6803 contain two ORFs, sll1629 and slr0854, which are highly similar photolyases. In another report, three UV-resistant bacteria isolated from distinct niches from HAALs: *Exiguobacterium* sp. S17 (stromatolite, Lake Socompa, 3570 m), *Acinetobacter* sp. Ver3 (water, Lake Verde; 4,400 m), and *Nesterenkonia* sp. Act20 (soil, Lake Socompa, 3570 m) also possess photolyases [195].

Collagenases isolated from *Vibrio alginolyticus*, *Clostridium histolyticum*, and *Bacillus cerus* were used for skin regeneration [196, 197]. Keratinases isolated from *Epidermophyton*, *Microsporum*, *Trichophyton*, *Scopulariopsis brevicaulis*, *Chrysosporiu*, *B. subtilis*, and *B. licheniform* were used for hair removal [198, 199, 200]. They are utilized to treat different parts like scar tissues and stretch marks and regenerate the epithelial cells to get better healing. Usually, keratin hydrolysates are utilized as topical ointments and creams for the heels, elbows, and knees, offering external smoothness and diminishing the damage of the skin. They also work in the treatment of enzymatic peeling, hair removal, and delaying hair growth. *Bacillus licheniformis* is an example of commercial keratinase-producing bacteria. SOD and several peroxidases, like glutathione peroxidase, catalase, and lactoperoxidase, work synergistically to remove dead skin cells from the surface of the skin as exfoliate. These enzymes also serve as scavengers for free radicals when applied to the skin surface to protect the skin from the harms of free radicals created by ultraviolet light [201].

Lactate dehydrogenase [LDH] is another enzyme able to catalyze the reduction of NADH and pyruvate leading to the production of NAD$^+$ and lactate as the end products. These reactions get diminished on the exposure of UV, while in the presence of this LDH, the subunits remain intact in the cells, permitting cells to bring out regular functioning [201]. Proteases are popular for hydrolyzing keratin, collagen, and elastin of the skin. Alkaline aspartic proteases obtained from a variety of alkaliphilic bacteria are used for the treatment of skin disorders, like ichthyosis (scaly skin), xerosis (dryness of the skin), and psoriasis (skin flaking and inflammation) [202].

Enzymes used in cosmetics have been increasing on increasing the demand continuously. These act as scavengers for free radicals in cream, mouthwash, toothpaste, and hair dye [13]. Proteases are also utilized in skin creams to clean and smoothen the skin by peeling off the deceased and injured skin [203]. Endoglycosidase and papain are used widely in toothpaste and mouthwash whiten teeth, remove plaque, and remove odor created in the deposition of teeth and gum tissue [204].

Various enzymes like peroxidases, laccases, oxidases, and polyphenol oxidases are used in hair dyeing [205]. In the normal wearing of contact lenses, tear proteinaceous films and debris are deposited in the surface of the lens, disturbing the optical clarity of the lens. The surface of the contact lens also accumulates a lot of pathogens, like *Pseudomonas aeruginosa* sticking together [206, 207].

Generally, solutions used for contact lens cleaning are prepared by using proteases of the plant-based papain and the animal-based trypsin, pancreatin, and chymotrypsin. Microbial proteases obtained from *Streptomyces* sp., *Bacillus* sp., and *Aspergillus* sp. were also reported for cleansing of tear films and debris [208, 209].

4.2.7 DETERGENTS INDUSTRY

Detergents symbolize the major utilization of enzymes amounting to 25–30% of the entire sales and are expected to grow faster at a CAGR of about 11.5% commencing 2015 to 2020 [14]. Enzymes have a great contribution toward the expansion and progress of the detergent industry that covers the major application area for enzymes in the recent era. Detergents have numerous applications, such as dishwashing, domestic cleaning, laundry, and cleaning of different industries and institutions [210].

The function of detergent enzymes is to remove starch, proteins, oils, and fats [211, 101]. Enzymes in laundry detergents have some excellent properties; they are weight-efficient, cleave off scratched cotton fibers, and whiten fabric or improve fabric color. Enzymes mainly present in detergents are of the hydrolase group, comprising amylase and protease. Sometimes a combination of enzymes, including proteases, cellulases, amylases, pectinases, and lipases, is used to increase efficiency on stain cleaning and fabric care [13]. Amylases are very effective in eliminating starchy food deposits, while lipases are for removing stains created as the result of fatty products [212].

The detergent industry is the most important consumer of enzymes in terms of both volume and value. Enzyme applications in detergents augment the ability to get rid of tough stains and also build detergent eco-friendly. Amylases are commonly used in the formulation of detergent; 90% of all liquid detergents contain these enzymes [213]. This plays an important role in laundry washing and automatic dishwashing to remove stains of starchy foods, chocolate, and other smaller oligosaccharides.

The growth of bacterial lipases in the detergent industry is an incredibly innovative thing that replaced harsh chlorine. Bleaching with lipases also minimizes different levels of pollution [214, 215]. Lipases in powder form are planned to dominate in markets, owing to the following properties they have: stability, easy handling, packaging, and suitable transportation chosen by the consumers [216, 217].

Because of the addition of lipases to laundry detergents and household dishwashers, hydrolytic lipases are considered very important in the detergent industry. Some details of lipase enzymes are mentioned in Table 4.3. Enzymes overall perform various functions, like reducing the amount of detergents thrown as environmental waste, saving energy by requiring a lower wash temperature, and lowering the use of chemicals, thus leaving no destructive residues in the environment, not impacting the sewage treatment processes, and not threatening aquatic life [218].

Protease eliminates various organic stains, like grass, blood, egg, and human sweat, while cellulases are used to brighten colors, soften fabrics, and eliminate small fibers without damaging the major fibers [211, 141]. Proteases and amylases are applied chiefly in dishwasher detergents to get rid of protein and carbohydrate food particles [219]. Enzyme application is considered extremely advantageous because enzymes contain some bleaching agents, phosphates, which ultimately benefit both the public and the environment [220, 221].

TABLE 4.3
Applications of Lipase in Detergent Industries

Source	Application	References
Acinetobacter radioresistens, Bacillus sp. FH5	Used in the detergent industry	[222]
Pseudomonas mendocina	Dishwashing or removal of fat strain laundry	[223]
Micrococcus sp.	Commonly used as detergents; augments the removal of oily stains from various types of fabrics	[224]
Pseudomonas plantarii	Solvay enzyme products; applicable for is a nonionic or anionic detergent formulation	[211]
Chromobacterium viscosum	Detergent formulations containing alkaline lipase used in laundry detergents	[225]
Bacillus sonorensis	Removal of corn oil stains fromundyed cotton fabric	[226, 227]
Pseudomonas aeruginosa strain BUP2	Produces an alkaline and thermotolerant lipase used in the detergent industry	[170]
Pseudomonas mendocina, P. glumae	Produces lipases with hightemperature; for commercial detergent formulations	[228]

4.2.8 WASTE TREATMENT

Enzymes utilized in waste management have a wide variety and degrade toxic pollutants. Effluents coming from industrial as well as domestic wastes contain numerous chemicals that hazard living beings and the ecosystem. Enzymes obtained from bacteria work alone or in combinations for the treatment of industrial effluents like phenols, aromatic amines, and nitriles by converting deadly chemical compounds to innocuous products [229, 230, 231, 232]. Different enzymes applied for waste management are lipases, amylases, cellulases, glucoamylases, pectinases, nitrile hydratases, and proteases [233, 234, 235].

Oxidoreductases detoxify the toxic organic compounds through oxidative coupling [233]. Several other enzymes, like laccase, peroxidase, and tyrosinase, breakdown the phenolic compounds from industrial effluents and remove the chlorinated molecules [42, 236, 237, 238, 239]. Bacterial enzymes are also applied to recycle the waste for use again (e.g., to get additional oil from oilseeds, to convert starch into sugar, to convert whey into a variety of practical products) [240, 241]. Various oxygenases, like monooxygenases and dioxygenases, have a broad substrate choice and active against different compounds, including the chlorinated aliphatics [242]. They are also work in the remediation of pollutants comprising halogenated organic compounds like herbicides, fungicides, and insecticides; hydraulic and heat transfer fluids; and intermediates for chemical synthesis [242, 233]. Some details of bacterial enzymes used in waste management are mentioned in Table 4.4.

TABLE 4.4
Applications of Some Enzymes in Waste Management

Source	Enzyme	Application	References
Klebsiella pneumoniae, *Bacillus safensis*	Laccase	Degrades diverse dyes used in industrial processes; decolorization of toluidine, malachite green, and reactive black 5	[12, 243]
Brevibacterium casei, *Bacillus* sp. VUS	Peroxidase	Decolorization of synthetic textile dyes; removal of chromate Cr (VI) and azo dye Acid Orange 7 [AO7]; have the capability for degrading a variety of dyes	[244, 245]
Ralstonia solanacearum, *Streptomyces espinosus* strain LK-4	Tyrosinase	Used for the improved catalytic efficiency toward D-tyrosine; removal of phenol and phenolic compounds from wastewater	[246, 247]

(Continued)

TABLE 4.4 (Continued)

Source	Enzyme	Application	References
Bacillus amyloliquefaciens, B. subtilis	Amylase	Under submerged fermentation using some agro-industrial by-products, likerice husk, wheat bran, and potato starchy waste	[248, 259]
Clostridium thermocellum, Bacillus subtilis	Cellulase	Bioconversion of agricultural waste to ethanol; cellulose degradation in mangrove environments	[250, 251]
Pseudomonas, Bacillus, Acinetobacter	Lipase	Useful in the industrial wastewaters treatment, such as food waste, dairy waste, grease from wool, manure, and wastewater from oil mills	[252, 253]
Rhodococcus rhodochrous strain J1, *Rhodococcus pyridinivorans* S85–2	Nitrile hydratase	Biotransformation of nitriles to valuable amides; convenient treatment of acetonitrile-containing wastes	[254, 255, 256]
Thermoactinomyces, B. amyloliquefaciens	Protease	Mitigates industrial pollution; featherdegradation	[257, 258]
Flavobacterium sp., *Bacillus* sp.	Glucoamylase	Degrading cyclodextrins	[259, 260]

REFERENCES

1. Aldridge S. Industry backs biocatalysis for greener manufacturing. Nature Biotechnology. 2013;31:95–96.
2. Piccolino M. Biological machines: From mills to molecules. *Nature Reviews Molecular Cell Biology.* 2000; 1:149–153.
3. Fersht A. *Enzyme Structure and Mechanism.* W.H. Freeman, New York. Wilay Online Library. Biochemical Education. 1985; 13, 3: 475.
4. Stryer L, Berg JM, Tymoczko JL. *Biochemistry* (5th ed.). W.H. Freeman, San Francisco. 2002. ISBN 0-7167-4955-6.
5. Soccol, C.R., Rojan, P.J., Patel, A.K., Woiciechowski, A.L., Vandenberghe, L.P., Pandey, A. 2006., Glucoamylase. In: Pandey, A., Webb, C., Soccol, C.R., Larroche, C. (eds) *Enzyme Technology.* Springer, New York, NY. https://doi.org/10.1007/978-0-387-35141-4_11.
6. Gurung N, Ray S, Bose S, et al. A broader view: Microbial enzymes and their relevance in industries, medicine, and beyond. *Hindawi Publishing Corporation BioMed Research International.* 2013;1–18. Article ID 329121.
7. Pandey A, Nigam P, Soccol CR, et al. Advances in microbial amylases. *Biotechnology and Applied Biochemistry.*2000; 31(Part 2):135–152.

8. DeSouza PM, Magalhães PO. Application of microbial α-amylase in industry—A review. *Brazilian Journal of Microbiology*. 2010; 41(4):850–861.

9. Abada, E.A., 2019. Application of microbial enzymes in the dairy industry. In: Kuddus, M., Enzymes in Food Biotechnology. *Production, Applications, and Future Prospects*. Elsevier Inc. Ch-5; 61–72.

10. Newton MS, Arcus VL, Gerth ML, et al. Enzyme evolution: Innovation is easy, optimization is complicated. *Current Opinion in Structural Biology*. Elsevier. 2018; 48:110–116.

11. Yang H, Li J, Du G, et al., 2017. Eds. *Biotechnology of Microbial Enzymes: duction, Biocatalysis and Industrial Applications*. Academic Press Books, Elsevier, Science direct. 2017; 151–165.

12. Liu X, Kokare C. Microbial enzymes of use in industry. In: Brahmachari G, Demain AL, Adrio JL (eds.), Biotechnology of microbial enzymes—production, biocatalysis and industrial applications. Elsevier, Academic Press, London. 2017; 267–298. https://doi.org/10.1016/b978-0 -12-803725-6.00011-x

13. Li S, Yang X, Yang S, et al. Technology prospecting on enzymes: Application, marketing and engineering. *Computational and Structural Biotechnology Journal*. 2012; 2: e201209017.

14. Singh R, Kumar M, Mittal A, et al. Microbial enzymes: Industrial progress in 21st century. *3 Biotech*. 2016; 6:174.

15. Dublin. Global industrial enzymes market growth, trends, and forecast 2019–2024: Competition for raw materials with other industries and price volatility restraining market growth. *Research and Markets*. www.researchandmarkets.com.

16. Industrial enzymes market to attain revenue of $12.84 bn by 2025, 2019; News-Transparency Market Research.

17. Mane P, Tale V. Overview of microbial therapeutic enzymes. *International Journal of Current Microbiology and Applied Sciences*. 2015; 4(4):17–26.

18. Ensor CM, Holtsberg FW, Bomalaski JS, et al. Pegylatedarginine deiminase (ADI-SS PEG20,000 mw) inhibits human melanomas and hepatocellular carcinomas in vitro and in vivo. *Cancer Research*. 2002; 62(19):5443–5450.

19. Blain F, Tkalec AL, Shao Z, et al., Expression system for high levels of GAG lyase gene. expression and study of the hepA upstream region in *Flavobacterium heparinum*. *Journal of Bacteriology*. 2002;184(12):3242–3252.

20. Bradbury EJ, Moon LDF, Popat RJ, et al. Chondroitinase ABC promotes functional recovery after spinal cord injury. *Nature*. 2002; 416:636–640.

21. Moon AF, Edavettal SC, Krahn JM. et al. Structural analysis of the sulfotransferase [3-OST-3] involved in the biosynthesis of an entry receptor for herpes simplex virus 1. *Journal of Biological Chemistry*. 2004; 279:45185–45193.

22. Jay JM. Production of lysozyme by staphylococci and its correlation with three other extracellular substances. *Journal of Bacteriology*. 1996; 91(5):1804–1810.

23. Huang SL, Huang PL, Sun Y, et al. Lysozyme and RNases as anti-HIV components in β-core preparations of human chorionic gonadotropin. *Proceedings of the National Academy of Sciences*. 1999; 96(6):2678–2681.

24. Okafor N., 2007. Biocatalysis: Immobilized enzymes and immobilized cells. In: *Book Modern Industrial Microbiology and Biotechnology*, Edition 1st. Taylor & Francis Group. 398. ISBN978-1-57808-434-0.

25. Feng W, Wang XQ, Zhou W, et al. Isolation and characterization of lipase-producing bacteria in the intestine of the silkworm, *Bombyx mori*, reared on different forage. *Journal of Insect Science*. 2011; 11(1):135.

26. Nikitina VE, Vetchinkina EP, Ponomareva EG, et al. Phenol oxidase activity in bacteria of the genus *Azospirillum. Microbiology*. 2010;79:327–333.

27. Faber K. Biotransformations in Organic Chemistry; A Textbook. Springer, Berlin. 1997.

28. Haq I, Ali S, Qadeer MA. Biosynthesis of l-DOPA by Aspergillus oryzae. *Bioresource Technology*. 2002; 85(1):25–29.

29. Zaidi KU, Ali AS, Ali SA, et al. Microbial tyrosinases: Promising enzymes for pharmaceutical, food bioprocessing, and environmental industry. Biochemistry Research International. *Hindawi Publishing Corporation.* 2014; 1–16. doi:10.1155/2014/ 854687.

30. Choi YJ, Kim EJ, Piao Z, et al. Purification and characterization of chitosanase from *Bacillus* sp. Strain KCTC 0377BP and its application for the production of chitosan oligosaccharides. 2004; 70(8):4522–4531.

31. Kim SK, Rajapakse N. Enzymatic production and biological activities of chitosan oligosaccharides [COS]: A review. 2005; 62(4):357–368.

32. Ming M, Kuroiwa T, Ichikawa S, et al. Production of chitosan oligosaccharides by chitosanase directly immobilized on an agar gelcoated multidisk impeller. 2006; 28(3):289–294.

33. Zhang H, Sang Q, Zhang W. Statistical optimization of chitosanase production by Aspergillus sp. QD-2 in submerged fermentation. 2012; 62(1):193–201.

34. Thadathil N, Velappan SP. Recent developments in chitosanase research and its biotechnological applications: A review. 20140; 150:392–399. doi:10.1016/j.foodchem.2013.10.083.

35. Devlin TM. *Textbook of Biochemistry: With Clinical Correlations.* Wiley, New York. 2010. 1240 Pages.

36. Vellard M. The enzyme as drug: Application of enzymes as pharmaceuticals. 2003; 14:444–450.

37. Sabu A. Sources, properties and applications of microbial therapeutic enzymes. 2003; 2:234–241.

38. Kaur R, Sekhon BS. Enzymes as drugs: An overview. 2012; 3(2):29–41.

39. Orth D, Grif K, Dierich MP, et al. Prevalence, structure and expression of urease genes in Shiga toxin-producing *Escherichia coli* from humans and the environment. 2006; 209(6):513–520.

40. Nanda P, Babu PEJ. Isolation, screening and production studies of uricase producing bacteria from poultry sources. 2014; 44(8):811–821.

41. Dordick JS. Biocatalysts for industry. 2013; ISBN 9781475745979 Encyclopedia of occupational health and safety (Beverage industry).

42. Hill MLR, Prins A. Biotechnological potential of oxidative enzymes from Actinobacteria. 2016. doi:10.5772/61321.

43. Tjoelker LW, Eberhardt C, Unger J, et al. Plasma platelet-activating factor acetylhydrolase is a secreted phospholipase A2 with a catalytic triad. 1995; 270(43):25481–25487.

44. Spiegel S, Foster D, Kolesnick R. Signal transduction through lipid second messengers. 1996; 8(2):159–167.

45. Vroom De E. an improved immobilized penicillin G Acylase. 1997; WO Patent WO 1997004086 A1.

46. Vroom De E. Penicillin G Acylase immobilized with a crosslinked mixture of gelled gelatin and amino polymer. 2000; US Patent 6060268.

47. Bianchi D, Bartolo R, Olini P, et al. Application of immobilised enzymes in the manufacture of f3-lactam antibiotics. 1998; 80:879–885.

48. Wedekind F, Daser A, Tischer W. Immobilization of penicillin G Amidase, Glutaryl 7-ACA Acylase or D-Amino acid oxidase on an amino functional organosiloxane. 1998; US Patent 5780260.

49. Parmar A, Kumar H, Marwaha SS, et al. Advances in enzymatic transformation of penicillins to 6-aminopenicillanic acid (6-APA). 2000; 18:289–301.

50. Zhang B, Wang J, Chen J, et al. Magnetic mesoporous microspheres modified with hyperbranched amine for the immobilization of penicillin G acylase. 2017; 127(15):43–52.

51. Lee G, Lee SS, Kay KY, Kim DW, et al. Isolation and characterization of a novel adenosine deaminase inhibitor, IADA-7, from *Bacillus* sp. J-89.2009; 24(1):59–64.

52. Aiuti A, Slavin S, Aker M, et al. Correction of ADA-SCID by stem cell gene therapy combined with nonmyeloablative conditioning. 2002; 296(5577):2410–2413.

53. Kumar G, Meghwanshi, Kaur N, Verma S, et al. Enzyme for pharmaceuticals and therapeutic application.2020; 67(4):586–601.
54. Gupte V., Luthra U. Analytical techniques for serratiopeptidase: A review. 2017; 7 (4):203–207.
55. Jadhav S, Shah N, Rathi A, et al. Serratiopeptidase: Insights into the therapeutic applications.2020; 28(3):e00544.doi:10.1016/j.btre.2020.e00544.
56. Metkar SK, Girigoswami A, Vijayashree R, et al. Attenuation of subcutaneous insulin induced amyloid mass in vivo using lumbrokinase and serratiopeptidase. 2020; 163:128–134.
57. Nakamura M, KonnoH, TanakaT, et al. Possible role of plasminogen activator inhibitor 2 in the prevention of the metastasis of gastric cancer tissue. 1992; 65(92):709–719.
58. Sumi H, Hamada H, Nakanishi K, et al. Enhancement of the fibrinolytic activity in plasma by oral administration of nattokinase. 1990; 84:139–143.
59. Vachher M, Sen A, Kapila R, et al. Microbial therapeutic enzymes: A promising area of biopharmaceuticals.2021; 3:195–208.
60. Weng Y, Yao J, Sparks S, et al. An oral antithrombotic agent for the prevention of cardiovascular disease. 2017; 18(3):523.
61. Cho YH, Song JY, Kim KM, et al. Production of nattokinase by batch and fed-batch culture of *Bacillus subtilis*. 2010; 27(4):341–346.
62. Hoffman RM, Tan Y, Li S., et al. Pilot phase I clinical trial of methioninase on high-stage cancer patients: Rapid depletion of circulating methionine. 2019; 1866:231–242.
63. Sharma B, Singh S, Kanwar SS. L-methionase: A therapeutic enzyme to treat malignancies. 2014; Article ID 506287 | https://doi.org/10.1155/2014/506287.
64. Jain R, Zaidi KU, Verma Y, et al. L-Asparaginase: A promising enzyme for treatment of acute lymphoblastic Leukiemia. 2012; 5(1):29–35.
65. Spiers AS, Wade HE. Bacterial glutaminase in treatment of acute Leukaemia. 1976; 1:1317–1319.
66. Singh P, Banik RM, Biochemical characterization and antitumor study of Lglutaminase from bacillus cereus MTCC 1305.2013; 171:522–531.
67. Babich OO, Pokrovsky VS, Anisimova NY, et al. Recombinant l-phenylalanine ammonia lyase from *Rhodosporidium toruloides* as a potential anticancer agent. 2013; 60:316–322.
68. Vakili B, Nezafat N, Negahdaripour M, et al. Staphylokinase enzyme: An overview of structure, function and engineered forms. 2017; 18:1026–1037.
69. Zaitsev S, Spitzer D, Murciano JC. Sustained thromboprophylaxis mediated by an rbc-targeted pro- urokinase zymogen activated at the site of clot formation. 2010; 115:5241 5248.
70. Banerjee A, Chisti Y, Banerjee UC, Streptokinase—a clinically useful thrombolytic agent. 2004; 22:287–307.
71. Gupta V, Kumarasamy K, Gulati N, et al. AmpC -lactamases in nosocomial isolates of Klebsiella pneumonia from India. 2012;136:237–241.
72. Dolynchuk K, Keast D, Campbell K. Best practices for the prevention and treatment of pressure ulcers. *Ostomy/Wound Manage*. 2000; 46:38–53.
73. Ozcan C, Ergun O, Celik A, et al. Enzymatic debridement of burn wound with collagenase in children with partial-thickness burns. 2002; 28:791–794.
74. Fitzgerald TJ, Gannon EM. Further evidence for hyaluronidase activity of Treponema pallidum. 1983; 29(11):1507–1513.
75. Nelapati AK, Ettiyappan JP. Computational analysis of therapeutic enzyme uricase from different source organisms. 2020; 17:59–77.
76. Hommelgaard T, Nielsen PH. Enzyme therapy in the treatmenta of sinusitis: Streptokinase-streptodornase (Varidase) buccal tablets used in the treatment of sinusitis maxillaris. 1964; 58:273–277.
77. Neidleman SL. Applications of biocatalysis to biotechnology. 1984; 1:1–38.

78. Qureshi MA, Khare AK, Pervez A. Enzymes used in dairy industries. 2015; 1(10):523–527.
79. Sharma R, Chisti Y, Banerjee UC. Production, purification, characterization and applications of lipases. 2001; 19:627–662.
80. Ghosh PK, Saxena RK, Gupta R, et al. Microbial lipases: Production and applications. 1996;79:119–157.
81. Chandra P, Enespa, Singh R, et al. Microbial lipases and their industrial applications: A comprehensive review. 2020;19(1):169.
82. Sardessai YN, Bhosle S. Industrial potential of organic solvent tolerant bacteria. 2004; 20(3):655–660.
83. Rossa PN, De Sa EMF, Burin VM, et al. Optimization of microbial transglutaminase activity in ice cream using response surface methodology. 2011; 44:29–34.
84. Kieliszek M, Misiewicz A. Microbial transglutaminase and its application in the food industry. A review. 2014; 59:241–250.
85. Soares I, Tavora Z, Patera R, et al. Microorganism-produced enzymes in the food industry. In: Valdez DB (ed.), *Food Industry, Scientific, Health and Social Aspects of the Food Industry*. 2012; 83–94. doi:10.5772/31256.
86. Forsgård RA. Lactose digestion in humans: Intestinal lactase appears to be constitutive whereas the colonic microbiome is adaptable. 2019; 110(2):273–279. https://doi.org/10.1093/ajcn/nqz104.
87. He T, VenemaK, Priebe MG, et al. The role of colonic metabolism in lactose intolerance. 2008; 38:541–547.
88. Geiger B, Nguyena H, Weniga S, et al. From by-product tovaluable components: Efficient enzymaticconversion of lactose in whey using β-galactosidase fromStreptococcusthermophilus. 2016; 116:45–53.
89. Saqib S, Akram A, Halim SA, et al. Sources of *β*-galactosidase and its applications in food industry. 2017; 7(1):79.
90. Kardel G, Furtado MM, Neto JPM. Lactase na induˊstria de laticıˊnios. Revista do Instituto de Laticıˊnios "Caˆndido Tostes" 1995; 50(294):15–17.
91. Pivarnik LF, Senegal AG, Rand AG. Hydrolytic and transgalactosil activities of commercial b-galactosidase (lactase) in food processing. 1995; 38:1–102.
92. Mahoney RR. Lactose: Enzymatic modification. In: Fox PF (ed.). *Advanced Dairy Chemistry-Lactose, Water, Salts and Vitamins* (2nd ed.). Chapman & Hall, London. 1997; 77–125.
93. Afroz QM, Khan KA, Ahmed P, et al. Enzymes used in dairy industries. 2015; 1(10):523–527.
94. www.marketsandmarkets.com/PressReleases/baking-enzymes.asp.2014a: Baking Enzymes Market.
95. Kuraishi C, Sakamoto J, Yamazaki K, et al. Production of restructured meat using microbial transglutaminase without salt or cooking. 1997; 62:488–490.
96. Moore MM, Heinbockel M, Dockery P, et al. Network formation in gluten-free bread with application of transglutaminase. 2006; 83:28–36.
97. Andreu P, Collar C, MartinezAnaya MA. Thermal properties of doughs formulated with enzymes and starters. 1999; 209:286–293.
98. Dauter Z, Dauter M, Brzozowski AM, et al. X-ray structure of Novamyl, the five-domain 'maltogenic' a-amylase from *Bacillus stearothermophilus*: Maltose and acarbose complexes at 1.7 A ° resolution. 1999; 38:8385–8392.
99. Monfort A, Blasco A, Sanz P, et al. Expression of LIP1 and LIP2 genes from Geotricum species in baker's yeast strains and their application to the bread-making process. 1999; 47:803–808.
100. Collar C, Martinez JC, Andreu P, et al. Effect of enzyme associations on bread dough performance: A response surface study. 2000; 6:217–226.
101. Kirk O, Borchert TV, Fuglsang CC. Industrial enzyme applications. 2002; 13:345–351.
102. Fernandes P. Enzymes in food processing: A condensed overview on strategies for better biocatalysis. 2010; 2010(1):862537.

103. Adrio JL, Demain AL. Microbial enzymes: Tools for biotechnological processes. 2014; 4(1):117–139.
104. Miguel ÂSM, Martins-Meyer TS, Veríssimoda CFE, et al. Enzymes in bakery: Current and future trends. In: Muzzalupo I (ed.), *Food Industry*. InTech, Rijeka, Croatia. 2013.
105. Tucker GA, Woods LFJ. Eds. *Enzymes in Food Processing*. Springer, Boston, MA. 1995. https://doi.org/10.1007/978-1-4615-2147-1.
106. Aruna K, Shah J, Birmole R. Production and partial characterization of alkaline protease from *Bacillus tequilensis* strains CSGAB 0139 isolated from spoilt cottage cheese. 2014;5:201–221.
107. Law BA. The nature of enzymes and their action in foods. In: Whitehurst RJ, Law BA (eds.), *Enzyme in Food Technology*. Sheffield Academic Press, Sheffileld. 2002; 1–18.
108. Kumar S. Role of enzymes in fruit juice processing and its quality enhancement. 2015; 6(6):114–124.
109. Garg G, Singh A, Kaur A, et al. Microbial pectinases: An ecofriendly tool of nature for industries. 2016; 6(1):47–59.
110. Mojsov K. Microbial alpha-amylases and their industrial applications: A review. 2012; 2(10):583–609. ISSN 2249–0558.
111. Grassin C, Fauquembergue P. Fruit juices. In: Godfrey T, West S (eds.), *Industrial Enzymology* (2nd ed.). MacMillan Press, London, UK. 1996; 226–264.
112. Kabir F, Sultana MS, Kurnianta H. Polyphenolic contents and antioxidant activities of underutilized grape (*Vitis vinifera* L.) pomace extracts. 2015; 20(3):210–214.
113. Humpf HU, Schreier P. Bound aroma compounds from the fruit and the leaves of blackberry (*Rubus laciniata*, L.). 1991; 39:1830–1832.
114. Sajith S, Priji P, Sreedevi S, et al. An overview on fungal cellulases with an industrial perspective. 2016; 6(1):461.
115. Vaillant F, Millan A, Dornier M, et al. Strategy for economical optimization of the clarification of pulpy fruit juices using crossflow microfiltration. 2001; 48:83–90.
116. Sivaramakrishnan S, Gangadharan D, Nampoothiri KM, et al. a-Amylases from microbial sources—an overview on recent developments. 2006; 44(2):173–184.
117. Hotchkis JH, Soares NFF. *The Use of Active Packaging to Improve Citrus Juice Quality*. (9th ed.). 2000. International Citrus Congress, Orlando, FL; 1202–1205.
118. Kashyap DR, Vohra PK, Chopra S. Applications of pectinases in the commercial sector: A review. 2001; 77:215–227.
119. Yadav S, Yadav PK, Yadav D, et al. Pectin lyase: A review. 2001; 44(1):1–10.
120. Karlund A, Moor U, Sandell M, et al. The impact of harvesting, storage and processing factors on health-promoting phytochemicals in berries and fruits. 2014; 2(3):596–624.
121. Bamforth CW. Current perspectives on the role of enzymes in brewing. 2009; 50(3):353–357.
122. Galante YM, DeConti A, Monteverdi R. Application of *Trichoderma* enzymes in food and feed industries. In Harman GF, Kubicek CP (eds.), *TrichodermaandGliocladium—Enzymes, Vol. 2 of Biological Control and Commercial Applications*. Taylor & Francis, London, UK. 1998; 311–326.
123. Oksanen J, Ahvenainen J, Home S. *Microbial Cellulose for Improving Filterability of Wort and Beer*. Proceedings of the 20th European Brewery Chemistry Congress, Helsinki, Finland. 1985.
124. Gallego CMV, Otamendi FP, Vidal DR, et al. Production and characterization of an aspergillus terteus α-L-rhamnosidase of oenological interest. 1996;203(6):522–527.
125. Steiner E, Becker T, Gast M. Turbidity and haze formation in beer—insights and overview. 2010;116(4):360–368.
126. Choct M. Enzymes for the feed industry: Past, present and future. 2006; 62:5–15.
127. Collection of information on enzymes European Communities. 2002.

128. Blackburn DM, Greiner R. Enzymes used in animal feed: Leading technologies and forthcoming developments. In: *Functional Polymers in Food Science* (1st ed.); Scrivener Publishing LLC4. 2015. 4:47–73.

129. Bhat MK. Cellulases and related enzymes in biotechnology. 2000; 18:355–383.

130. Walsh GA, Power RF, Headon DR. Enzymes in the animalfeed industry. 1993; 11(10):424–430.

131. Chesson A. Feed enzymes. 1993; 45:65–79. doi:10.1016/0377-8401(93)90072-R.

132. Frias J, Doblado R, Antezana JR, et al. Inositol phosphate degradation by the action of phytase enzyme in legume seeds. 2003; 81(2):233–239.

133. Lei XG, Stahl CH. Nutritional benefits of phytase and dietary determinants of its efficacy. 2000; 17:97–112.

134. Lei XG, Stahl CH. Biotechnological development of effective phytases for mineral nutrition and environmental protection. 2001; 257:474–481.

135. Ojha BK, Singh PK, ShrivastavaN. Chapter 7 — enzymes in the animal feed industry. Enzymes in food biotechnology. *Production, Applications, and Future Prospects.*2019; 93–109.

136. Srivastava N, Singh P. Degradation of toxic pollutants from pulp & paper mill effluent. 2015; 40(183):221–227.

137. Maijala P, Kleen M, Westin C, et al. Biomechanical pulping of softwood with enzymes and white-rot fungus *Physisporinus rivulosus*. 2008; 43:169–177.

138. Kohli U, Nigam P, Singh D, et al. Thermostable, alkalophilic and cellulase free xylanase production by *Thermoactinomyces thalophilus* subgroup C. 2001; 28:606–610. https://www.sciencedirect.com/science/article/abs/pii/S0141022901003209.

139. Dwivedi P, Vivekanand V, Pareek N, et al. Bleach enhancement of mixed wood pulp by xylanase—laccase concoction derived through co-culture strategy applied. 2010; 160(1):255–268.

140. Fu GZ, Chan AW, Minns DE. Preliminary assessment of the environmental benefits of enzyme bleaching for pulp and paper making. 2005; 10:136–142.

141. Kuhad RC, Gupta R, Singh A. Microbial cellulases and their industrial applications. 2011. doi:10.4061/2011/ 280696.

142. Nigam PS. Microbial enzymes with special characteristics for biotechnological applications. 2013; 3(3):597–611.

143. Kirk TK, Jeffries TW. Role of microbial enzymes for pulp and paper processing. ACS symposium series: Am Chem Soc Washington, DC. 1996; 2–14. doi:10.1021/bk-1996-0655.ch001.

144. Demuner BJ, Pereira JN, Antunes A. Technology prospecting on enzymes for the pulp and paper industry. 2011; 6(3):148–158.

145. Patrick K. Enzyme technology improves efficiency, cost, safety of stickies removal program. 2004; 120:22–25.

146. Clarke JH, Davidson K, Rixon JE, et al. A comparison of enzyme aided bleaching of softwood pulp using a combination of xylanase, mannanase and a-galactosidase. 2000; 53:661–667.

147. Ahuja SK, Ferreira GM, Moreira AR. Utilization of enzymes for environmental applications. 2004; 24:125–154.

148. Choi JM, Han SS, Kim HS. Industrial applications of enzyme biocatalysis: Current status and future aspect. 2015; 33:1443–1454.

149. Araujo R, Casal M, Cavaco-Paulo A. Application of enzymes for textiles fibers processing. 2008; 26:332–349.

150. Chen S, Su L, Chen J, et al. Cutinase: Characteristics, preparation, and application. 2013; 31(8):1754–1767.

151. Mojsov K. Applications of enzymes in the textile industry: A review. In: *2nd International Congress: Engineering, Ecology and Materials in the Processing Industry: Jahorina*. Bosnia and Herzegovina; Tehnoloski Fakultet, Zvornik. 2011; 230–239.

152. Ahlawat S, Dhiman SS, Battan B, et al. Pectinase production by *Bacillus subtilis* and its potential application in biopreparation of cotton and micropoly fabric.2009; 44:521–526.
153. https://infinitabiotech.com/blog/top-3-uses-of-enzymes-in-textile-industry.
154. Uygur A. An overview of oxidative and photooxidative decolorisation treatments of textile waste waters. 1997; 113:211–217.
155. Uygur A. Reuse of decoulorized wastewater of azo dyes containing dichlorotriazinyl reactive groups using an advanced oxidation method. 2001; 117:111–113.
156. Tzanov T, Costa S, Guebitz GM, et al. Dyeing in catalase-treated bleaching baths. 2001a; 117:1–5.
157. Amorim AM, Marcelo DG, Andreaus GJ, et al. 2002. The application of catalase for the elimination of hydrogen peroxide residues after bleaching of cotton fabrics. *Anais da Academia Brasileira de Ciências*. 2002; 74(3):433–436.
158. R. Doshi, V. Shelke. Enzymes in textile industry—An environment-friendly approach. 2001;26:202–205.
159. Campos R, Kandelbauer A, Robra K, et al. Indigo degradation with purified laccases from Trametes hirsuta and Sclerotium rolfsii. 2001; 89(2):131–139.
160. Zille A. Laccase reactions for textile applications, PhD Thesis, Universidade do Minho, Portugal. 1996.
161. Ryan S, Schnitzhofer W, Tzanov T, et al. An acid-stable laccase from *Sclerotium rolfsii* with potential for wool dye decolourization. 2003; 33:766–774.
162. Sharma S, Whiteside L, Kernaghan K. Enzymatic treatment of flax fibre at the roving stage for production of wet-spun yam. 2005; 37(4):386–394.
163. Mojsov K. Biotechnological applications of laccases in the textile industry. 2014; 3(1):76–79.
164. Coudhary RB, Jana AK, Jha MK. Enzyme technology applications in leather processesing. 2004; 11:659–671.
165. Covington AD. *Tanning chemistry: The science of leather*. RSC Publishing: Cambridge, UK. 2009.
166. Feigel T. Use of enzymes in the beam house possibilities and limitations. 1998; 5:54–55.
167. Mhya DH, Mankilik M. Bacterial enzymes: A good alternative to conventional chemicals in leather processing. 2015; 2:20–23.
168. Zambare VP, Nilegaonkar SS, Kanekar PP. Protease production and enzymatic soaking of salt-preserved buffalo hides for leather processing. 2013; 3:1–7.
169. Bienkewicz K. *Physical Chemistry of Leather Making*. Ed. Robert E. Krieger Publishing Co. Inc., Malabar, FL. 1983.
170. Unni KN, Priji P, Sajith S, et al. *Pseudomonas aeruginosa* strain BUP2, a novel bacterium inhabiting the rumen of Malabari goat, produces an efficient lipase. 2016; 71(4):378–387.
171. Iqbal SA, Rehman A. Characterization of lipase from *Bacillus subtilis* I-4 and its potential use in oil contaminated wastewater. 2015; 58(5):789–797.
172. Niyonzima FN, More SS. Coproduction of detergent compatible bacterial enzymes and stain removal evaluation. 2015; 55(10):1149–1158.
173. Fotouh DMA, Bayoumi RA, Hassan MA. Production of thermoalkaliphilic lipase from *Geobacillus thermoleovorans* DA2 and application in leather industry. 2016; Article ID 9034364. doi:10.1155/2016/9034364.
174. Ali NEH, Agrebi R, Ghorbel B, et al. Biochemical and molecular characterization of detergent stable alkaline serine protease from a newly *Bacillus lichenniformis* NH1.2007; 40(4):515–523.
175. Al-Abdalall AH, Al-Khaldi EM. Production of alkaline proteases by alkalophilic Bacillus subtilis during recycling animal and plant wastes. 2016; 15(47):2698–2702.
176. Gushterova A, Tonkova ESV, Dimova EY, et al. Keratinase production by newly isolated Antarctic *actinomycete* strains. 2005; 21(6):831–834.

177. Liu L, Yang H, Shin HD. How to achieve high-level expression of microbial enzymes strategies and perspectives. 2013; 4(4):212–223.

178. Sankaran S. *Five Decades of Leather: A Journey Down Memory Lane*. Indian Leather, Madras. 1995.

179. Bailey DG. Handling, greading and curing of hides and skins. In: Pearson AM, Dutson TR (eds.), *Inedible Meat by-Products*. Springer. 1992; 19–34. doi:10.1007/978-94-011-7933-1_2.

180. Raju AA, Chandrababu NK, Rose C, et al. Eco-friendly enzymatic dehairing using extracellular proteases from a *Bacillus* species isolate. 1996; 91:115–119.

181. De-Souza FR, Gutterres M. Application of enzymes in leather processing: A comparison between chemical and coenzymatic processes. 2012; 29(3):471–481.

182. Money CA. Unhairing and dewooling-requirement for quality and the environment. 1996; 80:175–186.

183. Green GH. Unhairing by means of enzymes. 1952; 36:127–134.

184. Rao MB, Tanksale AM, Ghatge MS, et al. Molecular and biotechnological aspects of microbial proteases. 1998; 62(3):597–635.

185. Bhatia SC. *Managing Industrial Pollution*. Macmillan India, New Delhi. 2003.

186. Gomes C, Silva AC, Marques AC, et al. Review biotechnology applied to cosmetics and aesthetic medicines. *Cosmetics*. 2020; 7:33.

187. Zappelli C, Barbulova A, Apone F, et al. Effective active ingredients obtained through biotechnology. 2016; 3:39.

188. Regulation (EC) No 1223/2009 of the european parliament and of the council of 30 November 2009 on cosmetic products.

189. Pandey A, Höfer R, Larroche C, et al. *Industrial Biorefineries and White Biotechnology* (1st ed.). Elsevier Science, New York. 2015; 608–640.

190. Wang Y, Branicky R, Noe A, et al. Superoxide dismutases: Dual roles in controlling ROS damage and regulating ROS signaling. 2018; 217:1915–1928.

191. Younus, H. Therapeutic potentials of superoxide dismutase. 2018; 12:88–93.

192. Seki T, Yajima I, Yabu T, et al. Examining an exfoliation-promoting enzyme for cosmetic applications. 2005; 120:87.

193. McNeil E, Melton D. The good and bad sides of DNA repair: DNA damage in the skin and melanoma. 2013; 35:25–29.

194. Hitomi K, Okamoto K, Daiyasu H, et al. Bacterial cryptochrome and photolyase: Characterization of two photolyase-like genes of *Synechocystis* sp. PCC6803.2000; 28(12):2353–2362.

195. Portero LR, Alonso-Reyes DG, Zannier F, et al. Photolyases and cryptochromes in UV-resistant bacteria from high-altitude Andean Lakes. 2018.https://doi.org/10.1111/php.13061.

196. Duarte AS, Correia A, Esteves AC, Bacterial collagenases—A review. 2016; 42:106.

197. Watanabe K. Collagenolytic proteases from bacteria. 2004; 63:520–526.

198. Gopinath SCB, Anbu P, Lakshmipriya T, et al. Biotechnological aspects and perspective of microbial keratinase production. 2015;Article ID 140726, 10 pages http://dx.doi.org/10.1155/2015/140726.

199. Brandelli A. Bacterial keratinases: Useful enzymes for bioprocessing agroindustrial wastes and beyond. 2008; 1:105–116.

200. Gherbawy YAMH, Maghraby TA, El-Sharony HM, et al. Diversity of keratinophilic fungi on human hairs and nails at four governorates in Upper Egypt. 2006; 34:4, 180–184.

201. Lods LM, Dres C, Johnson DB, et al. Brooks, the future of enzymes in cosmetics. 2000; 22(2):85–94.

202. Bernard D, Mehul BC. U.S. Patent 7521422B2 21, April 2009.

203. Cho SA, Cho JC, Han SH. Cosmetic composition containing enzyme and amino acid. 2007; 11/990:431.

204. Buckingham KWC. Methods for the treatment and prophylaxis of diaper rash and diaper dermatitis. *The Procter & Gamble Company*. 1985; Patent No US4556560 A.
205. Lang G, Cotteret J. Composition for the oxidation dyeing of keratinous fibres containing a laccase and dyeing method using this composition. US2004255401 A1.2004.
206. Bruinsma GM, Mei HCV, Busscher HJ. Bacterial adhesion to surface hydrophilic and hydrophobic contact lenses. 2001; 22:3217–3224.
207. Butrus SI, Klotz SA. Contact lens surface deposits increase the adhesion of *Pseudomonas aeruginosa*. 1990; 9:717–724.
208. Alfa MJ, Jackson M. A new hydrogen peroxide-based medical device detergent with germicidal properties: Comparison with enzymatic cleaners. 2001; 29(3):168–177.
209. Pawar R, Zambare V, Barve S, et al. Application of protease isolated from *Bacillus* sp. 158 in enzymatic cleansing of contact lenses. 2009; 8:276–280.
210. Schafer T, Kirk O, Borchert T, et al. Enzymes for technical applications. In: Fahnestock SR, Steinbü̈chel A (eds.), *Biopolymers*. Wiley-VCH, Cambridge. 2002; 377–437.
211. Hasan F, Shah AA, Javed S, et al. Enzymes used in detergents: Lipases. 2010; 9(31):4836–4844.
212. Masse L, Kennedy KJ, Chou S. Testing of alkaline and enzymatic pretreatment for fat particles in slaughterhouses wastewater. 2001; 77:145–155.
213. Mitidieri S, Martinelli S, Schrank AH, et al. Enzymatic detergent formulation containing amylase from *Aspergillus niger*: A comparative study with commercial detergent formulations. 2006; 97:1217–1224.
214. Elbrahim N, Ma K. Industrial applications of thermostable enzymes from extremophilic microorganisms. 2017; 4(2):75–98.
215. Hauthal HG. *Types and Typical Ingredients of Detergents. Handbook of Detergents, Part C*. CRC Press, Boca Raton. 2016; 19–118.
216. Varnam A, Sutherland JP. *Milk and Milk Products: Technology, Chemistry and Microbiology*. Springer, Berlin. 2001; 1(1):452.
217. Bano K, Kuddus M, Zaheer MR, Zia Q, et al. Microbial enzymatic degradation of biodegradable plastics. 2017;18(50):429–440.
218. Applications of Lipases. *AU-KBC Research Center, Life Sciences*, Anna University, Chennai, India.www.au-kbc.org/beta/bioproj2/introduction.html.
219. Keshwani A, Malhotra B, Kharkwal H. Natural polymer based detergents for stain removal. 2015; 4(4):490–508.
220. Olsen HS, Falholt P. The role of enzymes in modern detergency. 1998; 1(4):555–567.
221. Novozyme: Enzyme at Work. 2013 (4th ed.). http://novozymes.com/ en/about-us/brochures/Documents/Enzymes_at_work.pdf.
222. Kiamarsi Z, Soleimani M, Nezami A, et al. Biodegradation of n-alkanes and polycyclic aromatic hydrocarbons using novel indigenous bacteria isolated from contaminated soils. 2019;16(1):6805–6816.
223. Hasan F, Shah A, Hameed A. Purification and characterization of a mesophilic lipase from *Bacillus subtilis* FH5 stable at high temperature and pH. 2007; 58(1):115–132.
224. Chandra P, Singh DP. Removal of Cr (VI) by a halotolerant bacterium *Halomonas* sp. CSB 5 isolated from sāmbhar salt lake Rajastha (India). 2014; 60(5):64–72.
225. Wakelin NG, Forster CF. An investigation into microbial removal of fats, oils and greases. 1997; 59(1):37–43.
226. Fujii T, Tatara T, Minagawa M. Studies on applications of lipolytic enzyme in detergency I. Effect of lipase from *Candida cylindracea*on removal of olive oil from cotton fabric. 1986; 63(6):796–799.
227. Nerurkar M, Vaidyanathan J, Adivarekar R, et al. Use of a natural dye from *Serratia marcescens*subspecies *Marcescens*in dyeing of textile fabrics. 2013;1(2):129–135.
228. Bacha AB, Al-Assaf A, Moubayed NM, et al. Evaluation of a novel thermo-alkaline *Staphylococcus aureus*lipase for application in detergent formulations. 2018; 25(3):409–417.

229. Pandey D, Singh R, Chand D. An improved bioprocess for synthesis of acetohydroxamic acids using DTT (dithiothreitol) treated resting cells of *Bacillus sp.* APB-6.2011; 102(11):6579–6586.

230. Rubilar O, Diez MC, Gianfreda L. Transformation of chlorinated phenolic compounds by white rot fungi. 2008; 38:227–268.

231. Raj J, Prasad S, Bhalla TC. *Rhodococcus rhodochrous* PA-34: A potential biocatalyst for acrylamide synthesis. 2006; 41:1359–1363.

232. Klibanov AM, Tu TM, Scott KP. Enzymatic removal of hazardous pollutants from industrial aqueous effluents. 1982; 6:319–323.

233. Karigar CS, Rao SS. Role of microbial enzymes in the bioremediation of pollutants: A review. 2011; 1–11.doi:10.4061/ 2011/805187.

234. Riffaldi R, Levi-Minzi R, Cardelli R, et al. Soil biological activities in monitoring the bioremediation of diesel oilcontaminated soil. 2006;170:3–15.

235. Margesin R, Zimmerbauer A, Schinner F. Soil lipase activity-A useful indicator of oil biodegradation. 1999;13(12):859–863.

236. Piontek K, Smith AT, Blodig W. Lignin peroxidase structure and function. 2001; 29(2):111–116.

237. Have RT, Teunissen PJM. Oxidative mechanisms involved in lignin degradation by white-rot fungi. 2001;101(11):3397–3413.

238. Mai C, Schormann W, Milstein O, et al. Enhanced stability of laccase in the presence of phenolic compounds. 2000; 54(4):510–514.

239. Gianfreda L, Xu F, Bollag JM. Laccases: A useful group of oxidoreductive enzymes. 1999; 3(1):1–25.

240. Kalia VC, Lal SR, et al. Using enzymes for oil recovery from edible seeds. 2001; 60:298–310.

241. https://www.un.org/en/development/desa/news/population/2015-report.html.

242. Fetzner S, Lingens F. Bacterial dehalogenases: Biochemistry, genetics, and biotechnological applications. 1994; 58(4):641–685.

243. Siroosi M, Amoozegar MA, Khajeh K, et al. Decolorization of dyes by a novel sodium azide-resistant spore laccase from a halotolerant bacterium, *Bacillus safensis* sp. strain S31.2018; 77(12):2867–2875.

244. Ng TW, Cai Q, Wong C, et al. Simultaneous chromate reduction and azo dye decolourization by *Brevibacterium casei*: Azo dye as electron donor for chromate reduction. 2010; 182(1–3):792–800.

245. Dawkar VV, Jadhav UU, Jadhav SU, et al. Biodegradation of disperse textile dye Brown 3REL by newly isolated *Bacillus* sp. VUS. 2008; 105(1):14–24.

246. Molloy S, Nikodinovic-Runic J, Martin LB, et al. Engineering of a bacterial tyrosinase for improved catalytic efficiency towards D-tyrosine using random and site directed mutagenesis approaches. 2013; 110(7):1849–1857.

247. Roy S, Das I, Munjal M, et al. Isolation and characterization of tyrosinase produced by marine actinobacteria and its application in the removal of phenol from aqueous environment. 2014; 9:306–316.

248. Abd-Elhalem BT, Rawia MS, Khadiga FG, et al. Production of amylases from *Bacillus amyloliquefaciens* under submerged fermentation using some agro-industrial by-products. 2015; 60(2):193–202.

249. Asgher M, Asad MJ, Rahman SU, et al. A thermostable α-amylase from a moderately thermophilic *Bacillus subtilis* strain for starch processing. 2007; 79:950–955.

250. Mutreja R, Das D, Goyal D, et al. Research article bioconversion of agricultural waste to ethanol by SSF using recombinant cellulase from *clostridium thermocellum*. 2011; Article ID 340279, 6 pages. doi:10.4061/2011/340279.

251. Reka V, Ananthi T. Isolation and characterization of extracellular cellulase using *Bacilluls subtilis* from mangrove soil. 2013; 8(2):67–71.

252. Agobo KU, Arazu VA, Uzo K, et al. Microbial lipases: A prospect for biotechnological industrial catalysis for green products: A review. 2017; 6(2):1–12.
253. Porwal HJ, Mane AV, Velhal SG. Biodegradation of dairy effluent by using microbial isolates obtained from activated sludge. 2015; 9:1–15.
254. Asano Y, Tani Y, Yamada H. A new enzyme "Nitrile Hydratase" which degrades acetonitrile in combination with amidase. 1980; 44:2251–2252. doi:10.1271/bbb1961.44.2251.
255. Chen J, Zheng RC, Zheng YG, et al. Microbial transformation of nitriles to high-value acids or amides. In: Zhong JJ, Bai FW, Zhang W (ed.), *Biotechnology in China I: From Bioreaction to Bioseparation and Bioremediation.* Springer Berlin Heidelberg, Berlin. 2009; 33–77.
256. Kohyama E, Yoshimura A, Aoshima D, et al. Convenient treatment of acetonitrile-containing wastes using the tandem combination of nitrile hydratase and amidase-producing microorganisms. 2006; 72(3):600–606.
257. Verma A, Ansari MW, Anwar MS., et al. Alkaline protease from *Thermoactinomyces* sp. RS1 mitigates industrial pollution.2014; 251(3):711–718.
258. Cortezi M, Cilli EM, Contiero J. *Bacillus amyloliquefaciens*: A new keratinolytic feather-degrading bacteria. 2008; 2:170–177.
259. Bender H. A bacterial glucoamylase degrading cyclodextrins, Partial purification and properties of the enzyme from a *Flavobacterium*species. 1981; 115:287–291.
260. Gill RK, Kaur J. A thermostable glucoamylase from a thermophilic *Bacillus* sp.: Characterization and thermostability. 2004; 11:540–543.

5 A Quick Look-Around of Microbial Enzymes in Modern Food Industries and Dietary Research

Vineet Singh, Anjali Pande, and Jae-Ho Shin

ABSTRACT

The use of microbial enzymes in food production started way before the discovery of the enzyme itself. In modern times microbial enzymes are rigorously utilized in the production and processing of food, especially in bakery, brewery, dairy and animal feed industries. Commercially, these enzymes are less expensive and easy to produce on a large scale, which makes them suitable for their industrial utilization. In food industries, microbial enzymes are used to improve the nutritional value and food processing as they provide essential metabolites, color, aroma, consistency, and texture to the final product. According to the industrial requirement, specific microbial enzymes are used based on their suitability in the food industry, such as transglutaminase is used to prepare gluten-free bread, lactase is used to prepare dairy products for milk-intolerant people, naringinase controls the bitterness in citrus juices, invertase is utilized to prepare probiotic drinks like "Kombucha", and proteases from *Bacillus subtilis* and *Aspergillus oryzae* are used to tenderize the meat products, etc. Additionally, microbial enzymes also assist to maintain healthy livestock by increasing the nutrient availability in the host gut, such as *Aspergillus niger*-derived phytase degrades the anti-nutritional phytic acid, while xylanase and β-glucanase ease the digestion of starch. At present, many of the microbial enzymes used in food industries are obtained from manipulated recombinant microbes with exogenous inserted sequences, as they can provide a better yield of the required enzyme. Moreover, microbial enzymes such as amylase, lactase, etc., can be used for dietary research involving study on biochemical properties of different dietary fibers and studies mimicking the murine digestion by in-vitro model. Furthermore, the industrialization of microbial enzymes is highly advantageous as it neither involves any animal cruelty nor it overburdens natural resources.

Keywords: Microbial enzyme, Food production, Food processing, Dietary research, In-vitro digestion studies

DOI: 10.1201/9781003202998-5

CONTENTS

5.1   Introduction.. 92
5.2   Industrial Production of Microbial Enzymes.. 93
5.3   Enzymes in the Food Industry 93
5.4   Microbial Enzymes in the Bakery Industry 96
5.5   Microbial Enzymes in the Dairy Industry 96
5.6   Microbial Enzymes in the Beverage Industry.. 97
5.7   Microbial Enzymes in the Meat and Animal Feed Industry 97
5.8   Microbial Enzymes in Food and Dietary Research 98
5.9   Conclusion and Future Prospects... 98
5.10 Acknowledgment... 99
References.. 99

5.1   INTRODUCTION

Enzymes are high-molecular-weight proteinaceous catalysts that are highly specific to their substrate and have a higher rate of reaction compared to catalysts. Unlike catalysts, enzymes are biological in origin (i.e., synthesized in living organisms), and they are highly sensitive to various factors, such as temperature, pH, and amount of substrate. Since ancient times, humankind has been using microbes in preparing food such as cheese, curd, and fermented soy-based products, but it took us a little longer to understand that, in most cases, it was actually the microbial enzymes that were involved in food production. Presently, we are totally aware of the importance of microbial enzymes in various fields of the food industry, such as the milk industry, meat industry, fruit/juice industry, and bakery industry, and all these industries actively utilize various enzymes of microbial origin, such as amylase, rennin, and papain. For instance, the enzyme rennin, which was earlier extracted from the stomach of young ruminant animals like calves and lambs, is now obtained from the microbes.

The global food industry supports the food demand of the world, but as the world population is continuously growing and expected to be around 9.1 billion by 2050, the food demand will be 70% higher than today, thus overburdening the food industry. In this scenario, microbial enzymes can be a great asset to address the huge future food demands, as these biomolecules are involved in fast production, higher yield, and quality improvement in the food industry. Enzymes can be obtained from animals, plants, and microbes, but compared to microbial enzymes, animal and plant enzymes have higher production costs, along with other manufacturing complexities. Additionally, microbial cultures are easy to maintain and can be manipulated for better yield, and there are no such animal cruelty issues that are especially associated with animal enzymes.

Moreover, present advancement in biotechnology, industrial microbiology, and enzyme technology has also made it possible to produce high-quality microbial enzymes on a commercial scale [1]. According to reports, the global food enzyme market is estimated to be $2.2 billion in 2021 and projected to reach $3.1 billion by 2026 (www.marketsandmarkets.com/Market-Reports/food-enzymes-market-800. html). Such great economic potential and relevance of enzymes in food industries are expected to generate a huge demand for microbial enzymes in the future. Even at present, more than 80% global enzyme market is captured by the enzymes used in the industrial requirements, in which enzymes related to the food industry dominates [2]. Microbial enzymes are quite suitable for the food industry as their bulk production is cheaper and less time-consuming, which makes them much more economical compared to enzymes obtained from other sources. In the food industry, microbial enzymes are used in various food sectors, such as the bakery industry, dairy industry, beverage industry, and animal feed industry.

5.2 INDUSTRIAL PRODUCTION OF MICROBIAL ENZYMES

Enzymes are valuable bioactive compounds available in all living organisms, but for industrial purposes, enzymes are usually obtained from microbial sources, such as bacteria, yeast, actinomycetes, and fungi. Though enzymes can be obtained from plants and animals, but microbes can be manipulated very easily to get the desired product with a better yield. For example, initially, rennin was obtained from calves, but nowadays, most of the rennin is obtained from the microbe *Rhizomucor miehei* [3]. Additionally, microbial productivity can be enhanced further by inserting multiple copies of a specific gene and by adding a strong promoter. Moreover, a suitable foreign gene can be inserted into the microbial genome to create a heterologous expression system that also enhances enzyme production [4]. Other than that, screening of selected microbes of interest can be simplified by adding any selectable marker, such as antibiotic resistance genes. After successful manipulation, a high-yielding microbial strain is obtained that can be used in downstream processing for the commercial production of enzymes. Presently, most of the enzymes utilized in the food and feed industry are obtained from such recombinant microorganisms.

Generally, commercial production of enzymes occurs in a large batch of microbes allowed to grow in an industrial fermenter in optimized controlled conditions. Mainly, there are two commonly operated fermentation methods for enzyme production; the first is solid-state fermentation, and the second is submerged fermentation. The production of enzymes through solid-state fermentation is carried out on the solid-substrate surface itself. Solid-state fermentation is considered to be more suitable for the production of fungal enzymes than bacterial ones and has better productivity over submerged fermentation [5]. On the other hand, in the submerged fermentation method, microbes are allowed to grow in a liquid broth media enriched with nutrients to optimize their growth [6]. Commercially, submerged fermentation is the most used strategy in the production of different fermented products, including enzymes for the food industry [7–8]. Further, based on the oxygen requirement of microbes, the fermentation process can be broadly classified into aerobic fermentation and anaerobic fermentation.

5.3 ENZYMES IN THE FOOD INDUSTRY

Enzymes have become an essential part of the food industry as they are used to hydrolyze complex compounds into their simpler forms. Along with that, they also provide additional advantages, such as enhanced taste, texture, and added metabolites. On the basis of industrial requirements, various enzymes are frequently used in the food industry (Table 5.1).

TABLE 5.1

Various Microbial Enzymes Used in Food Industry along with Their Detailed Information

Class	Enzyme	Role	Application	Microbes
Oxidoreductases	Lipoxygenase	Catalyzes the oxygenation, by molecular oxygen, of fatty acids containing a cis, cis-1,4-pentadiene system to produce conjugated hydroperoxydiene derivatives	Dough strengthening, bread whitening	Lasiodiplodia theobromae
	Laccases	Oxidizes the phenolic and nonphenolic compounds to produce dimers, oligomers, and polymers	Clarifies the beverages; enhances flavor	Aspergillus niger, Aspergillus oryzae, and E. coli BL21
	Glucose oxidase	Oxidation of β-D-glucose into D-glucono-δ-lactone at its first hydroxyl group	Dough strengthening	Aspergillus niger ZGL528–72, Aspergillus oryzae Mtl-72
Transferases	Fructosyltransferase	Microbial fructan biosynthesis	Biosynthesis of fructose	Aureobasidium pullulans, Aspergillus japonicus
	Transglutaminase	Catalyzes the formation of isopeptide bonds between proteins	Dough and meat processing	Streptoverticillium mobaraense S-8112
Hydrolases	Amylases	Hydrolyzes the glycosidic bonds in starch molecules to convert them into simple sugar	Bread softening, beverages, and beer processing	Bacillus subtilis, Bacillus licheniformis
	Esterases	Catalyzes the splitting of esters into acid and alcohol	Modification of oil and fat in fruit juices; produces fragrances and flavor	Bacillus licheniformis
	Glycosidases	Catalyzes the hydrolysis of the β-glucosidic linkages	Enhances aroma and flavor of sparkling wine	Saccharomyces cerevisiae

Class	Enzyme	Action	Application	Source
	Glucoamylases	Catalyzes the hydrolysis of polysaccharide starch, releasing β-glucose	Reduces dough staling; improves bread crust color and the quality of baked products	*Rhizopus oryzae*
	Invertase	Catalyzes the hydrolysis of the disaccharide sucrose into glucose and fructose	Sugar syrup	*Aspergillus japonicus, Saccharomyces sp.*
	Lactase	Cleaves lactose into its two components: glucose and galactose	Ice cream, condensed milk, and baby food production	*Bacillus licheniformis PP3930, Kluyveromyces fragilis*
	Lipase	Breaks down the triglycerides into free fatty acids and glycerol	Enhances flavor in dairy products and fatty food	*Aspergillus niger, Aspergillus oryzae*
	Proteases	Hydrolyzes peptide bonds of proteins	In baking, regulates the gluten strength, texture, and flavor	*Bacillus stearothermophilus*
	Phospholipases	Catalyzes the hydrolysis of acyl esters and phosphate esters	Degumming of vegetable oils, cheese, and bread	*Fusarium oxysporum*
	Pectinase	Breaks the glycosidic bonds present between the galacturonic acid monomers	Clarification and filtration of beverages (e.g., juices)	*Aspergillus niger, Rhizopus oryzae*
Lyases	Acetolactate decarboxylase	Catalyzes the decarboxylation of α-acetolactate to acetoin	Beer maturation	*Bacillus subtilis ToC46*
Isomerases	Xylose (glucose) isomerase	Catalyzes the reversible isomerization of glucose to fructose and that of xylose to xylulose	Industrial production of high-fructose corn syrup	*Microbacterium arborescens NRRL B-11022, Streptomyces murinus DSM 3252*

5.4 MICROBIAL ENZYMES IN THE BAKERY INDUSTRY

The bakery industry involves the manufacturing of different breads, buns, cakes, and various other food products, to provide daily nourishment. Globally, the bakery industry is the single largest utilizer of microbial enzymes in the food industry and is growing at the rate of 8.2% from 2013 to 2019 [9]. In the bakery industry, microbial enzymes, such as amylase, proteinase, and lipolytic, are routinely used for efficient mass-scale production and quality improvement of baking products, which are used for better dough texture and color. Especially in bread-making, fungal α-amylase is used to degrade the damaged starch molecule into respective sugars, which can be further utilized by yeast during fermentation.

These microbial enzymes, mainly amylase and lipase, are also used to enhance the shelf life of baked products by reducing starch crystallization. Lipases also enhance the health-benefiting short-chain fatty acids in the dough, which is known to improve gut health [10]. Similarly, enzymes of lactic acid bacteria and other external microbial enzymes (lipase, protease, glucose-oxidase) also improve the nutritional value of bakery products during fermentation, such as in the manufacturing of sourdough bread [2, 11]. Similarly, while transglutaminase is widely used for its cross-linking property to make the dough firm, elastic, and of desired viscosity, other enzymes like cellulose and hemicelluloses are used to enhance the crumb-softness and quality of fresh bread, respectively, mainly by degrading starch into sugar. Additionally, low-fiber cereal products containing gluten (a nonnutritive protein causing gluten allergy characterized by blotting and stomach pain) are processed with microbial enzymes such as transglutaminase and laccase to prepare gluten-free bakery products for health-conscious people and those with celiac disease (an autoimmune disease, with severe gluten sensitivity) [12].

5.5 MICROBIAL ENZYMES IN THE DAIRY INDUSTRY

The global dairy industry was of whopping $718.9 billion in the year 2019 and expected to be $1.0327 trillion by 2024. As a major contributor to the dairy industry, the demand for fermented milk products and microbial enzymes is also expected to rise within this time span. Fermented milk products are one of the ancient products prepared using enzymes, and to date, they are equally popular as they were before. The dairy industry offers a lot of products, such as curd, cheese, yogurt, and kefir, which commonly utilize microbial enzymes.

The microbial enzymes, such as lactases, lipases, proteases, and esterases, are used to enhance the aroma, texture, flavor, and shelf life of dairy products. In the cheese industry, traditionally, animal rennet is used for the coagulation of milk, but now approximately 33% of global cheese demand is fulfilled through microbial rennet. Other than rennet, microbial lipase and protease are also used, which enhance the flavor with quicker ripening of the cheese, thus making them useful for accelerated commercial production of cheese. These enzymes also enhance the nutritional value of the final product. For instance, in most fermented dairy products, milk protein (casein) is converted into simpler peptides and amino acids, which makes it easier to digest and enhances its nutrition content. Microbial enzymes such as lactase are also used to prepare predigested milk products for lactose-intolerant people, which break down the lactose into glucose and fructose.

5.6 MICROBIAL ENZYMES IN THE BEVERAGE INDUSTRY

The beverage industry is divided into two separate commercial segments: the first is alcoholic beverages, and the second is nonalcoholic beverages. The alcoholic beverage industry includes breweries and distilleries, where microbial enzymes are consistently used for manufacturing high-quality products such as wine and beer. In the brewing industry, microbial enzymes are used to dissolve the cell wall of raw plant material to get a higher yield, desired aroma, and flavor. While for commercial liquor production, starch is converted to simple sugar molecules by α-amylase and amyloglucosidase, which is fermented by yeast to give a raw liquor, that is further purified and distilled to get the desired ethanol concentration [13–14]. In the liquor industry, nowadays, malt enzymes are completely replaced by enzymes of microbial origin because, as compared to traditional plant enzymes, they have higher activity, with a wide range of temperatures and pH [14]. Moreover, these enzymes also provide liquor with a distinct individual taste because β-amylase converts starch to maltose and amyloglucosidase hydrolyzes the product of α-amylase into glucose, along with the generation of several fragrant metabolites.

On the other hand, the nonalcoholic beverage segment includes fruit juices, vegetable juices, probiotic drinks, and tea. According to reports, the global fruit juice and vegetable market was $172 billion in 2020. In the fruit juice industry, microbial enzymes are mainly used for the extraction of fruit pulp from the fruit, digestion of the plant tissues, and lowering the viscosity of and clarifying the final product. For these purposes, amylase, cellulase, and pectinase are used more often, among others, to regulate the quantity and quality of juice products, while some specific enzymes are also used depending on certain requirements; naringinase and limoninase, for example, are used to control the bitterness in citrus juices [9]. Microbial enzymes are also used in tea manufacturing, where *Aspergillus*-derived pectinase is used for the tea leaves fermentation to improve the taste and color [15]. In the case of the coffee industry, enzyme mixtures containing pectinase, hemicellulose, and cellulose are used to produce some special types of coffees [16]. Additionally, microbial pectinase may be used during the fermentation of coffee beans to efficiently remove the mucilage coat from them [17]. The commercial production of health-benefiting probiotic products also utilizes microbial enzymes; kombucha, for example, is prepared using invertase.

5.7 MICROBIAL ENZYMES IN THE MEAT AND ANIMAL FEED INDUSTRY

In commercial meat processing, proteolytic enzymes are commonly used to tenderize the hard meat fibers, to prepare different meat products, where bacterial proteases are specifically used to digest the tough elastin and collagen fibers present in meat [18]. Proteases from *Aspergillus oryzae* and *Bacillus subtilis* are recognized as safe to be used in meat processing by the US department of agriculture [19]. Proteolytic subtilisin, neutral proteases, alkaline elastase, and aspartic proteases can be obtained from the *Bacillus subtilis* and *Aspergillus oryzae*, respectively [18]. Studies also found that certain microbial proteases can even work on a range of pH and low temperature, which makes them suitable for industrial meat production. For example,

alkaline elastase obtained from *Bacillus* spp. has optimum enzymatic activity at
10–15°C. Similarly, fungal proteases have also been used for the industrial production of proteases.

As we discussed above, the world population largely depends on animal or different animal products, such as milk, meat, and eggs, for their daily dietary requirements, and this worldwide demand is continuously growing. As one of the measures
to counter such a crisis, microbial enzymes can be added to animal feed to support
their healthy, balanced growth, which can ensure the continuous supply of animal
products. These feed enzymes enhance animals' ability to digest feed, availability of
nutrients, and feed consumption by animals, which ensures better mass gain, reduction in feed cost, and improvement in meat quality. Additionally, these enzymes are
also added with animal feed to achieve desired results, such as in pig feed xylanase,
and β-glucanase is added, which degrades the higher amount of starch in feed.
Microbial proteases and phytase are also used to degrade the anti-nutritional factors
present in animal feed to support their health [20–21]. Especially in the poultry and
pig industry, *Aspergillus niger*–derived phytase is used to increase the availability
of phosphorus by releasing the phosphorus from the phytate-phosphorus couple [21].

5.8   MICROBIAL ENZYMES IN FOOD AND DIETARY RESEARCH

Food and dietary science is a branch of science where the dietary impact of various
food substances over the host is studied using various models, which may be either
in vitro or in vivo. The in vivo model involves real animals, so the administration
of any external enzyme is not required, but in the case of the in vitro model, there
is a requirement for external digestive enzymes [22]. For the in vitro dietary study,
we need to digest the dietary substance outside the body, which is similar to a host's
digestive system, and for that purpose, there is a requirement of various digestive
enzymes. Certainly, there is an activity difference between the microbial enzymes
and human enzymes; therefore, for human dietary studies, human-origin enzymes
or animal enzymes having activity near to human enzymes are preferred, but for
primary evaluation or mice-based studies, microbial enzymes of similar enzymatic
activity can be used.

Thus, different in vitro models use microbial amylase to mimic the salivary digestion of dietary substances, while microbial lactase is used in other dietary analyses,
as their activity is similar to respective mice enzymes [23]. Similarly, fungal and
bacterial amylase is also used to hydrolyze the different dietary fibers to compare
their biochemical structure and properties. Moreover, enzymes of microbial origin
are also used to study carbohydrate absorption by the gut-epithelial layer in the in
vitro model using microbial galactosidase [23].

5.9   CONCLUSION AND FUTURE PROSPECTS

Microbial enzymes play a significant role in the modern food industry; being microbial in origin, these enzymes are easy to produce at a mass scale and can be manipulated according to any requirement, which makes them highly suitable for food
production. Further, the role of microbial enzymes is not limited to food production only, but they are also used to improve the nutrient content and production of a

variety of healthy commercial foods, such as kombucha and certain expensive types of coffee. In the animal feed industry, microbial enzymes are used to provide quality nutrition to the livestock, which supports the global food supply chain. Microbial enzymes hold a great prospect in the food industry, as these enzymes are immensely capable of producing various new commercial food products of economic and nutritional importance. Additionally, these enzymes have a huge potential to be exploited in the field of nutraceuticals and the production of new functional food.

5.10   ACKNOWLEDGMENT

This work was supported by the project to train professional personnel in biological materials by the Ministry of Environment, South Korea.

REFERENCES

1. Sabu, A., *Sources, Properties and Applications of Microbial Therapeutic Enzymes.* Indian Journal of Biotechnology, 2003. **2**(3): p. 334–341.
2. Fallahi, P., H.-M. Habte-Tsion, and W. Rossi, *Chapter 10 — Depolymerizating Enzymes in Human Food: Bakery, Dairy Products, and Drinks*, in *Enzymes in Human and Animal Nutrition*, C.S. Nunes and V. Kumar, Editors. 2018, Academic Press: Cambridge. p. 211–237.
3. Singh, R., A. Singh, and S. Sachan, *Chapter 48 — Enzymes Used in the Food Industry: Friends or Foes?* in *Enzymes in Food Biotechnology*, M. Kuddus, Editor. 2019, Academic Press: Cambridge. p. 827–843.
4. Demain, A.L., and P. Vaishnav, *Production of Recombinant Proteins by Microbes and Higher Organisms.* Biotechnology Advances, 2009. **27**(3): p. 297–306.
5. Nighojkar, A., M.K. Patidar, and S. Nighojkar, *8 — Pectinases: Production and Applications for Fruit Juice Beverages*, in *Processing and Sustainability of Beverages*, A.M. Grumezescu and A.M. Holban, Editors. 2019, Woodhead Publishing: Sawston. p. 235–273.
6. Doriya, K., et al., *Chapter Six—Solid-State Fermentation vs Submerged Fermentation for the Production of l-Asparaginase*, in *Advances in Food and Nutrition Research*, S.-K. Kim and F. Toldrá, Editors. 2016, Academic Press: Cambridge. p. 115–135.
7. Sharma, R., H.S. Oberoi, and G.S. Dhillon, *Chapter 2 — Fruit and Vegetable Processing Waste: Renewable Feed Stocks for Enzyme Production*, in *Agro-Industrial Wastes as Feedstock for Enzyme Production*, G.S. Dhillon and S. Kaur, Editors. 2016, Academic Press: San Diego: California. p. 23–59.
8. Martínez-Medina, G.A., et al., *Chapter 14 — Fungal Proteases and Production of Bioactive Peptides for the Food Industry*, in *Enzymes in Food Biotechnology*, M. Kuddus, Editor. 2019, Academic Press: Cambridge. p. 221–246.
9. Singh, R., et al., *Microbial Enzymes: Industrial Progress in 21st Century.* 3 Biotech, 2016. **6**(2): p. 174.
10. Singh, V., Y.-C. Ryu, and T. Unno, *Dietary Intervention Induced Distinct Repercussions in Response to the Individual Gut Microbiota as Demonstrated by the in Vitro Fecal Fermentation of Beef.* Applied Sciences, 2021. **11**(15).
11. Rollán, G.C., C.L. Gerez, and J.G. LeBlanc, *Lactic Fermentation as a Strategy to Improve the Nutritional and Functional Values of Pseudocereals.* Frontiers in Nutrition, 2019. **6**(98).
12. Renzetti, S., F. Dal Bello, and E.K. Arendt, *Microstructure, Fundamental Rheology and Baking Characteristics of Batters and Breads from Different Gluten-Free Flours Treated with a Microbial Transglutaminase.* Journal of Cereal Science, 2008. **48**(1): p. 33–45.
13. Raveendran, S., et al., *Applications of Microbial Enzymes in Food Industry.* Food Technology and Biotechnology, 2018. **56**(1): p. 16–30.

14. Pielech-Przybylska, K., et al., *The Effect of Different Starch Liberation and Sacchari-fication Methods on the Microbial Contaminations of Distillery Mashes, Fermentation Efficiency, and Spirits Quality.* Molecules (Basel, Switzerland), 2017. **22**(10): p. 1647.

15. Angayarkanni, J., et al., *Improvement of Tea Leaves Fermentation with Aspergillus spp. Pectinase.* Journal of Bioscience and Bioengineering, 2002. **94**(4): p. 299–303.

16. Haile, M., and W.H. Kang, *The Role of Microbes in Coffee Fermentation and Their Impact on Coffee Quality.* Journal of Food Quality, 2019. **2019**: p. 4836709.

17. Suhaimi, H., et al., *Fungal Pectinases: Production and Applications in Food Industries.* Fungi in Sustainable Food Production. 2021, Springer: Cham. p. 85.

18. Arshad, M.S., et al., *Plant and Bacterial Proteases: A Key Towards Improving Meat Ten-derization, a Mini Review.* Cogent Food & Agriculture, 2016. **2**(1): p. 1261780.

19. Ashie, I.N.A., T.L. Sorensen, and P.M. Nielsen, *Effects of Papain and a Microbial Enzyme on Meat Proteins and Beef Tenderness.* Journal of Food Science, 2002. **67**(6): p. 2138–2142.

20. Samtiya, M., R.E. Aluko, and T. Dhewa, *Plant Food Anti-Nutritional Factors and Their Reduction Strategies: An Overview.* Food Production, Processing and Nutrition, 2020. **2**(1): p. 6.

21. Selle, P.H., and V. Ravindran, *Microbial Phytase in Poultry Nutrition.* Animal Feed Sci-ence and Technology, 2007. **135**(1): p. 1–41.

22. Singh, V., et al., *Effects of Digested Cheonggukjang on Human Microbiota Assessed by in Vitro Fecal Fermentation.* Journal of Microbiology, 2021. **59**(2): p. 217–227.

23. Hernandez-Hernandez, O., et al., *In Vitro Digestibility of Dietary Carbohydrates: Toward a Standardized Methodology Beyond Amylolytic and Microbial Enzymes.* Fron-tiers in Nutrition, 2019. **6**(61).

6 Fungal Enzymes in Organic Pollutants Bioremediation

Adam Grzywaczyk, Wojciech Smułek,
Jakub Zdarta, and Ewa Kaczorek

ABSTRACT

The problem of environmental contamination with chemical compounds continues to be a serious, large-scale, global problem. Petroleum compounds, pesticides, and pharmaceuticals are some of the most emerging pollutants in the world. Bioremediation methods using enzymes for the degradation of pollutants are distinguished from physicochemical and biological methods by their high efficiency achieved in a short time and their lack of harmful effects on the environment and living organisms. The aim of this work is to present possibilities of the application of different types of fungal enzymes, which are used in the bioremediation of environmental pollutants. In recent years, much attention has been paid to laccases, which effectively degrade synthetic dyes and pharmaceuticals. Nitrogenous compounds, in turn, can be treated, among others, by nitroreductases. Other oxidoreductases, transferases, and hydrolases also have the ability to catalyze the degradation of stubborn pollutants of anthropogenic origin. The review of literature on the subject clearly indicates that the application of fungal enzymes, also in the immobilized form, is going to develop intensively in the near future.

CONTENTS

6.1   Introduction.. 102
6.2   Fungi as Enzymes' Sources 103
6.3   Fungal Enzymes in Bioremediation.. 108
6.4   Laccases... 109
6.5   Monooxygenases 115
6.6   Nitroreductases 118
6.7   Other Groups of Enzymes 119
6.8   Conclusions and Perspectives... 121
6.9   Acknowledgments 122
References... 123

DOI: 10.1201/9781003202998-6

6.1 INTRODUCTION

Various chemical compounds used in many industries and areas of everyday life very often cause contamination of the environment. Persistent organic pollutants (POPs) are found in water, soil, and air (Ashraf, 2017). Particularly hazardous groups of organic compounds include

(a) petroleum hydrocarbons, including polycyclic aromatic hydrocarbons (PAHs) (Varjani, 2017);
(b) plant protection chemicals, including herbicides, fungicides, insecticides, and other biocide groups (Mahmood et al., 2016);
(c) synthetic surfactants (Xiaolin Li et al., 2018);
(d) plastics, including microplastics (W. Luo et al., 2019);
(e) halogenated hydrocarbons, in particular chlorinated derivatives of aromatic compounds, such as chlorophenols (Lin et al., 2020);
(f) dioxins and other products emitted to the atmosphere during incomplete combustion of wastes and fuels (Rathna et al., 2018); and
(g) pharmaceuticals, including antibiotics and compounds with endocrine activity (Cunha et al., 2019; Liu et al., 2018).

The harmful effects of mentioned compounds are observed in many fields. First, they have a direct toxic effect, causing the deaths of living organisms, contributing to the disappearance of biodiversity in ecosystems. Some of these compounds interact with genetic material, which makes them mutagenic (Pandey et al., 2019). Pharmaceuticals, even when they occur in trace concentrations, may have endocrine-disrupting effects on the animal and human hormone system (Hameed et al., 2021). On the other hand, the presence of antibiotics in the environment results in antibiotic resistance in environmental microorganisms, including pathogens (Tacconelli et al., 2018). It should also be emphasized that not only toxic compounds are a problem and pollute the environment. Many waste compounds, although considered safe, present in high concentrations, and not degrading quickly, pose a great challenge for organisms, disrupting the balance of nutrients and the food chains. Such compounds include waste from the cellulose industry (e.g., lignin).

It should also be emphasized that the durability of the mentioned groups of chemical compounds in the environment results from their strong sorption on the organic matrix and the slow processes of their natural degradation, which occurs as a result of environmental factors (light, oxidants) and microorganisms' action. In the light of these facts, numerous actions are taken to remove pollution from the environment (Ashraf, 2017). The basic methods are, first, physicochemical techniques based on the sorption of contaminants, their separation on semipermeable membranes, their evaporation, or other forms of physical desorption and, second, chemical neutralization or oxidation. However, the disadvantage of this group of methods is their low selectivity, as well as their negative impact on the living and abiotic parts of the environment (Singh et al., 2009; Sun et al., 2018).

As a result, biological methods, in which bacterial and fungal cells (sometimes microalgae as well [Pandey et al., 2019]) are the degrading agents of pollutants, are being developed. Many strains of microorganisms are able to incorporate chemical compounds, such as hydrocarbons, pesticides, and pharmaceuticals into their metabolic pathways. Consequently, cells can use them as a source of carbon and energy in aerobic and anaerobic processes. Despite their low cost and low environmental impact, biological methods based on the use of whole microbial cells have their drawbacks. The first one is the relatively long process time, which is conditioned by the growth kinetics of the microorganism and its metabolism under given environmental conditions. Another disadvantage is the difficulty in predicting in which direction the pollutants will be metabolized and whether the microbial cell will not produce intermediates with toxicity higher or comparable to that of the original compound (da Silva et al., 2020) (Zhang et al., 2019) (da Silva et al., 2020; S. Zhang et al., 2019).

This lack of complete control over the biodegradation process prompts the use of enzymes since they are directly responsible in cells for the biotransformation process of organic pollutants (Karigar & Rao, 2011; Shakerian et al., 2020). With a clearly defined enzyme or group of enzymes, it is much easier to plan the process, and it runs several orders of magnitude faster because it is not limited by the physiology of the cells. What limits the use of enzymes is certainly their relatively high price, associated with the cost of their isolation from living cells. Hence, the greatest attention is paid to high-performance enzymes, whose efficiency of action compensates for their high price (Mukherjee, 2019). The production of enzymes by fungi possesses several benefits, like the low cost of the substrates for growth, relatively fast production, and the ease with which the enzymes can be modified. Moreover, many of the fungal enzymes are extracellular and then easy to recover from the culture media (Vishwanatha et al., 2010).

6.2   FUNGI AS ENZYMES' SOURCES

The chemical industry is one of the crucial branches of the world economy. This statement cannot be denied due to the fact that gas, oil, pesticides, and many other chemicals are widely used in every country throughout the world. This is, let's say, a part of the world globalization process; on the one hand, it is beneficial for the country's GDP (for example, for countries such as the United Arab Emirates, Bahrain, or even Azerbaijan), guarantees jobs, and drives the economy, but on the other hand, oil and gases are the cause of conflicts and generate pollutants harmful to human health and natural environment (Alharbi et al., 2018; Carpenter, 2011; Chung, 2016). Moreover, fuel production is not the only industry responsible for the production of chemicals. Organic pollutants include substances such as phenolic compounds (dyes, bisphenol A, 2,4,6-trichlorophenol), many different hydrocarbons, heavy metals, polycyclic aromatic compounds PAHs, and POPs such as polychlorinated biphenyls, pesticides (e.g., dichlorodiphenyltrichloroethane [DDT]), furans, and dioxins (Guo et al., 2019). POPs are so-called forever chemicals due to the fact that they are resistant to biological, chemical, and photolytic degradation and pose a serious health risk (Ritter et al., 2011). Fortunately, according to European Environment Agency, the emission trends of POPs in Europe have decreased since the beginning of the 21st century by almost 80% in 2017 compared to the emissions in 1990. Despite the emission decline, the aforementioned chemicals still pose a risk to the environment and human health. This forces to search for novel, effective methods of degradation of organic and inorganic pollutants. One possible method is the bioremediation process, which uses microorganisms to degrade or transform hazardous substances. This includes, for example, composting, bioventing, phytoremediation, and mycoremediation.

The mycoremediation process (from the Greek *mukēs*, "fungus") uses the enzymes present in fungus for decontamination processes. Fungi, which are present all over the world, are the most effective agents for the decomposition of matter. This includes basically every substance which possesses a carbon-carbon bond in its structure. That causes fungi to be the only organism on earth that is able to decompose wood (Rhodes, 2014). The Fungi kingdom includes eukaryotic organisms, such as single-celled yeast, mold (which grows in the form of multicellular filaments called hyphae), and mushrooms. A detailed division of fungi is presented in Table 6.1.

Fungi have accompanied man from the beginning of human history. The evidence shows that beer was produced as early as 8500 BC to 5500 BC (Dietrich et al., 2012). This information is not accidental, as it is well known that spontaneous fermentation by yeast is used in the production of beer. Since that time, humans have used fungi

TABLE 6.1
Division of Fungi Kingdom

Phylum	Description	Source
Ascomycota	Largest phylum; contains 150,000 described species; comprises saprotrophs, lichens, pathogens, and singled-cell and multicelled organisms	(Hill et al., 2021)
Basidiomycota	Together with the *Ascomycota*, constitutes the subkingdom Dikarya; includes mushrooms, puffballs, boletes, and chanterelles	(Hibbett et al., 2007)
Blastocladiomycota	Microscopic fungi of a very diverse structure; the most famous species are parasites	(Wijayawardene et al., 2020)
Chytridiomycota	Has retained many features of the original fungi, including flagella gametes; contains 600 described species	(Hibbett et al., 2007)
Entomophthoromycota	Includes over 250 species; mostly arthropod pathogens or saprobionts living in soil and litter	(Humber, 2012)
Glomeromycota	Widespread soil mycorrhizal fungi; 230 described species	(Błaszkowski & Czerniawska, 2011)
Microsporidia	Parasites that infect insects, crustaceans, and fish; 1,500 described species	(Didier, 2005)
Neocallimastigomycota	Wicked anaerobic organisms that live in the digestive system of herbivorous mammals and (possibly) also in other anaerobic conditions; encompasses only one family	(Hibbett et al., 2007)
Zygomycota	About 1,060 species of mostly saprophytic fungi; can cause rare and deadly human diseases—zygomycoses	(Spatafora et al., 2016)

for many purposes. In Europe, mushroom hunting is a popular pastime, mainly in the fall. It is related to the use of mushrooms for culinary purposes. The fruiting bodies of Blastomycota mushrooms are usually picked and then cooked or otherwise processed. It is recommended to pick mushrooms with a hymenophore in the form of folds and tubes due to the fact that most of them are edible. Some mushrooms with lamellae hymenophore are also edible, but there are also deadly toxic species, so picking is recommended for experienced mushroom pickers (Valverde et al., 2015). In Figure 6.1, *Boletus edulis* Bull and, in Figure 6.2, *Lactarius deterrimus* Gröger, are presented; both are edible and are found in Wielkopolska, Poland.

The use of fungi in the production of food also includes the production of beers, wine, and bread, basic food products that are produced almost everywhere in the world. For each product, the baker yeast *Saccharomyces cerevisiae* is used for fermentation processes (Ali et al., 2012). During these processes, yeast produces ethanol and carbon dioxide from carbohydrates, which are the basis for loosening the bread dough or the alcohol content in wine and beer. Moreover, many by-products related to yeast metabolism are formed, which significantly affect the taste, aroma, and properties of beer (Lodolo et al., 2008). What is worth mentioning, fungi are also used for the production of cheese. Cheeses with *Penicilum camemberti* on its outer surface are called Camembert and Brie. Roquefort, Stilton, and Gorgonzola—these cheeses belong to the group of blue cheese, created by inoculating cheese with *Penicillium roqueforti* or *P. glaucum* (Hymery et al., 2014). In Figure 6.3 and Figure 6.4, the homemade production of bread and beer are presented.

FIGURE 6.1 *Boletus edulis* Bull, example of mushroom with tubular hymenophore, own source.

FIGURE 6.2 *Lactarius deterrimus*, example of mushroom with lamellae hymenophore, own source.

FIGURE 6.3 Bread dough after formation process, own source.

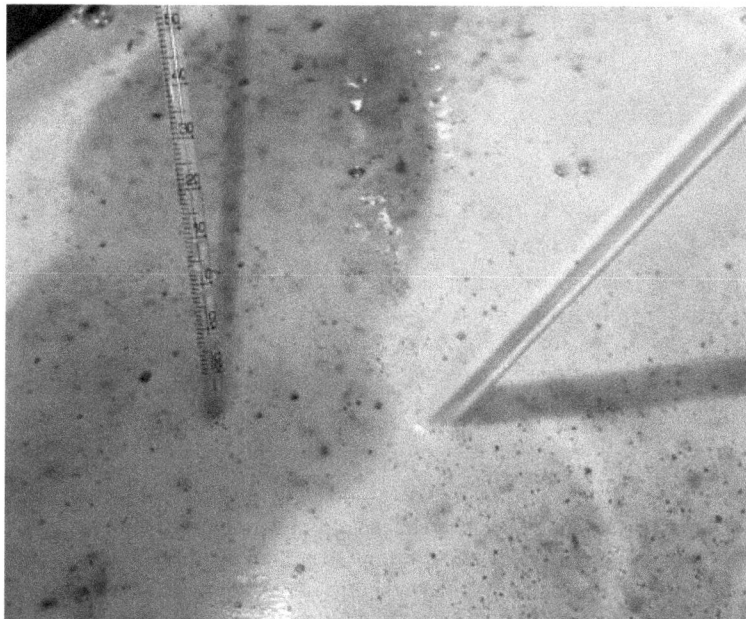

FIGURE 6.4 Process of beer fermentation, own source.

When discussing the uses of fungi, the production of antibiotics must be mentioned. Penicillin was the first isolated antibiotic, by Alexander Fleming, almost 100 years ago, in 1928, from *Penicillium notatum*. The mechanism of action of penicillins as antibiotics is based on blocking the activity of bacterial enzymes called transpeptidases (PBP) involved in the final stage of peptidoglycan synthesis in the bacterial cell wall (Soares et al., 2012). The source of another important antibiotic, cephalosporin, was the fungus *Cephalosporium acremonium*. The action of this antibiotic is similar to that of penicillin; they covalently bind to the active center of bacterial enzymes: carboxypeptidases and transpeptidases, blocking their action. Thus, they inhibit the process of bacterial cell wall synthesis (Bhide et al., 2020; Yotsuji et al., 1988). Both penicillin and cephalosporin belong to the group of β-lactam antibiotics. Another example is griseofulvin, obtained in 1939 from *Penicillium griseofulvum*. Interestingly, it has antifungal properties against *Trichophyton*, *Microsporum*, and *Epidermophyton* species and was the first oral antifungal drug used in the treatment of dermatomycoses (Bai et al., 2019; Gupta et al., 2018). Structures of penicillin and cephalosporin are presented in Figure 6.5.

The possible uses of mushrooms mentioned above do not exhaust the topic but constitute an appropriate basis for further considerations. The multitude of the species of fungi described so far means that the fungi have found a wide spectrum of applications in many areas of life. Of course, possible opportunistic species that pose a direct threat to life (*Candida albicans*, causing candidiasis) or molds that cause allergic reactions should also be mentioned (Mendell et al., 2011; Pappas et al., 2018).

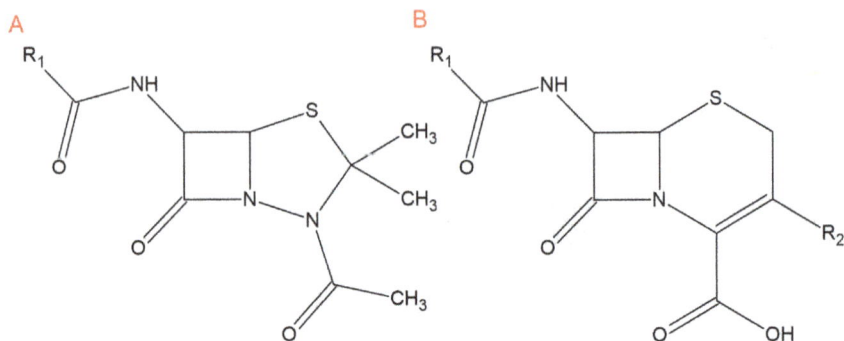

FIGURE 6.5 Structures of (A) penicillin and (B) cephalosporin, created with ChemSketch, based on Chaudhry & Veve (2019).

Enzymes (i.e., macromolecular catalysts), usually of a protein nature, are responsible for the action of fungi, such as fermentation, decomposition of organic matter, and other functions. Their most important structural feature is the presence of a region in the molecule called the active site in which catalytic transformation takes place (Robinson, 2015; Weng et al., 2011). The role of the active center is to directly bind the substrate with the enzyme through the action of such van der Waals forces or electrostatic forces. Depending on the enzyme, a protein catalyst requires the presence of an organic or inorganic molecule called a coenzyme to function properly. This usually determines the type of reaction being catalyzed. The apoenzyme is the remaining protein part of the enzyme that is responsible for the recognition of the substrate as well as the direction of the catalytic protein (Fruk et al., 2009; Weng et al., 2011). Enzymes are divided into six classes according to the mechanism of the reactions they catalyze. These are EC1 oxidoreductases, EC2 transferases, EC3 hydrolases, EC4 lyases, EC5 isomerases, and EC6 ligases (Robinson, 2015). Fungal biocatalysts include mostly oxidoreductases, which catalyze the electron transfer from donor to acceptor, and hydrolases, which catalyze the breakdown of chemical bond in the hydrolysis process.

Enzymes present in fungi have found application in the removal of organic chemicals present in soil, water bodies, sewage, and other places. It is enzymes that are responsible for the ability of the discussed group of organisms to decompose organic compounds and thus constitute saprotrophic nutrition. This chapter will be devoted to the issue of the use of enzymes present in fungi in the process of removing organic compounds. The mechanism of action and methods of obtaining enzymes will also be presented.

6.3 FUNGAL ENZYMES IN BIOREMEDIATION

Nowadays, biological methods of removal of pollutants from industry seem to be an alternative for physical or chemical methods. It is caused by the fact that they can effectively remove contaminants without additional chemicals and nontoxic products and without the generation of sludge. The biological methods can be divided into two

TABLE 6.2
Industrial Enzymes and Their Source (Adapted from McKelvey & Murphy, 2011)

Enzyme	Source
α-amylase	*Aspergillus niger, A. oryzae*
Chymosin	*A. niger*
Cellulase	*Trichoderma viride, T. reesei*
Glucoamylase	*A. phoenicis, Rhizopus delemar*
Invertase	Yeast
Laccase	*Trametes versicolor*
Lipases	*A. niger, A. oryzae*
Proteases	*A. niger, R. delemar*
Phytase	*A. niger, A. oryzae*

major ways of reduction of pollutants from the textile industry—treatment by microorganisms and enzymes. The most popular fungi used for the process are usually white rot fungi species, such as *Trametes* spp., and *Pleurotus* spp. (dos Santos et al., 2007). It should be noted that due to the variety of different microbial species, the process can occur in an aerobic, anaerobic, or mixed aerobic-anaerobic way (Kumar et al., 2012). Besides microorganisms as tools for the removal of organic contaminants, special attention should be paid to enzymes. Fungi cells (like any living cell) produced many different biocatalysts, beneficial from the economic point of view. Such enzymes are presented in Table 6.2.

Among the many enzymes produced by fungi, not all of them are used in bioremediation processes. Moreover, not all biocatalysts used in industry are used in environmental cleaning processes. This is, of course, due to the different mechanisms of action of individual enzyme classes. Perhaps there is not yet, and may never be, a need to use some of them in remediation processes. Nevertheless, in the search for green alternative methods, we should focus our attention on enzymes such as laccases, tyrosinases, manganese and lignin peroxidases and others, chloroperoxidases, monooxygenases, nitroreductases, reductive dehalogenases, quinone reductases, and transferases (Harms et al., 2011). In these considerations, we will focus on selected enzymes.

6.4 LACCASES

Laccases (EC. 1.10.3.2) belong to the oxidoreductase group of enzymes. They catalyze the oxidation and reduction reactions of a wide variety of compounds. Oxidases are enzymes that, in the presence of molecular oxygen, catalyze oxidation and reduction reactions by electron transfer. Laccases have four copper atoms in their active center and are built from 220–800 amino acids. This widespread in nature enzymes is found in fungi, microorganisms, and also plants. Arregui et al. indicate that laccase can be found in more than 1,000 bacteria and 6,200 eukaryotes, according to

UniProtKB base (Arregui et al., 2019). Laccases' broad spectrum of activity, and thus the variety of compounds they can oxidize, makes this group of enzymes the most widely used in recent years in bioremediation processes. One of their particular advantages is conducting oxidation processes with the participation of atmospheric oxygen, which ensures the high efficiency of catalyzed processes, often without the need to use cofactors or, for example, additional oxidants. At the same time, the frequency of their occurrence, also in fungal cells, significantly affects their low acquisition cost (Arregui et al., 2019). Hence, most of the literature includes studies on the use of this particular group of enzymes in bioremediation.

Among fungal species, laccases can be found in *Trametes* spp., *Nectriella pironi*, *Cladosporium* spp., *Agaricus bisporus*, and *Pleurotus* spp. (de Freitas et al., 2017; Góralczyk-Binkowska et al., 2020; Jankowska et al., 2020; Nikolaivits et al., 2021; Othman et al., 2018; Zheng et al., 2017). The structure of laccase derived from the fungus *Trametes versicolor* is presented in Figure 6.6, and the photo of the fungus *Pleurotus ostreatus* is in Figure 6.7.

Laccase finds application in various industries, such as the food industry or paper industry. It is used as a stabilizer during the production of wines and juices and also during the removal of chlorophenols and chlorolignins from the paper industry. However, the most interesting application of laccase is to use it in bioremediation processes.

These enzymes have been widely used for dye degradation. The largest group of synthetic dyes used in the textile industry are azo dyes (Sarkar et al., 2017). More than 80% of azo dyes are used for the coloring of materials, such as fur or natural or synthetic textile fibers. Due to inefficient binding between dye and material, dyes are washed out from the fabrics. This causes the release of huge amounts, up to 10,000 tons of dyes annually, to wastewater and surface water, which causes major

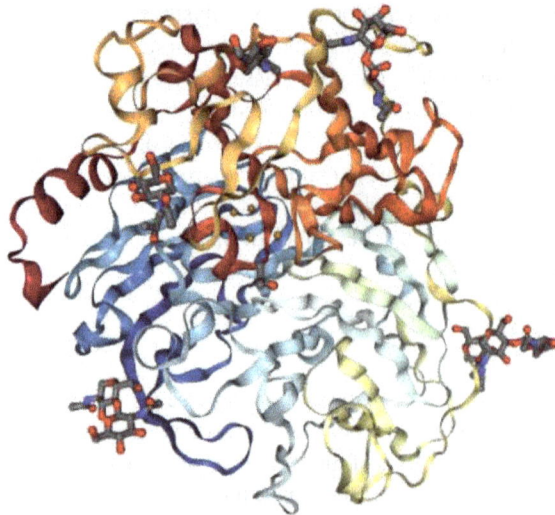

FIGURE 6.6 The example of a 3D structure of the laccase molecule from the fungus *Trametes versicolor*, adapted from Piontek et al. (2002).

FIGURE 6.7 *Pleurotus ostreatus*, own source.

environmental pollution (S. Wong et al., 2020). It is worth mentioning that textile wastewater consists relatively high concentration of dyes—up to 200 mg/L. What is more, even a relatively small concentration, such as 1 mg/L, can be visible in water solutions and might lead to limitations in sunlight and oxygen transmission to each water reservoirs level (Alinsafi et al., 2007). Nowadays, many different studies show that even a small amount of dye and the products of its degradation can be cancerogenic and possess mutagenic activity; hence, waters with dyes are a serious environmental problem (De Aragão Umbuzeiro et al., 2005; S. Wong et al., 2020).

During the enzymatic hydrolysis of dyes, the enzyme binds to the dye particle. Five mechanisms of this binding can be distinguished—covalent, acid-base catalysis, substrate distortion and induced-fit, approximation and orientation effect, synergistic effect, and multiple catalyses (Zhou et al., 2021). Each mechanism aims to achieve the same goal—to degrade dyes to colorless products. Many different dyes were tried to undergo enzymatic degradation, presented in Table 6.3.

The presented table is only a fragment of a huge number of examples of the effective use of laccase in the processes of dye degradation. The presented table is only a fragment of a huge number of examples of the effective use of laccase in the processes of dye degradation. It should be emphasized here not only the multitude of dyes subjected to the degradation processes, and even potentially mineralized, but also the effectiveness of this process. The effectiveness of the decolorization process of the Reactive Black 5 dye has been checked by Antecka et al. and Tavares et al. In the first case, the laccase derived from *Trametes versicolor* was immobilized on hybrid oxide systems TiO_2-ZrO_2 and TiO_2-ZrO_2-SiO_2 with an efficiency of 83% and 96%, respectively. During the RB 5 dye decolorization process, 100% efficiency was achieved for the concentration of 1 mg/L for both systems; for the concentration of 5 mg/L and 10 mg/L for the TiO_2-ZrO_2-SiO_2-laccase system, about 80% efficiency

TABLE 6.3

Examples of Dyes That Undergo Enzymatic Degradation

Degraded Dye	Laccase Source	References
Congo red	*Oudemansiella canarii*	(Iark et al., 2019)
Remazol Brilliant Blue R, Coomassie Brilliant Blue G-250	*Trametes versicolor*	(Jankowska et al., 2020)
Reactive Black 5, Alizarin Red S	*Trametes versicolor*	(Antecka et al., 2018)
Malachite green, bromophenol blue, crystal violet	*Trametes trogii*	(Yang et al., 2020)
Malachite green, methylene blue, Congo red	*Trichoderma harzianum*	(Bagewadi et al., 2017)
Safranine, methylene blue Azure blue	*Cerrena unicolor*	(J. Zhang et al., 2018)
Reactive Blue 220, Acid Green 2	*Marasmiellus palmivorus*	(Cantele et al., 2017)
Acid Orange 156, Acid Red 52, Coomassie Brilliant Blue, methyl violet, malachite green	*Myceliophthora thermophila*	(Salami et al., 2018)
Mauveine	*Trametes versicolor*	(Tišma et al., 2020)
Reactive Blue 220, Reactive Black 5, Remazol Brilliant Blue R	*Lentinus crinitus*	(Tavares et al., 2020)

and 47% were achieved, respectively. In the case of the TiO_2-ZrO_2-laccase system, in both cases, the efficiency oscillated around 15%. Researchers suggest that the higher the concentration of the dye, the lower the degradation rate is (Antecka et al., 2018). Tavares et al. obtained similar results. For a concentration of 1 mg/L dye, they achieved 100% decolorization of Reactive Black 5 dye using laccase with an activity of about 6,000 U/g dry mass (Tavares et al., 2020).

The abovementioned examples and other studies show that achieving more than 90% efficiency of the decolorization process is possible with the appropriate selection of process conditions and appropriate dye dilution (which is reflected in the conditions in the wastewater), and with the appropriate selection of the carrier, it can be enhanced by the probable process absorption and adsorption of the dye on the surface of the carrier (Antecka et al., 2018; Bagewadi et al., 2017; Cantele et al., 2017; Jankowska et al., 2020; Peng et al., 2021; Salami et al., 2018; Tavares et al., 2020; Tišma et al., 2020; Yang et al., 2020; J. Zhang et al., 2018).

Another area of action of laccase is the degradation of polychlorinated biphenyls (PCBs) (Nikolaivits et al., 2021). These compounds, banned in 1970s, are responsible for the pollution of both the environment and pollution of foods, such as animal-derived foods fishery products. These compounds are not currently of practical use. They are a by-product of the combustion of wood, fuels, fires, natural disasters, and chemical disasters. Previously, they were used as transformer oils, high-temperature lubricants, and plasticizers. The human body absorbs the most PCBs from food, much less from air and water (Leong et al., 2019; Weber et al., 2018). They are potentially carcinogenic and neurotoxic compounds and are also considered endocrine disruptors (Jung et al., 2017; Klenov et al., 2021; Klocke & Lein, 2020; Marotta et al., 2020; Onozuka et al., 2020). Moreover, Leong et al. suggest a relationship between a decrease in serum testosterone levels and an increase in PCB153 levels (Leong et al., 2019). It is worth mentioning that so far, more than 200 different PCBs have been described, so Figure 6.8 shows only the structure diagram.

The effective degradation of PCBs is associated with the formation of new compounds. These products vary in structure depending on the starting compound and time. Nevertheless, gas chromatography coupled with mass spectrometry (GCMS) analyses by Šrédlová et al. made it possible to obtain the structures of compounds formed during the enzymatic decomposition of PCBs by the enzyme laccase. These structures are presented in Figure 6.9.

Laccases for the degradation of PCBs were used by the team of Li et al. The enzyme was loaded onto cellulose/cellulose fibril (CF) beads (MCCBs). Ten mg of MCCBs were incubated in 10 mL of laccase solution containing 0.5–32 g/L of enzyme. This system was used to degrade PCBs; the degradation efficiency was determined by adding 50 mg immobilized laccase in 10 mL of PCBs aqueous solution and 10 mL of pH 4 phosphate buffer at room temperature. In this way, process efficiency of 85% is achieved after 8 hours (N. Li, Xia, Li, et al., 2018). Similar performance after the same process time was obtained by Li et al. Laccase immobilized on biocarbon was used to remove PCBs from sewage. Two kinds of wood biochar were used as a type of support for immobilizing. They achieved a yield of 71.4% degradation after 8 hours of running the process (N. Li, Xia, Niu, et al., 2018).

The use of laccase does not end with the two compounds mentioned above. An attempt to degrade bisphenol A was undertaken by BarriosEstrada et al. They used oxidoreductase enzymes in the free system and immobilized system based on ceramic membranes. Two laccase forms from *Pycnoporus sanguineus* and *Trametes versicolor* achieved 100% degradation efficiency in less than 24 hours of the process. The enzymes showed optimal activity at pH 5 and 5.4 with a degradation rate of 204.8 ± 1.8 and 79 ± 0.1 μmol/min/U for *P. sanguineus* and *T. versicolor*, respectively (Barrios-Estrada et al., 2018). Another example of the degradation of organic compounds using laccase is the degradation of perfluorooctanoic acid. Research by Luo et al. aimed at investigating the possibility of degradation of PFOA present in sandy loam soil using the discussed enzyme. In the presence of soybean meal as a mediator and laccase, PFOA was degraded 24% in water after 36 days and 40% in a soil slurry system after 140 days. Moreover, the authors identified 12 possible enzymatic degradation products of PFOA, most of which are primarily partially fluorinated compounds containing perfluoroalkyl moieties (Q. Luo et al., 2018). An interesting and very promising example of laccase use is the degradation of polyurethane in the form of polycaprolactone-based (PCL-based) and polytetrahydrofuran-based

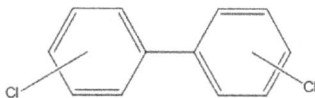

FIGURE 6.8 General chemical structure of PCBs, based on Horwat et al. (2015).

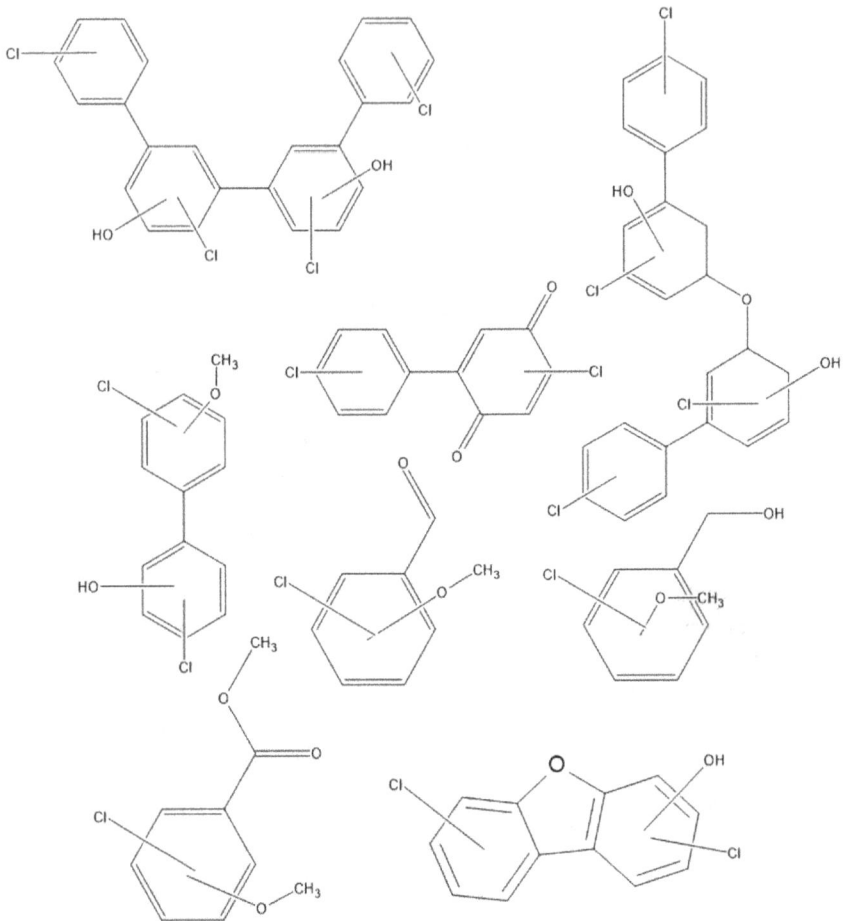

FIGURE 6.9 Compounds formed during the enzymatic decomposition; structures proposed by Šrédlová et al. (2021).

(PTHF-based) foam and thermoplastic. SEM observations, weight loss of PCL-based PU foam, and yellowing of the samples indicate successful PU degradation. However, researchers indicate that after 18 days, the weight loss is small. For PCL-based foam, it is 12%; for PTHF-based foam, 2%; and for TPU, less than 20%. Nevertheless, these results are promising and worth quoting (Magnin et al., 2021). The last example cited in this chapter is the removal of aniline from simulated wastewater by Yang et al. Scientists indicate that for the removal of aniline, the optimum conditions were 50? aniline concentration 80 mg/L and laccase concentration 1 g/L. The total removal of aniline reached 97.1% after 8 hours. Additionally, the o-phenylenediamine removal process was checked. The optimal process conditions were similar to aniline, with the concentration of o-phenylenediamine changing to 100 mg/L. The test compound was completely removed from the simulated sewage after only 60 minutes.

The presented examples are only a fragment of the efforts of scientists to search for newer and more effective methods of using laccase in the processes of remediation of organic compounds. At the moment, laccase is a flagship example of the use of a fungal enzyme in such processes due to the widespread presence of fungi capable of producing it. Moreover, the use of this biocatalyst is effective and is a promising alternative to other less green methods of removing organic compounds from the environment. However, considerations cannot focus solely on laccase.

6.5   MONOOXYGENASES

Overall, monooxygenases (EC 1.13.x.x and EC 1.14.x.x) catalyze the transfer of an oxygen atom to the substrate, which is an organic molecule. In order for this enzyme to function properly, the oxygen molecule O_2 must be activated, which depends on the type of cofactor present in the biocatalyst. Therefore, we distinguish heme-dependent monooxygenases (EC 1.14.13.x, EC 1.14.14.x and EC 1.14.15.x), flavin-dependent monooxygenases (EC 1.13.12.x and EC 1.14.13.x), copper-dependent monooxygenases (EC 1.14.17.x), nonheme iron-dependent monooxygenases, pterin-dependent monooxygenases, cytochrome P450 moonooxygenases, and cofactor-independent monooxygenases (Torres Pazmiño et al., 2010). An example of the structure of monooxysgenase is shown in Figure 6.10.

As mentioned by Torres Pazmino et al., the use of monooxygenases, in addition to the discussed remediation processes, is limited to the use of these biocatalysts in chemical, electrochemical, enzymatic, and photochemical processes of coenzyme regeneration (Torres Pazmiño et al., 2010). Of course, this enzyme has also found wide application in the degradation of organic compounds. Nevertheless, several examples of such compounds, commonly found and commonly used in various industries, deserve attention. The first example will be the degradation of nonylphenol conducted by Subramanian et al. During their research, they used P450 monooxygenases derived from the fungus *Phanerochaete chrysosporium*. As the researchers themselves inform, this relationship is one of the so-called endocrine-disrupting chemical to which the surfactant they are testing belongs. *Phanerochaete chrysosporium* extensively degraded the nonylphenol (100% of 100 ppm) in both nutrient-limited cultures and nutrient-sufficient cultures (Subramanian & Yadav, 2009). Monooxygenase is also capable of degrading salicylic acid. Rocheleau et al. assessed the ability to degrade the salicylic acid compounds by mutated *Fusarium graminearum* cells. The ability to utilize supplemented SA in a liquid culture assay was compared quantitatively between three replacement mutants and one ectopic insertion mutant strain from the same transformation batch. Most or all of the salicylic acid had been utilized after 4 hours of incubation. Moreover, there was no significant change in the level of remaining SA was observed between 0 and 6 hours (Rocheleau et al., 2019). Their proposed metabolic pathway for salicylic acid decomposition includes 2,3-dihydroxybenzoic acid, catechol, cis-muconate, and β-ketoadipate. The structures of these compounds are shown below in Figure 6.11.

Polycyclic aromatic hydrocarbons PAHs are a group of compounds naturally occurring in fossil fuels. They are substances composed only of carbon and hydrogen, containing condensed aromatic rings. They are released during incomplete

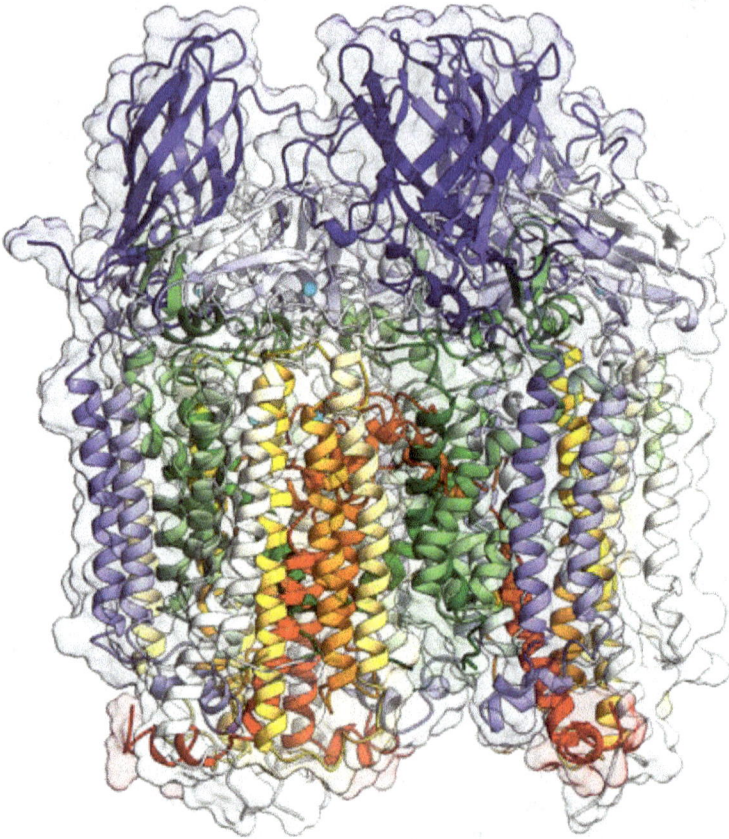

FIGURE 6.10 The structure of copper-dependent monooxygenases, adapted from Musiani et al. (2020).

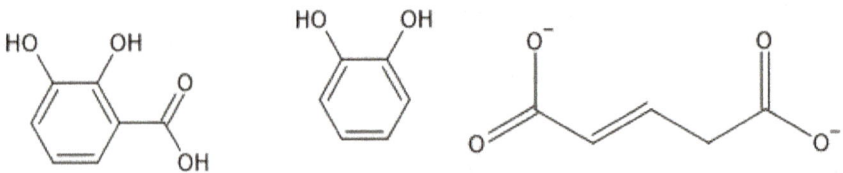

FIGURE 6.11 Structures of salicylic acid breakdown products, left to right: 2,3-dihydroxybenzoic acid, catechol, and *cis*-muconate; based on Rocheleau et al. (2019).

combustion of all hydrocarbons, so they can also be found in food as a result of their processing in processes such as frying or grilling. These compounds are also a component of smog, and many of them are suspected or have proven strong geno-toxic, mutagenic, and carcinogenic properties. All this makes the entire humanity vulnerable to their negative effects (Jia et al., 2018; Mihankhah et al., 2020; Stading et al., 2021; Y. Zhang et al., 2019). The structures of a few selected PAH compounds are shown in Figure 6.12.

As reported by Elyamine et al., aerobic oxidation of one of PAHs—pyrene, by monooxygenase-containing fungi begins with the configuration of dihydrodiol through the enzymatic cascade of oxygenases, leading to the formation of intermediates, such as catechol and protocatechuate. Fungi capable of producing monooxygenases are *Aspergillus* spp., *Penicilium* spp., *Phanaerochate* spp., *Pleurotus* spp., and *Tramates* spp., so these species may find potential use in PAH removal (Elyamine et al., 2021). Monooxygenases have also found potential use together with laccase to remove phenanthrene and benzanthracene—a cyclic, foreign carcinogenic compound, inter alia, in tobacco smoke (Talhout et al., 2011). Li et al., in their experiment, used the fungus *Pycnoporus sanguineus*, able to form laccase in large amounts, and fungi of the species—*Pleurotus ostreatus*, *Coriolus versicolor*, and *Trametes versicolor*. After 14 days of incubation in the medium containing the tested PAHs, at the conditions of 28?, 160 rpm, the concentration of PAHs was 20 mg/L. For phenanthrene, *Pleurotus ostreatus* showed the strongest biodegradability (54.5%), followed by *Coriolus versicolor* (48.9%), *Pycnoporus sanguineus* (45.6%), and *Trametes versicolor* (33.5%). In the case of benzanthracene, *Pycnoporus sanguineus*, the degradation was 90.1%, much higher than the other rates, which ranged around 20%. Moreover, scientists also proposed possible degradation products of phenanthrene and benzanthracene (Xuanzhen Li et al., 2018). They are presented in Figure 6.13.

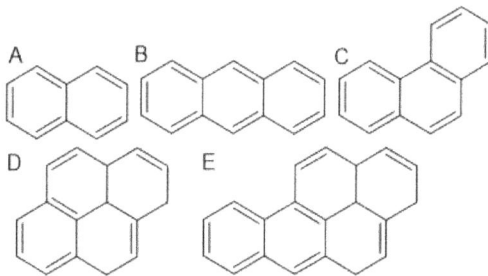

FIGURE 6.12 Structures of example PAHs: A—naphthalene, B—anthracene, C—phenanthrene, D—pyrene, E—benzo[a]piren.

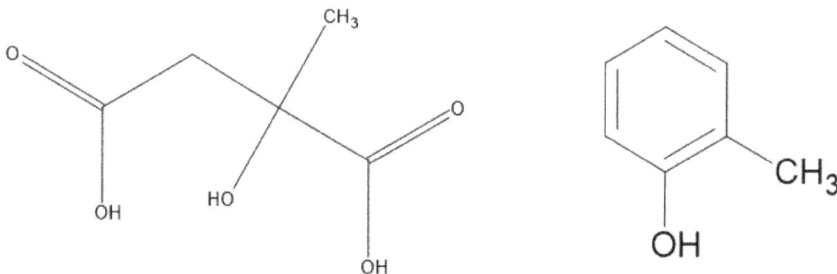

FIGURE 6.13 Possible degradation products of phenanthrene and benzanthracene, proposed by Xuanzhen Li et al. (2018).

It is also worth quoting the use of monooxygenases in lignin and cellulose degradation processes (Beeson et al., 2015; Elyamine et al., 2021; F. Li et al., 2019; Xuanzhen Li et al., 2018; Morgenstern et al., 2014; R. Zhang et al., 2019). As this is not a substance responsible for environmental pollution, and it is even an indispensable part of it, it will be omitted in further considerations. Nevertheless, the use of the described biocatalysts may cause significant progress in seeking effective methods for removing carcinogenic compounds as multi-circular aromatic hydrocarbons. They are not only a threat to human life but also for the environment, and through the universality of their occurrence (e.g., ease of creation, even when preparing food), this problem does not apply only to selected areas, but it is a global problem.

6.6   NITROREDUCTASES

Reductases, as the name suggests, catalyze reduction reactions. This particular group of enzymes is capable of reducing organic nitro compounds. Representatives of such compounds include nitrotoluenes, furazones, and other aromatic compounds with a nitro substituent. This group of biocatalysts can be divided into two groups—the first and the second type. The mechanism of action of type 1 nitroreductases is the reduction of organic nitro compounds using a two-electron transfer. The operation of type 2, on the other hand, is based on one-electron transfer and the formation of an anionic nitro radical (Song et al., 2015). There is also a further division already in types into groups A and B depending on the reducing agent (NADH or NADPH) and similarities to *E. coli* nitroreductases NfsA and NfsB. An example of ribbon nitroreductase structure, divided into two monomers, is shown in Figure 6.14.

Nitroreductases are cell-bound enzymes that are used to reduce, among others, trinitrotoluene to hydroxylamino- and amino-dinitrotoluenes and explosives containing N-heterocyclic structures into their respective mononitroso derivatives (Harms et al., 2011). Fungi of the genus *Phanerochate* and *Stropharia* are capable

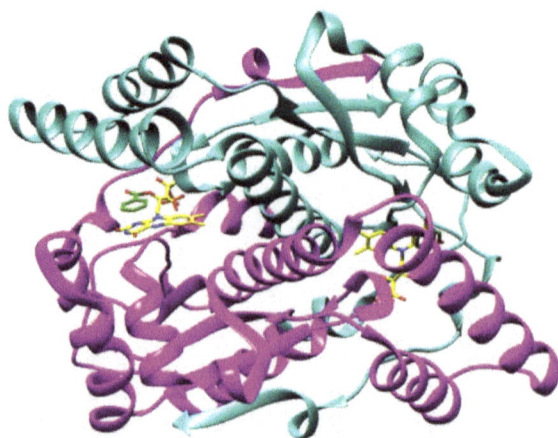

FIGURE 6.14 An exemplary structure of nitroreductase divided into two monomers; created with the participation of *PyMOL* program by Pitsawong et al. (2014)

of degrading or even mineralizing TNT and thus contain enzymes from the group of nitroreductases. The products of this decomposition may undergo further decomposition processes by other enzymes, such as oxygenases or laccases, in further stages (Harms et al., 2011; Serrano-González et al., 2018). The mechanism of this process is not fully understood. Nevertheless, the researchers led by Serrano et al. suggest that the first step in TNT degradation is the reduction of the molecule to 4-amino-2,6-di-nitrotoluene, 2-amino-4,6-di-nitrotoluene, hydroxylamino-dinitrotoluene, azoxytetranitrotoluenes, phenolic, and acylated derivatives. The 4-amino-2,6-di-nitrotoluene is degraded to 4-formamido-2,6-dinitrotoluene, then this is reduced to 2-amino-4-formamido-6-nitrotoluene and transformed to 2,4- diamino-6-nitrotoluene (Serrano-González et al., 2018). One type of fungus capable of decomposing nitro derivatives of organic compounds is *Trichoderma* sp. It has been used in trinitrotoluene (TNT) bioremediation studies by Alothman et al. *Trichoderma viride* was raised on media containing TNT at a concentration of 50 and 100 ppm. An increase in the diameter of the colony inoculated on agar medium to 85 mm after 8 and 11 days of growth suggests that *T. viride* is able to decompose TNT explosives and uses them as a nitrogenous source for growth. Researchers identified the decomposition products of 2,4,6-TNT using an analytical method of gas chromatography coupled with mass spectrometry. The main product was 5-(hydroxymethyl)-2-furancarboxaldehyde and 4-propyl benzaldehyde as the minor compound. The conclusions from the research suggest that it is still necessary to conduct appropriate studies on the degradation of nitro compounds, but they constitute an appropriate example and basis for further considerations on the use of nitroreductases in bioremediation processes (Alothman et al., 2020).

Microorganisms can quickly adapt to very variable environmental conditions. This sentence is not mentioned by accident. The example of the degradation of trinitrotoluene by fungi (and bacteria) confirms the previous thesis. This chemical compound does not occur naturally in nature; it is the work of man, and its presence is a potential threat to the environment and humans (Alothman et al., 2020). The use of nitroreductases can significantly contribute to reducing the risk posed by the presence of TNT.

6.7   OTHER GROUPS OF ENZYMES

Among the multitude of enzymes, it is worth paying attention to a few more of them. The first group is the peroxidases (EC 1.11.1.x), which catalyze the oxidation of substrates with hydrogen peroxide as a final electron acceptor. They perform a number of functions, both in the human body (glutathione and ascorbate) and in the degradation processes of organic compounds (e.g., manganese peroxidase and lignin peroxidase).

It is worth mentioning here that the two mentioned peroxidases, as well as laccases, belong to the so-called group of ligninolytic enzymes (i.e., those capable of decomposing lignin) (Fernández-Fueyo et al., 2015; Sen et al., 2016). Peroxidases are mainly used for dyes degradation. As reported by Zhang et al. manganese peroxidase, with a molecular mass of 45 kDa, was purified from the white rot fungus *Cerrena unicolor* and used for decolorizing dyes, such as Congo red, Remazol Brilliant Blue R (RBBR), bromophenol blue, methyl orange, and crystal violet. Manganese peroxidase had strong decolorizing activity on many dyes; crystal violet obtained degradation at level 85.1% in 14 hours, RBBR 83% in 5 hours, methyl orange 79.2% in 13 hours, and bromophenol blue 62.3% in 8 hours. Furthermore, the addition of gallic acid as the redox mediator enabled manganese peroxidase to decolorize azure blue by 63% in 18 hours. The process was carried out at 65? and pH 4.5, which was the most optimal value for activity peroxidase (H. Zhang et al., 2018).

Rekik et al., instead, used manganese peroxidase from *Trametes pubescens* strain i8 with an activity of 221 U/mg. The decolorization efficiencies of the dyes Acid Blue 158 (AB), polymeric dye (Poly R-478), Remazol Brilliant Violet 5R (RBV5R), Direct Red 5B (DR5B), indigo carmine (IC), methyl green (MG), Cibacet Brilliant Blue BG (CBB), and Remazol Brilliant Blue R (RBBR) were 95%, 88%, 76%, 66%, 64%, 50%, 46%, and 42%, respectively. Compared to the manganese peroxidase from *Bjerkander adusta*, the manganese peroxidase from *Trametes pubescens* showed higher catalytic efficiency and dye-decolorization ability. In the case of manganese peroxidase from *Bjerkander adusta*, degradation values of 90%, 77%, 75%, 55%, 49%, 14%, 88%, and 35% were achieved for the dyes AB, Poly R-478, RBV5R, DR5B, IC, MG, CBB, and RBBR, respectively (Rekik et al., 2019).

Peroxidases can also be used to efficiently degrade PAHs, presented earlier in this chapter. Manganese peroxidase from white rot fungus *Trametes* sp. 48424 for such a process was used by Zhang et al. The degradation of 10 mg/L fluorene (FLR), fluoranthene (FLA), pyrene (PYR), phenanthrene (PHE), and anthracene (ANT) was studied over a 24-hour period. The degradation of FLR, FLA, PYR, PHE, and ANT reached 93%, 96%, 98%, 97%, and 98%, respectively. Moreover, degradation of PAHs mixtures was carried out during the research, where over 90% was achieved in each case.

Transferases are enzymes that catalyze the reaction of transferring a chemical group or atom from one molecule (donor) to another (acceptor). They constitute the so-called phase II enzyme of the biotransformation of xenobiotics in the cell (glycosyltransferases, sulfotransferases, and glutathione S-transferases) (Bulkan et al., 2020). They are most likely responsible for the interaction with monooxygenases and the conversion of PAHs to aryl sulfates and hydroxyl aryl sulfates, anthraquinone, and 1,4-dihydroxyanthraquinone in *Aspergillus terreus* fungi (Marco-Urrea et al., 2015). Hydroxy, dihydroxy, dihydrodiol, and quinone derivatives are formed from polycyclic aromatic hydrocarbons in the first phase of decomposition. Subsequently, coupling of the oxidized metabolites to sulfate, methyl, or glucose acid groups takes place by phase II transferases (Bulkan et al., 2020).

Tyrosinase (EC 1.14.18.1) is a metalloenzyme (contains copper) glycoprotein found in fungal, animal, bacterial, and plant cells. Tyrosinase's ability to react with phenols makes it participate in the process of melanogenesis; as a result of which, melanin is produced. This ability has been exploited for the degradation of compounds such as cyanide, phenol, chlorophenols, cresols, and dyes (Ba & Vinoth Kumar, 2017; Nawaz et al., 2017; Osuoha et al., 2019). Martínková et al. report almost complete degradation of free cyanide (8.3–520 mg/L) in the model and actual coking the wastewater obtained by using recombinant cyanide hydratase in the first step, and tyrosinase degradation as the second step of the process. Phenol (1,552 mg/L) was completely removed from the actual coke wastewater in 20 hours, and the cresols (540 mg/L) were removed by 66% under the same conditions (Martínková & Chmátal, 2016).

Polyphenol oxidases (PPOs) are metalloenzymes that, similar to laccases, possess copper atoms in their active center (Marusek et al., 2006). In the presence of an oxygen molecule, PPO catalyzes the hydroxylation of monophenol and produces diphenol, followed by an oxidation reaction that removes the hydrogen molecule from diphenol. The sources of polyphenol oxidase are fungi of the genus *Lactarius* sp., *Entomocorticium* sp., *Trameres* sp., and *Polyporus* sp. These organisms produce PPO as an extracellular moiety (Panadare & Rathod, 2018).

An example of the use of polyphenol oxidase to degrade compounds responsible for environmental pollution is an attempt by Gouma et al., during which they used white rot fungi from the species *Pleurotus ostreatus*, *Pycnoporus coccineus*, *Phlebiopsis gigatea*, and *Trametes versicolor* to degrade linuron, metribuzin, and chlorpyrifos pesticides.

The addition of these compounds to the fungal culture increased the production of the said enzyme in *T. vesricolor* and thus the possibility of using these compounds as a carbon source for growth. Moreover, each of the tested mushrooms produces high levels of polyphenol oxidase and has the capacity to tolerate and degrade mixtures of pesticides (Gouma et al., 2019).

Moreover, this biocatalyst shows enzymatic activity against compounds such as phenol, mono- and di-chlorophenols, catechol and its derivatives, benzenetriol derivatives, phenol derivatives, other pesticides (e.g., chlorpyrifos) (A. Kumar et al., 2021; Nikolaivits et al., 2018; Valmas et al., 2020).

6.8 CONCLUSIONS AND PERSPECTIVES

The prevalence of many species of fungi around the world is an important starting point for considering the effectiveness of mycoremediation. This availability and practically every possible habitat or growing condition of various species of fungi make them have many applications, also in the so-called bioremediation of the natural environment.

Specialized proteins called enzymes, which catalyze a multitude of chemical reactions, are responsible for enabling the degradation of organic environmental pollutants. The examples of the use of different classes of enzymes, such as oxidoreductases, transferases, or hydrolases presented in this chapter, are only an overview of the multitude of their potential applications in the processes of remediation, degradation, and even mineralization of pollutants.

Among the listed pollutants, there are dyes, responsible for the deterioration of the quality of surface waters; polycyclic aromatic hydrocarbons, which pose a potential threat to human health; and other compounds, such as aromatic nitro derivatives or pharmaceuticals. The use of enzymes of fungal origin may constitute the so-called green alternative to commonly used methods due to the rational cost of the degradation process and the lack of the need to use additional chemicals.

In light of the continuing problem of environmental contamination with various types of chemicals, the challenges associated with their effective removal from the environment will not diminish in the near future. In particular, environmental pollution from pharmaceuticals is a growing problem that will become even worse as global drug consumption increases (M. Kumar et al., 2019). Moreover, the increase in pesticide use in agricultural production, especially observed in developing countries, will intensify the demand for the removal of agrochemicals from soils and surface waters (de Castro et al., 2017). As highlighted earlier, fungal enzymes have many potential applications in remediation processes, and due to their high selectivity, efficiency, and lack of toxic effects on the environment, their use will increase.

Current trends in environmental protection direct particular attention to concentrating remediation processes for hazardous pollutants at the source of emissions, at the point of generation of industrial effluents and waste sites. Therefore, it is observed that fungal enzymes will be particularly desirable in industrial wastewater treatment systems—for example, the textile industry, which emits large amounts of dyes (J. K. H. Wong et al., 2019), or the pharmaceutical industry and hospitals, where highly bioactive compounds are discharged into wastewaters (Stadlmair et al., 2018).

In order to ensure the possibility of continuous operation, it is necessary to use enzymes on immobilized beds. An appropriately selected media will maintain not only the expected effluent flow rates but also a high reduction ratio of pollutant concentrations over a longer period of time. Immobilization of enzymes reduces their susceptibility to degrading factors, such as extreme pH or increased temperature. Additionally, an appropriate carrier can also be a sorbent supporting the removal of pollutants by means of adsorption (Jesionowski et al., 2014).

However, the challenge in the process of enzyme immobilization is the choice of suitable carrier material and immobilization method, which do not decrease the efficiency of the enzyme by degradation or deactivation. In addition, leaching of the enzyme during contact with the wastewater stream must be prevented (Zdarta et al., 2018).

The use of fungal enzymes will also likely increase with the identification of both new fungal strains and new enzymes produced by them. As with bacteria, special attention is being paid to extremophilic fungi inhabiting specific ecological niches (Giovanella et al., 2020). Enzymes isolated from them may prove useful in the case of contamination with organic compounds that occur in specific places (e.g., in regions contaminated with heavy metals or in polar regions). In such places, the use of enzymes in the remediation of pollution will have a particular advantage over other bioremediation methods—that is, those that use live microorganisms—because their metabolism will be slowed down by adverse environmental conditions, such as low water activity or temperatures below 0? (Berde et al., 2019; Duarte et al., 2018).

Growing needs and growing knowledge about enzymes will make it possible to use them isolated from fungi cells in bioremediation processes with high efficiency, regardless of the type of environmental contamination. Then, it can be expected that the next years will bring many valuable and necessary solutions using fungal enzymes in environmental protection.

6.9   ACKNOWLEDGMENTS

This work was supported by the Ministry of Education and Science (Poland).

REFERENCES

Alharbi, O. M. L., Basheer, A. A., Khattab, R. A., & Ali, I. (2018). Health and environmental effects of persistent organic pollutants. *Journal of Molecular Liquids*, *263*(2017). Elsevier B.V. https://doi.org/10.1016/j.molliq.2018.05.029.

Ali, A., Shehzad, A., Khan, M., Shabbir, M., & Amjid, M. (2012). Yeast, its types and role in fermentation during bread making process-A. *Pakistan Journal of Food Sciences*, *22*(3), 171–179.

Alinsafi, A., Evenou, F., Abdulkarim, E. M., Pons, M. N., Zahraa, O., Benhammou, A., Yaacoubi, A., & Nejmeddine, A. (2007). Treatment of textile industry wastewater by supported photocatalysis. *Dyes and Pigments*, *74*(2), 439–445. https://doi.org/10.1016/j.dyepig.2006.02.024.

Alothman, Z. A., Bahkali, A. H., Elgorban, A. M., Al-Otaibi, M. S., Ghfar, A. A., Gabr, S. A., Wabaidur, S. M., Habila, M. A., & Hadj Ahmed, A. Y. B. (2020). Bioremediation of explosive TNT by trichoderma viride. *Molecules*, *25*(6), 1–13. https://doi.org/10.3390/molecules25061393.

Antecka, K., Zdarta, J., Siwińska-Stefańska, K., Sztuk, G., Jankowska, E., Oleskowicz-Popiel, P., & Jesionowski, T. (2018). Synergistic degradation of dye wastewaters using binary or ternary oxide systems with immobilized laccase. *Catalysts*, *8*(9). https://doi.org/10.3390/catal8090402.

Arregui, L., Ayala, M., Gómez-Gil, X., Gutiérrez-Soto, G., Hernández-Luna, C. E., Herrera De Los Santos, M., Levin, L., Rojo-Domínguez, A., Romero-Martínez, D., Saparrat, M. C. N., Trujillo-Roldán, M. A., & Valdez-Cruz, N. A. (2019). Laccases: Structure, function, and potential application in water bioremediation. *Microbial Cell Factories*, *18*(1), 1–33. https://doi.org/10.1186/s12934-019-1248-0.

Ashraf, M. A. (2017). Persistent organic pollutants (POPs): A global issue, a global challenge. *Environmental Science and Pollution Research*, *24*(5), 4223–4227. https://doi.org/10.1007/s11356-015-5225-9.

Ba, S., & Vinoth Kumar, V. (2017). Recent developments in the use of tyrosinase and laccase in environmental applications. *Critical Reviews in Biotechnology*, *37*(7), 819–832. https://doi.org/10.1080/07388551.2016.1261081.

Bagewadi, Z. K., Mulla, S. I., & Ninnekar, H. Z. (2017). Purification and immobilization of laccase from trichoderma harzianum strain HZN10 and its application in dye decolorization. *Journal of Genetic Engineering and Biotechnology*, *15*(1), 139–150. https://doi.org/10.1016/j.jgeb.2017.01.007.

Bai, Y. B, Gao, Y. Q., Nie, X. D., Tuong, T. M. L., Li, D., & Gao, J. M. (2019). Antifungal activity of griseofulvin derivatives against phytopathogenic fungi in vitro and in vivo and three-dimensional quantitative structure-activity relationship analysis [research-article]. *Journal of Agricultural and Food Chemistry*, *67*(22), 6125–6132. https://doi.org/10.1021/acs.jafc.9b00606.

Barrios-Estrada, C., Rostro-Alanis, M. de J., Parra, A. L., Belleville, M. P., Sanchez-Marcano, J., Iqbal, H. M. N., & Parra-Saldívar, R. (2018). Potentialities of active membranes with immobilized laccase for Bisphenol A degradation. *International Journal of Biological Macromolecules*, *108*, 837–844. https://doi.org/10.1016/j.ijbiomac.2017.10.177.

Beeson, W. T., Vu, V. V., Span, E. A., Phillips, C. M., & Marletta, M. A. (2015). Cellulose degradation by polysaccharide monooxygenases. *Annual Review of Biochemistry*, *84*(March), 923–946. https://doi.org/10.1146/annurev-biochem-060614-034439.

Berde, C. P., Berde, V. B., Sheela, G. M., & Veerabramhachari, P. (2019). *Discovery of New Extremophilic Enzymes from Diverse Fungal Communities* (Vol. 3, Issue October). https://doi.org/10.1007/978-3-030-10480-1.

Bhide, A., Datar, S., & Stebbins, K. (2020). Case histories of significant medical advances: Tamoxifen. *SSRN Electronic Journal*. https://doi.org/10.2139/ssrn.3679645.

Błaszkowski, J., & Czerniawska, B. (2011). Arbuscular mycorrhizal fungi (Glomeromycota) associated with roots of Ammophila arenaria growing in maritime dunes of Bornholm (Denmark). *Acta Societatis Botanicorum Poloniae, 80*(1), 63–76. https://doi.org/10.5586/asbp.2011.009.

Bulkan, G., Ferreira, J. A., & Taherzadeh, M. J. (2020). Chapter 15 — removal of organic micro-pollutants using filamentous fungi. *Current Developments in Biotechnology and Bioengineering.* https://doi.org/10.1016/b978-0-12-819594-9.00015-2.

Cantele, C., Fontana, R. C., Mezzomo, A. G., da Rosa, L. O., Poleto, L., Camassola, M., & Dillon, A. J. P. (2017). Production, characterization and dye decolorization ability of a high level laccase from Marasmiellus palmivorus. *Biocatalysis and Agricultural Biotechnology, 12*(August), 15–22. https://doi.org/10.1016/j.bcab.2017.08.012.

Carpenter, D. O. (2011). Health effects of persistent organic pollutants: The challenge for the pacific basin and for the world. *Reviews on Environmental Health, 26*(1), 61–69. https://doi.org/10.1515/REVEH.2011.009.

Chaudhry, S. B., & Veve, M. P. (2019). Cephalosporins : A focus on side chains and β-Lactam cross-reactivity. *Pharmacy, 7*(103), 1–16.

Chung, K.-T. (2016). Azo dyes and human health: A review. *Journal of Environment and Health Science, 34*(4), 1–60.

Cunha, D. L., Mendes, M. P., & Marques, M. (2019). Environmental risk assessment of psychoactive drugs in the aquatic environment. *Environmental Science and Pollution Research, 26*(1), 78–90. https://doi.org/10.1007/s11356-018-3556-z.

da Silva, I. G. S., de Almeida, F. C. G., da Rocha e Silva, N. M. P., Casazza, A. A., Converti, A., & Sarubbo, L. A. (2020). Soil bioremediation: Overview of technologies and trends. *Energies, 13*(18). https://doi.org/10.3390/en13184664.

De Aragão Umbuzeiro, G., Freeman, H. S., Warren, S. H., De Oliveira, D. P., Terao, Y., Watanabe, T., & Claxton, L. D. (2005). The contribution of azo dyes to the mutagenic activity of the Cristais River. *Chemosphere, 60*(1), 55–64. https://doi.org/10.1016/j.chemosphere.2004.11.100.

de Castro, A. A., Prandi, I. G., Kuca, K., & Ramalho, T. C. (2017). Organophosphorus degrading enzymes: Molecular basis and perspectives for enzymatic bioremediation of agrochemicals. *Ciência e Agrotecnologia, 41*(5), 471–482. https://doi.org/10.1590/1413-70542017415000417.

de Freitas, E. N., Bubna, G. A., Brugnari, T., Kato, C. G., Nolli, M., Rauen, T. G., Peralta Muniz Moreira, R. de F., Peralta, R. A., Bracht, A., de Souza, C. G. M., & Peralta, R. M. (2017). Removal of bisphenol a by laccases from Pleurotus ostreatus and Pleurotus pulmonarius and evaluation of ecotoxicity of degradation products. *Chemical Engineering Journal, 330*, 1361–1369. https://doi.org/10.1016/j.cej.2017.08.051.

Didier, E. S. (2005). Microsporidiosis: An emerging and opportunistic infection in humans and animals. *Acta Tropica, 94*(1), 61–76. https://doi.org/10.1016/j.actatropica.2005.01.010.

Dietrich, O., Heun, M., Notroff, J., Schmidt, K., & Zarnkow, M. (2012). The role of cult and feasting in the emergence of Neolithic communities. New evidence from Göbekli Tepe, south-eastern Turkey. *Antiquity, 86*(333), 674–695. https://doi.org/10.1017/S0003598X00047840.

dos Santos, A. B., Cervantes, F. J., & van Lier, J. B. (2007). Review paper on current technologies for decolourisation of textile wastewaters: Perspectives for anaerobic biotechnology. *Bioresource Technology, 98*(12), 2369–2385. https://doi.org/10.1016/j.biortech.2006.11.013.

Duarte, A. W. F., dos Santos, J. A., Vianna, M. V., Vieira, J. M. F., Mallagutti, V. H., Inforsato, F. J., Wentzel, L. C. P., Lario, L. D., Rodrigues, A., Pagnocca, F. C., Pessoa, A., & Durães Sette, L. (2018). Cold-adapted enzymes produced by fungi from terrestrial and marine Antarctic environments. *Critical Reviews in Biotechnology, 38*(4), 600–619. https://doi.org/10.1080/07388551.2017.1379468.

Elyamine, A. M., Kan, J., Meng, S., Tao, P., Wang, H., & Hu, Z. (2021). Aerobic and anaerobic bacterial and fungal degradation of pyrene: Mechanism pathway including biochemical reaction and catabolic genes. *International Journal of Molecular Sciences*, *22*(15). https://doi.org/10.3390/ijms22158202.

Fernández-Fueyo, E., Linde, D., Almendral, D., López-Lucendo, M. F., Ruiz-Dueñas, F. J., & Martínez, A. T. (2015). Description of the first fungal dye-decolorizing peroxidase oxidizing manganese(II). *Applied Microbiology and Biotechnology*, *99*(21), 8927–8942. https://doi.org/10.1007/s00253-015-6665-3.

Fruk, L., Kuo, C. H., Torres, E., & Niemeyer, C. M. (2009). Apoenzyme reconstitution as a chemical tool for structural enzymology and biotechnology. *Angewandte Chemie—International Edition*, *48*(9), 1550–1574. https://doi.org/10.1002/anie.200803098.

Giovanella, P., Vieira, G. A. L., Ramos Otero, I. V., Pais Pellizzer, E., de Jesus Fontes, B., & Sette, L. D. (2020). Metal and organic pollutants bioremediation by extremophile microorganisms. *Journal of Hazardous Materials*, *382*(July 2019), 121024. https://doi.org/10.1016/j.jhazmat.2019.121024.

Góralczyk-Binkowska, A., Jasinska, A., Dlugonski, A., Plocinski, P., & Dlugonski, J. (2020). Laccase activity of the ascomycete fungus Nectriella pironii and innovative strategies for its production on leaf litter of an urban park. *PLoS One*, *15*(4), 1–18. https://doi.org/10.1371/journal.pone.0231453.

Gouma, S., Papadaki, A. A., Markakis, G., Magan, N., & Goumas, D. (2019). Studies on pesticides mixture degradation by white rot fungi. *Journal of Ecological Engineering*, *20*(2), 16–26. https://doi.org/10.12911/22998993/94918.

Guo, W., Pan, B., Sakkiah, S., Yavas, G., Ge, W., Zou, W., Tong, W., & Hong, H. (2019). Persistent organic pollutants in food: Contamination sources, health effects and detection methods. *International Journal of Environmental Research and Public Health*, *16*(22), 10–12. https://doi.org/10.3390/ijerph16224361.

Gupta, A. K., Mays, R. R., Versteeg, S. G., Piraccini, B. M., Shear, N. H., Piguet, V., Tosti, A., & Friedlander, S. F. (2018). Tinea capitis in children: A systematic review of management. *Journal of the European Academy of Dermatology and Venereology*, *32*(12), 2264–2274. https://doi.org/10.1111/jdv.15088.

Hameed, A., Farooq, T., & Shabbir, S. (2021). Role of pharmaceuticals as EDCs in metabolic disorders. In M. S. H. Akash, K. Rehman, & M. Z. Hashmi (Eds.), *Endocrine Disrupting Chemicals-Induced Metabolic Disorders and Treatment Strategies* (pp. 357–365). Springer International Publishing. https://doi.org/10.1007/978-3-030-45923-9_21.

Harms, H., Schlosser, D., & Wick, L. Y. (2011). Untapped potential: Exploiting fungi in bioremediation of hazardous chemicals. *Nature Reviews Microbiology*, *9*(3), 177–192. https://doi.org/10.1038/nrmicro2519.

Hibbett, D. S., Binder, M., Bischoff, J. F., Blackwell, M., Cannon, P. F., Eriksson, O. E., Huhndorf, S., James, T., Kirk, P. M., Lücking, R., Thorsten Lumbsch, H., Lutzoni, F., Matheny, P. B., McLaughlin, D. J., Powell, M. J., Redhead, S., Schoch, C. L., Spatafora, J. W., Stalpers, J. A., . . . Zhang, N. (2007). A higher-level phylogenetic classification of the Fungi. *Mycological Research*, *111*(5), 509–547. https://doi.org/10.1016/j.mycres.2007.03.004.

Hill, R., Leitch, I. J., & Gaya, E. (2021). Targeting Ascomycota genomes: What and how big? *Fungal Biology Reviews*, *36*, 52–59. https://doi.org/10.1016/j.fbr.2021.03.003.

Horwat, M., Tice, M., & Kjellerup, B. (2015). Biofilms at work: Bio-, phyto- and rhizoremediation approaches for soils contaminated with polychlorinated biphenyls. *AIMS Bioengineering*, *2*(4), 324–334. https://doi.org/10.3934/bioeng.2015.4.324.

Humber, R. A. (2012). Entomophthoromycota: A new phylum and reclassification for entomophthoroid fungi. *Mycotaxon*, *120*, 477–492. https://doi.org/10.5248/120.477.

Hymery, N., Vasseur, V., Coton, M., Mounier, J., Jany, J. L., Barbier, G., & Coton, E. (2014). Filamentous fungi and mycotoxins in Cheese: A review. *Comprehensive Reviews in Food Science and Food Safety*, *13*(4), 437–456. https://doi.org/10.1111/1541-4337.12069.

Iark, D., Buzzo, A. J. dos R., Garcia, J. A. A., Côrrea, V. G., Helm, C. V., Corrêa, R. C. G., Peralta, R. A., Peralta Muniz Moreira, R. de F., Bracht, A., & Peralta, R. M. (2019). Enzymatic degradation and detoxification of azo dye Congo red by a new laccase from Oudemansiella canarii. *Bioresource Technology, 289*(April), 121655. https://doi. org/10.1016/j.biortech.2019.121655.

Jankowska, K., Zdarta, J., Grzywaczyk, A., Kijeńska-Gawrońska, E., Biadasz, A., & Jesion-owski, T. (2020). Electrospun poly(methyl methacrylate)/polyaniline fibres as a support for laccase immobilisation and use in dye decolourisation. *Environmental Research, 184*(October 2019). https://doi.org/10.1016/j.envres.2020.109332.

Jesionowski, T., Zdarta, J., & Krajewska, B. (2014). Enzyme immobilization by adsorption: A review. *Adsorption, 20*(5–6), 801–821. https://doi.org/10.1007/s10450-014-9623-y.

Jia, J., Bi, C., Zhang, J., Jin, X., & Chen, Z. (2018). Characterization of polycyclic aromatic hydrocarbons (PAHs) in vegetables near industrial areas of Shanghai, China: Sources, exposure, and cancer risk. *Environmental Pollution, 241*, 750–758. https://doi. org/10.1016/j.envpol.2018.06.002.

Jung, J., Hah, K., Lee, W., & Jang, W. (2017). Meta-analysis of microarray datasets for the risk assessment of coplanar polychlorinated biphenyl 77 (PCB77) on human health. *Toxicology and Environmental Health Sciences, 9*(2), 161–168. https://doi.org/10.1007/ s13530-017-0317-1.

Karigar, C. S., & Rao, S. S. (2011). Role of microbial enzymes in the bioremediation of pollutants: A review. *Enzyme Research, 2011*(1). https://doi.org/10.4061/2011/805187.

Klenov, V., Flor, S., Ganesan, S., Adur, M., Eti, N., Iqbal, K., Soares, M. J., Ludewig, G., Ross, J. W., Robertson, L. W., & Keating, A. F. (2021). The Aryl hydrocarbon receptor mediates reproductive toxicity of polychlorinated biphenyl congener 126 in rats. *Toxicology and Applied Pharmacology, 426*(May), 115639. https://doi.org/10.1016/j.taap.2021.115639.

Klocke, C., & Lein, P. J. (2020). Evidence implicating non-dioxin-like congeners as the key mediators of polychlorinated biphenyl (Pcb) developmental neurotoxicity. *International Journal of Molecular Sciences, 21*(3), 1–39. https://doi.org/10.3390/ijms21031013.

Kumar, A., Sharma, A., Chaudhary, P., & Gangola, S. (2021). Chlorpyrifos degradation using binary fungal strains isolated from industrial waste soil. *Biologia, 76*(10), 3071–3080. https://doi.org/10.1007/s11756-021-00816-8.

Kumar, M., Jaiswal, S., Sodhi, K. K., Shree, P., Singh, D. K., Agrawal, P. K., & Shukla, P. (2019). Antibiotics bioremediation: Perspectives on its ecotoxicity and resistance. *Environment International, 124*(October 2018), 448–461. https://doi.org/10.1016/j. envint.2018.12.065.

Kumar, P., Agnihotri, R., Wasewar, K. L., Uslu, H., & Yoo, C. K. (2012). Status of adsorptive removal of dye from textile industry effluent. *Desalination and Water Treatment, 50*(1–3), 226–244. https://doi.org/10.1080/19443994.2012.719472.

Leong, J. Y., Blachman-Braun, R., Patel, A. S., Patel, P., & Ramasamy, R. (2019). Association between polychlorinated biphenyl 153 exposure and serum testosterone levels: Analysis of the national health and nutrition examination survey. *Translational Andrology and Urology, 8*(6), 666–672. https://doi.org/10.21037/tau.2019.11.26.

Li, F., Ma, F., Zhao, H., Zhang, S., Wang, L., Zhang, X., & Yu, H. (2019). A lytic polysaccharide monooxygenase from a white-rot fungus drives the degradation of lignin by a versatile peroxidase. *Applied and Environmental Microbiology, 85*(9). https://doi. org/10.1128/AEM.02803-18.

Li, N., Xia, Q., Li, Y., Hou, X., Niu, M., Ping, Q., & Xiao, H. (2018). Immobilizing laccase on modified cellulose/CF Beads to degrade chlorinated biphenyl in wastewater. *Polymers, 10*(7). https://doi.org/10.3390/polym10070798.

Li, N., Xia, Q., Niu, M., Ping, Q., & Xiao, H. (2018). Immobilizing laccase on different species wood biochar to remove the chlorinated biphenyl in wastewater. *Scientific Reports, 8*(1), 1–9. https://doi.org/10.1038/s41598-018-32013-0.

Li, Xuanzhen, Pan, Y., Hu, S., Cheng, Y., Wang, Y., Wu, K., Zhang, S., & Yang, S. (2018). Diversity of phenanthrene and benz[a]anthracene metabolic pathways in white rot fungus Pycnoporus sanguineus 14. *International Biodeterioration and Biodegradation*, *134*(May), 25–30. https://doi.org/10.1016/j.ibiod.2018.07.012.

Lin, X., Xu, C., Zhou, Y., Liu, S., & Liu, W. (2020). A new perspective on volatile halogenated hydrocarbons in Chinese agricultural soils. *Science of the Total Environment*, *703*(xxxx), 134646. https://doi.org/10.1016/j.scitotenv.2019.134646.

Liu, X., Lu, S., Guo, W., Xi, B., & Wang, W. (2018). Antibiotics in the aquatic environments: A review of lakes, China. *Science of the Total Environment*, *627*, 1195–1208. https://doi.org/10.1016/j.scitotenv.2018.01.271.

Lodolo, E. J., Kock, J. L. F., Axcell, B. C., & Brooks, M. (2008). The yeast saccharomyces cerevisiae—the main character in beer brewing. *FEMS Yeast Research*, *8*(7), 1018–1036. https://doi.org/10.1111/j.1567-1364.2008.00433.x.

Luo, Q., Liang, S., & Huang, Q. (2018). Laccase induced degradation of perfluorooctanoic acid in a soil slurry. *Journal of Hazardous Materials*, *359*, 241–247. https://doi.org/10.1016/j.jhazmat.2018.07.048.

Luo, W., Su, L., Craig, N. J., Du, F., Wu, C., & Shi, H. (2019). Comparison of microplastic pollution in different water bodies from urban creeks to coastal waters. *Environmental Pollution*, *246*, 174–182. https://doi.org/10.1016/j.envpol.2018.11.081.

Magnin, A., Entzmann, L., Pollet, E., & Avérous, L. (2021). Breakthrough in polyurethane bio-recycling: An efficient laccase-mediated system for the degradation of different types of polyurethanes. *Waste Management*, *132*(January), 23–30. https://doi.org/10.1016/j.wasman.2021.07.011.

Mahmood, I., Imadi, S. R., Shazadi, K., Gul, A., & Hakeem, K. R. (2016). Effects of pesticides on environment. In K. R. Hakeem, M. S. Akhtar, & S. N. A. Abdullah (Eds.), *Plant, Soil and Microbes: Volume 1: Implications in Crop Science* (pp. 253–269). Springer International Publishing. https://doi.org/10.1007/978-3-319-27455-3_13.

Marco-Urrea, E., García-Romera, I., & Aranda, E. (2015). Potential of non-ligninolytic fungi in bioremediation of chlorinated and polycyclic aromatic hydrocarbons. *New Biotechnology*, *32*(6), 620–628. https://doi.org/10.1016/j.nbt.2015.01.005.

Marotta, V., Malandrino, P., Russo, M., Panariello, I., Ionna, F., Chiofalo, M. G., & Pezzullo, L. (2020). Fathoming the link between anthropogenic chemical contamination and thyroid cancer. *Critical Reviews in Oncology/Hematology*, *150*(April), 102950. https://doi.org/10.1016/j.critrevonc.2020.102950.

Martínková, L., & Chmátal, M. (2016). The integration of cyanide hydratase and tyrosinase catalysts enables effective degradation of cyanide and phenol in coking wastewaters. *Water Research*, *102*, 90–95. https://doi.org/10.1016/j.watres.2016.06.016.

Marusek, C. M., Trobaugh, N. M., Flurkey, W. H., & Inlow, J. K. (2006). Comparative analysis of polyphenol oxidase from plant and fungal species. *Journal of Inorganic Biochemistry*, *100*(1), 108–123. https://doi.org/10.1016/j.jinorgbio.2005.10.008.

McKelvey, S. M., & Murphy, R. A. (2011). Biotechnological use of fungal enzymes. *Fungi: Biology and Applications: Second Edition*, 179–204. https://doi.org/10.1002/9781119976950.ch7.

Mendell, M. J., Mirer, A. G., Cheung, K., Tong, M., & Douwes, J. (2011). Respiratory and allergic health effects of dampness, mold, and dampness-related agents: A review of the epidemiologic evidence. *Environmental Health Perspectives*, *119*(6), 748–756. https://doi.org/10.1289/ehp.1002410.

Mihankhah, T., Saeedi, M., & Karbassi, A. (2020). Contamination and cancer risk assessment of polycyclic aromatic hydrocarbons (PAHs) in urban dust from different land-uses in the most populated city of Iran. *Ecotoxicology and Environmental Safety*, *187*(October 2019), 109838. https://doi.org/10.1016/j.ecoenv.2019.109838.

Morgenstern, I., Powlowski, J., & Tsang, A. (2014). Fungal cellulose degradation by oxidative enzymes: Fromdysfunctional GH61 family to powerful lytic polysaccharide monooxygenase family. *Briefings in Functional Genomics*, *13*(6), 471–481. https://doi.org/10.1093/bfgp/elu032.

Mukherjee, S. (2019). Isolation and purification of industrial enzymes: Advances in enzyme technology. In: *Biomass, Biofuels, Biochemicals: Advances in Enzyme Technology* (pp. 41–70). https://doi.org/10.1016/B978-0-444-64114-4.00002-9.

Musiani, F., Broll, V., Evangelisti, E., & Ciurli, S. (2020). The model structure of the copper-dependent ammonia monooxygenase. *Journal of Biological Inorganic Chemistry*, *25*(7), 995–1007. https://doi.org/10.1007/s00775-020-01820-0.

Nawaz, A., Shafi, T., Khaliq, A., Mukhtar, H., & ul Haq, I. (2017). Tyrosinase: Sources, structure and applications. *International Journal of Biotechnology and Bioengineering*, *3*(5), 135–141. https://doi.org/10.25141/2475-3432-2017-5.0135.

Nikolaivits, E., Dimarogona, M., Karagiannaki, I., Chalima, A., Fishman, A., & Topakasa, E. (2018). Versatile fungal polyphenol oxidase with chlorophenol bioremediation potential: Characterization and protein engineering. *Applied and Environmental Microbiology*, *84*(23). https://doi.org/10.1128/AEM.01628-18.

Nikolaivits, E., Siaperas, R., Agrafiotis, A., Ouazzani, J., Magoulas, A., Gioti, A., & Topakas, E. (2021). Functional and transcriptomic investigation of laccase activity in the presence of PCB29 identifies two novel enzymes and the multicopper oxidase repertoire of a marine-derived fungus. *Science of the Total Environment*, *775*. https://doi.org/10.1016/j.scitotenv.2021.145818.

Onozuka, D., Nakamura, Y., Tsuji, G., & Furue, M. (2020). Mortality in Yusho patients exposed to polychlorinated biphenyls and polychlorinated dibenzofurans: A 50-year retrospective cohort study. *Environmental Health: A Global Access Science Source*, *19*(1), 1–10. https://doi.org/10.1186/s12940-020-00680-0.

Osuoha, J. O., Abbey, B. W., Egwim, E. C., & Nwaichi, E. O. (2019). Production and characterization of tyrosinase enzyme for enhanced treatment of organic pollutants in petroleum refinery effluent. *Society of Petroleum Engineers—SPE Nigeria Annual International Conference and Exhibition 2019, NAIC 2019*. https://doi.org/10.2118/198791-MS.

Othman, A. M., Elsayed, M. A., Elshafei, A. M., & Hassan, M. M. (2018). Purification and biochemical characterization of two isolated laccase isoforms from Agaricus bisporus CU13 and their potency in dye decolorization. *International Journal of Biological Macromolecules*, *113*, 1142–1148. https://doi.org/10.1016/j.ijbiomac.2018.03.043.

Panadare, D., & Rathod, V. K. (2018). Extraction and purification of polyphenol oxidase: A review. *Biocatalysis and Agricultural Biotechnology*, *14*(March), 431–437. https://doi.org/10.1016/j.bcab.2018.03.010.

Pandey, A., Singh, M. P., Kumar, S., & Srivastava, S. (2019). Phycoremediation of persistent organic pollutants from wastewater: Retrospect and prospects. In S. K. Gupta & F. Bux (Eds.), *Application of Microalgae in Wastewater Treatment: Volume 1: Domestic and Industrial Wastewater Treatment* (pp. 207–235). Springer International Publishing. https://doi.org/10.1007/978-3-030-13913-1_11.

Pappas, P. G., Lionakis, M. S., Arendrup, M. C., Ostrosky-Zeichner, L., & Kullberg, B. J. (2018). Invasive candidiasis. *Nature Reviews Disease Primers*, *4*(May), 1–20. https://doi.org/10.1038/nrdp.2018.26.

Peng, J., Wu, E., Lou, X., Deng, Q., Hou, X., Lv, C., & Hu, Q. (2021). Anthraquinone removal by a metal-organic framework/polyvinyl alcohol cryogel-immobilized laccase: Effect and mechanism exploration. *Chemical Engineering Journal*, *418*(March). https://doi.org/10.1016/j.cej.2021.129473.

Piontek, K., Antorini, M., & Choinowski, T. (2002). Crystal structure of a laccase from the fungus trametes versicolor at 1.90-Å resolution containing a full complement of coppers. *Journal of Biological Chemistry*, *277*(40), 37663–37669. https://doi.org/10.1074/jbc.M204571200.

Pitsawong, W., Hoben, J. P., & Miller, A. F. (2014). Understanding the broad substrate repertoire of nitroreductase based on its kinetic mechanism. *Journal of Biological Chemistry*, *289*(22), 15203–15214. https://doi.org/10.1074/jbc.M113.547117.

Rathna, R., Varjani, S., & Nakkeeran, E. (2018). Recent developments and prospects of dioxins and furans remediation. *Journal of Environmental Management*, *223*(April), 797–806. https://doi.org/10.1016/j.jenvman.2018.06.095.

Rekik, H., Zaraî Jaouadi, N., Bouacem, K., Zenati, B., Kourdali, S., Badis, A., Annane, R., Bouanane-Darenfed, A., Bejar, S., & Jaouadi, B. (2019). Physical and enzymatic properties of a new manganese peroxidase from the white-rot fungus trametes pubescens strain i8 for lignin biodegradation and textile-dyes biodecolorization. *International Journal of Biological Macromolecules*, *125*, 514–525. https://doi.org/10.1016/j.ijbiomac.2018.12.053.

Rhodes, C. J. (2014). Mycoremediation (bioremediation with fungi)—growing mushrooms to clean the earth. *Chemical Speciation and Bioavailability*, *26*(3), 196–198. https://doi.org/10.3184/095422914X14047407349335.

Ritter, L., Solomon, K., & Forget, J. (2011). Persistent organic pollutants—an assessment report on: DDT-aldrin-dieldrin-endrin-chlordane-heptachlor-heptachlorobenzene-mirex-to-zapene-polychlorinated biphenyls-dioxins and furans. *Chemosphere*, 43 pp. www.ncbi.nlm.nih.gov/pubmed/22018961.

Robinson, P. K. (2015). Enzymes: Principles and biotechnological applications. *Essays in Biochemistry*, *59*, 1–41. https://doi.org/10.1042/BSE0590001.

Rocheleau, H., Al-harthi, R., & Ouellet, T. (2019). Degradation of salicylic acid by Fusarium graminearum. *Fungal Biology*, *123*(1), 77–86. https://doi.org/10.1016/j.funbio.2018.11.002.

Salami, F., Habibi, Z., Yousefi, M., & Mohammadi, M. (2018). Covalent immobilization of laccase by one pot three component reaction and its application in the decolorization of textile dyes. *International Journal of Biological Macromolecules*, *120*, 144–151. https://doi.org/10.1016/j.ijbiomac.2018.08.077.

Sarkar, S., Banerjee, A., Halder, U., Biswas, R., & Bandopadhyay, R. (2017). Degradation of synthetic azo dyes of textile industry: A sustainable approach using microbial enzymes. *Water Conservation Science and Engineering*, *2*(4), 121–131. https://doi.org/10.1007/s41101-017-0031-5.

Sen, S. K., Raut, S., Bandyopadhyay, P., & Raut, S. (2016). Fungal decolouration and degradation of azo dyes: A review. *Fungal Biology Reviews*, *30*(3), 112–133. https://doi.org/10.1016/j.fbr.2016.06.003.

Serrano-González, M. Y., Chandra, R., Castillo-Zacarias, C., Robledo-Padilla, F., Rostro-Alanis, M. de J., & Parra-Saldivar, R. (2018). Biotransformation and degradation of 2,4,6-trinitrotoluene by microbial metabolism and their interaction. *Defence Technology*, *14*(2), 151–164. https://doi.org/10.1016/j.dt.2018.01.004.

Shakerian, F., Zhao, J., & Li, S. P. (2020). Recent development in the application of immobilized oxidative enzymes for bioremediation of hazardous micropollutants—a review. *Chemosphere*, *239*, 124716. https://doi.org/10.1016/j.chemosphere.2019.124716.

Singh, A., Kuhad, R. C., & Ward, O. P. (2009). Biological remediation of soil: An overview of global market and available technologies. In A. Singh, R. C. Kuhad, & O. P. Ward (Eds.), *Advances in Applied Bioremediation* (pp. 1–19). Springer Berlin Heidelberg. https://doi.org/10.1007/978-3-540-89621-0_1.

Soares, G. M. S., Figueiredo, L. C., Faveri, M., Cortelli, S. C., Duarte, P. M., & Feres, M. (2012). Mechanisms of action of systemic antibiotics used in periodontal treatment and mechanisms of bacterial resistance to these drugs. *Journal of Applied Oral Science*, *20*(3), 295–305. https://doi.org/10.1590/S1678-77572012000300002.

Song, H. N., Jeong, D. G., Bang, S. Y., Paek, S. H., Park, B. C., Park, S. G., & Woo, E. J. (2015). Crystal structure of the fungal nitroreductase Frm2 from Saccharomyces cerevisiae. *Protein Science*, *24*(7), 1158–1163. https://doi.org/10.1002/pro.2686.

Spatafora, J. W., Chang, Y., Benny, G. L., Lazarus, K., Smith, M. E., Berbee, M. L., Bonito, G., Corradi, N., Grigoriev, I., Gryganskyi, A., James, T. Y., O'Donnell, K., Roberson, R. W., Taylor, T. N., Uehling, J., Vilgalys, R., White, M. M., & Stajich, J. E. (2016). A phylum-level phylogenetic classification of zygomycete fungi based on genome-scale data. *Mycologia*, *108*(5), 1028–1046. https://doi.org/10.3852/16-042.

Šrédlová, K., Šírová, K., Stella, T., & Cajthaml, T. (2021). Degradation products of polychlorinated biphenyls and their in vitro transformation by ligninolytic fungi. *Toxics*, *9*(4). https://doi.org/10.3390/toxics9040081.

Stading, R., Gastelum, G., Chu, C., Jiang, W., & Moorthy, B. (2021). Molecular mechanisms of pulmonary carcinogenesis by polycyclic aromatic hydrocarbons (PAHs): Implications for human lung cancer. *Seminars in Cancer Biology*, *June*. https://doi.org/10.1016/j.semcancer.2021.07.001.

Stadlmair, L. F., Letzel, T., Drewes, J. E., & Grassmann, J. (2018). Enzymes in removal of pharmaceuticals from wastewater: A critical review of challenges, applications and screening methods for their selection. *Chemosphere*, *205*, 649–661. https://doi.org/10.1016/j.chemosphere.2018.04.142.

Subramanian, V., & Yadav, J. S. (2009). Role of P450 monooxygenases in the degradation of the endocrine-disrupting chemical nonylphenol by the white rot fungus Phanerochaete chrysosporium. *Applied and Environmental Microbiology*, *75*(17), 5570–5580. https://doi.org/10.1128/AEM.02942-08.

Sun, S., Sidhu, V., Rong, Y., & Zheng, Y. (2018). Pesticide pollution in agricultural soils and sustainable remediation methods: A review. *Current Pollution Reports*, *4*(3), 240–250. https://doi.org/10.1007/s40726-018-0092-x.

Tacconelli, E., Carrara, E., Savoldi, A., Harbarth, S., Mendelson, M., Monnet, D. L., Pulcini, C., Kahlmeter, G., Kluytmans, J., Carmeli, Y., Ouellette, M., Outterson, K., Patel, J., Cavaleri, M., Cox, E. M., Houchens, C. R., Grayson, M. L., Hansen, P., Singh, N., . . . Zorzet, A. (2018). Discovery, research, and development of new antibiotics: The WHO priority list of antibiotic-resistant bacteria and tuberculosis. *The Lancet Infectious Diseases*, *18*(3), 318–327. https://doi.org/10.1016/S1473-3099(17)30753-3.

Talhout, R., Schulz, T., Florek, E., van Benthem, J., Wester, P., & Opperhuizen, A. (2011). Hazardous compounds in tobacco smoke. *International Journal of Environmental Research and Public Health*, *8*(2), 613–628. https://doi.org/10.3390/ijerph8020613.

Tavares, M. F., Avelino, K. V., Araújo, N. L., Marim, R. A., Linde, G. A., Colauto, N. B., & do Valle, J. S. (2020). Decolorization of azo and anthraquinone dyes by crude laccase produced by Lentinus crinitus in solid state cultivation. *Brazilian Journal of Microbiology*, *51*(1), 99–106. https://doi.org/10.1007/s42770-019-00189-w.

Tišma, M., Šalić, A., Planinić, M., Zelić, B., Potočnik, M., Šelo, G., & Bucić-Kojić, A. (2020). Production, characterisation and immobilization of laccase for an efficient aniline-based dye decolourization. *Journal of Water Process Engineering*, *36*(May), 101327. https://doi.org/10.1016/j.jwpe.2020.101327.

Torres Pazmiño, D. E., Winkler, M., Glieder, A., & Fraaije, M. W. (2010). Monooxygenases as biocatalysts: Classification, mechanistic aspects and biotechnological applications. *Journal of Biotechnology*, *146*(1–2), 9–24. https://doi.org/10.1016/j.jbiotec.2010.01.021.

Valmas, A., Dedes, G., & Dimarogona, M. (2020). Structural studies of a fungal polyphenol oxidase with application to bioremediation of contaminated water. *Proceedings*, *66*(1), 10. https://doi.org/10.3390/proceedings2020066010.

Valverde, M. E., Hernández-Pérez, T., & Paredes-López, O. (2015). Edible mushrooms: Improving human health and promoting quality life. *International Journal of Microbiology*, *2015*(January). https://doi.org/10.1155/2015/376387.

Varjani, S. J. (2017). Microbial degradation of petroleum hydrocarbons. *Bioresource Technology*, *223*, 277–286. https://doi.org/10.1016/j.biortech.2016.10.037.

Vishwanatha, K. S., Rao, A. G. A., & Singh, S. A. (2010). Acid protease production by solid-state fermentation using Aspergillus oryzae MTCC 5341: Optimization of process parameters. *Journal of Industrial Microbiology and Biotechnology*, *37*(2), 129–138. https://doi.org/10.1007/s10295-009-0654-4.

Weber, R., Herold, C., Hollert, H., Kamphues, J., Blepp, M., & Ballschmiter, K. (2018). Reviewing the relevance of dioxin and PCB sources for food from animal origin and the need for their inventory, control and management. *Environmental Sciences Europe*, *30*(1). https://doi.org/10.1186/s12302-018-0166-9.

Weng, Y. Z., Chang, D. T., Huang, Y. F., & Lin, C. W. (2011). A study on the flexibility of enzyme active sites. *BMC Bioinformatics*, *12*(SUPPL. 1), S32. https://doi.org/10.1186/1471-2105-12-S1-S32.

Wijayawardene, N. N., Hyde, K. D., Al-Ani, L. K. T., Tedersoo, L., Haelewaters, D., Rajeshkumar, K. C., Zhao, R. L., Aptroot, A., Leontyev, D. V., Saxena, R. K., Tokarev, Y. S., Dai, D. Q., Letcher, P. M., Stephenson, S. L., Ertz, D., Lumbsch, H. T., Kukwa, M., Issi, I. V., Madrid, H., . . . Thines, M. (2020). Outline of Fungi and fungus-like taxa. *Mycosphere*, *11*(1), 1060–1456. https://doi.org/10.5943/mycosphere/11/1/8.

Wong, J. K. H., Tan, H. K., Lau, S. Y., Yap, P. S., & Danquah, M. K. (2019). Potential and challenges of enzyme incorporated nanotechnology in dye wastewater treatment: A review. *Journal of Environmental Chemical Engineering*, *7*(4), 103261. https://doi.org/10.1016/j.jece.2019.103261.

Wong, S., Ghafar, N. A., Ngadi, N., Razmi, F. A., Inuwa, I. M., Mat, R., & Amin, N. A. S. (2020). Effective removal of anionic textile dyes using adsorbent synthesized from coffee waste. *Scientific Reports*, *10*(1), 1–13. https://doi.org/10.1038/s41598-020-60021-6.

Yang, X., Wu, Y., Zhang, Y., Yang, E., Qu, Y., Xu, H., Chen, Y., Irbis, C., & Yan, J. (2020). A thermo-active laccase isoenzyme from trametes trogii and its potential for dye decolorization at high temperature. *Frontiers in Microbiology*, *11*(February), 1–12. https://doi.org/10.3389/fmicb.2020.00241.

Yotsuji, A., Mitsuyama, J., Hori, R., Yasuda, T., Saikawa, I., Inoue, M., & Mitsuhashi, S. (1988). Mechanism of action of cephalosporins and resistance caused by decreased affinity for penicillin-binding proteins in bacteroides fragilis. *Antimicrobial Agents and Chemotherapy*, *32*(12), 1848–1853. https://doi.org/10.1128/AAC.32.12.1848.

Zdarta, J., Meyer, A. S., Jesionowski, T., & Pinelo, M. (2018). A general overview of support materials for enzyme immobilization: Characteristics, properties, practical utility. *Catalysts*, *8*(2). https://doi.org/10.3390/catal8020092.

Zhang, H., Zhang, J., Zhang, X., & Geng, A. (2018). Purification and characterization of a novel manganese peroxidase from white-rot fungus Cerrena unicolor BBP6 and its application in dye decolorization and denim bleaching. *Process Biochemistry*, *66*(December), 222–229. https://doi.org/10.1016/j.procbio.2017.12.011.

Zhang, J., Sun, L., Zhang, H., Wang, S., Zhang, X., & Geng, A. (2018). A novel homodimer laccase from Cerrena unicolor BBP6: Purification, characterization, and potential in dye decolorization and denim bleaching. *PLoS One*, *13*(8). https://doi.org/10.1371/journal.pone.0202440.

Zhang, R., Liu, Y., Zhang, Y., Feng, D., Hou, S., Guo, W., Niu, K., Jiang, Y., Han, L., Sindhu, L., & Fang, X. (2019). Identification of a thermostable fungal lytic polysaccharide monooxygenase and evaluation of its effect on lignocellulosic degradation. *Applied Microbiology and Biotechnology*, *103*(14), 5739–5750. https://doi.org/10.1007/s00253-019-09928-3.

Zhang, S., Lin, T., Chen, W., Xu, H., & Tao, H. (2019). Degradation kinetics, byproducts formation and estimated toxicity of metronidazole (MNZ) during chlor(am)ination. *Chemosphere*, *235*, 21–31. https://doi.org/10.1016/j.chemosphere.2019.06.150.

Zhang, Y., Zheng, H., Zhang, L., Zhang, Z., Xing, X., & Qi, S. (2019). Fine particle-bound polycyclic aromatic hydrocarbons (PAHs) at an urban site of Wuhan, central China: Characteristics, potential sources and cancer risks apportionment. *Environmental Pollution*, *246*, 319–327. https://doi.org/10.1016/j.envpol.2018.11.111.

Zheng, F., An, Q., Meng, G., Wu, X. J., Dai, Y. C., Si, J., & Cui, B. K. (2017). A novel laccase from white rot fungus Trametes orientalis: Purification, characterization, and application. *International Journal of Biological Macromolecules*, *102*, 758–770. https://doi.org/10.1016/j.ijbiomac.2017.04.089.

Zhou, Q., Guo, M., Ni, K., & Kerton, F. M. (2021). Construction of supramolecular laccase enzymes and understanding of catalytic dye degradation using multispectral and molecular docking approaches. *Reaction Chemistry & Engineering*, *6*(10), 1940–1949. https://doi.org/10.1039/d1re00111f.

7 Enzymes Involved in the Bioremediation of Pesticides

Sajjad Ahmad, Pankaj Bhatt, Hafiz Waqas Ahmad, Dongming Cui, Jiatai Guo, Guohua Zhong, and Jie Liu

ABSTRACT

The imprudent use of xenobiotics to control pest infestation for the high yield of crops causes dangerous impacts on diverse living organisms in their respective environment. The toxic residues of these highly hazardous compounds have created a route into the underground water system and polluted drinking water. Hence, extensive use of pesticides causes severe threats to all living and non-living organisms in different ways. The biodegradation method is considered as one of the efficient, inexpensive, and valuable methods compared to the other ones in which hazardous pollutants are eliminated from the ecosystem by applying different microbial species and their biologically active products, such as proteins, enzymes, and genes. The bacterial and fungal enzymes can degrade the hazardous substances existing in the agroecosystem and transform them into less toxic products via their catalytic reaction mechanisms. Laccase, esterase, monooxygenase, peroxidase, hydrolases, and carboxylesterases are the foremost microbial enzymes that actively take part in the remediation of the majority of the hazardous substances accumulated in the ecosystem. Recently, various immobilizations and genetic engineering practices have been established to improve enzyme efficiency and reduce the process cost for pesticides degradation. This book chapter highlighted the enzymatic degradation of xenobiotics in the agroecosystem. Through enzymatic degradation, current developments and further expansion for effective degradation of pesticides and other harmful pollutants, such as dyes, heavy metals, and plastics, would be efficiently achieved.

Keywords: Pesticides, toxicity, biodegradation, nano enzymes, genetic engineering

DOI: 10.1201/9781003202998-7

CONTENTS

7.1   Introduction.. 134
7.2   Classification of Enzymes and Biodegradation of Pesticides....................... 135
     7.2.1   Laccase 135
     7.2.2   Esterase.. 148
     7.2.3   Monooxygenase... 149
     7.2.4   Peroxidase.. 150
     7.2.5   Hydrolases 151
     7.2.6   Carboxylesterases 152
7.3   Innovative Enzyme-Based Tools and Technologies...................................... 153
     7.3.1   Genetic Engineering 153
     7.3.2   Enzymatic Engineering.. 155
     7.3.3   Immobilization Enzyme Technology... 156
     7.3.4   Nano Enzymes.. 157
7.4   Conclusion and Future Perspectives 158
References... 159

7.1   INTRODUCTION

The tremendous use of pesticides to prevent the proliferation of unwanted insects and plants in agriculture is increasing day by day, which may comprise and lead to pollution in the agroecosystem (Gonçalves et al., 2021). The unwise use of pesticides leads to diffuse contamination by spray drift, runoff, and the alteration of various phases (adsorption and desorption) in their distribution into the air, soil, and water (Avila et al., 2021). Furthermore, their toxic residues are accumulated in food chains, which further transfer to the non-target organisms, constitute a risk to human health, spread chronic diseases, and are the source of pollution in the ecosystem (Cimino et al., 2017; Zhu et al., 2019). Improper handling and application of pesticides cause severe problems to the aquatic environment, disturb their habitats, and decrease water quality (Materu et al., 2021). Many studies reported that the excessive use of pesticides alters the climate and causes negative impacts on the ecosystem (Daam et al., 2019) (Figure 7.1).

However, the presence of excessive residues of pesticides seriously threatens non-target organisms, including humans. It brings potential health risks, such as cancer, hormonal disorders, asthma, allergies, reduced birth rate, increased mortality rate, leukemia, Parkinson disease, Hodgkin disease, non-Hodgkin lymphoma, Burkitt lymphoma, and respiratory, reproductive, and endocrine disorders (Sabarwal et al., 2018; Brouwer et al., 2017).

Nowadays, many researchers are prompted to solve the present issue of pesticide residues and their toxic metabolites in the food chain and the environment (Li et al., 2018). Several conventional techniques and methods are used for environmental safety, such as physical, chemical, and biological ones. In comparison, the biological approach is ecofriendly and less expensive and has emerged nowadays, which plays a vital role in the degradation of pesticides (Zhan et al., 2020; Mustapha et al., 2019). Using the biological method, many enzymes from microbes were purified, which is

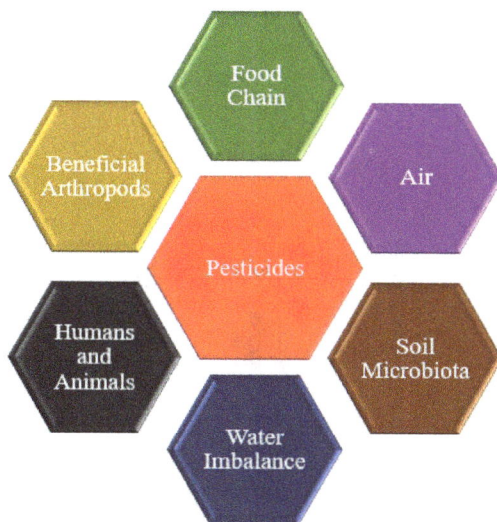

FIGURE 7.1 Pollution of pesticides in the environment.

helpful for the extensive and straightforward depollution of xenobiotic wastes and is considered more effective and superior (Mishra et al., 2020) (Table 1). Among microbes, bacteria are considered a more effective and efficient source of these enzymes than other microbial species due to the manipulation and cultured of bacterial cells (Jamwal et al., 2017). These enzymes transform the hazardous compounds into less-toxic particles, which blend naturally with the environment (Kumar et al., 2017). Besides this, these enzymes also play a vital role in most soil biochemical processes such as detection of soil quality, decomposition of organic matters, and involvement of nutrient recycling (Zhang et al., 2017b). The overall impact of the enzymes to clean the environment is gaining great attention. Therefore, this book chapter focuses on the hidden areas of the enzymes for the removal of xenobiotics from the environment. Finally, new horizons, including modern biological tools, have been discussed to treat the environment.

7.2 CLASSIFICATION OF ENZYMES AND BIODEGRADATION OF PESTICIDES

7.2.1 LACCASE

Laccase enzymes are extensively secreted by various living organisms, such as different types of plants, microbes, and arthropods. The structure of these enzymes comprises four copper ions involved to catalyze the oxidation reaction in the existence of oxygen. This reaction openly breaks down different phenolic compounds (like phenolic dye, substituted phenol, chlorophenol and sulfur phenol, and bisphenol A) and some aromatic amines. Due to their prominent characters like oxygen, an open environment enhances catalytic oxidation and is known as a "green catalyst" (Rashtbari

TABLE 7.1
Different Types of Enzymes Their Source and Degradation Conditions for Bioremediation of Pesticides

Enzyme	Type of Pesticide	Source	Degradation Condition	References
Laccase	Chlorpyrifos, Phoskill	*Bacillus* sp.	*Bacillus* sp. was used to produce laccase in the existence of solid-state fermentation and lignocellulosic agricultural byproducts (pear millet and finger millet husks). After 4 days of cultivation at pH 7 with pearl millet (2 g/L) and finger millet (1.5 g/L), highest laccase activity, 402 U/mL, was recorded. At the temperature of 40°C, neutral pH, and continuous cultivation of five days, a small quantity of purified laccase was rapidly removed (chlorpyriphos [71.8±3.5%] and Phoskill [77.3±3.4%]).	Srinivasan et al., 2019
	2,4-dichlorophenol (2,4-DCP)	*Trametes versicolor*	After the purification of laccase, their immobilization was done by the covalent bonding method in porous glass to improve the activity. The immobilized enzymes removed 90% with 5 mM concentration of 2,4-DCP within 2 h.	Jia et al., 2012
	Chlorpyrifos	*Trametes versicolor*	Immobilization of laccase was carried out on magnetic iron nanoparticles with the dimension of 10–15 nm and extra coated with chitosan and carbodiimide. In 12 h at optimal conditions (pH 7, temperature 60°C), 99% removal was gained.	Das et al., 2017
	Imazalil	*Trametes* sp.	Laccase enzymes were investigated to remove imazalil in the existence of a mediator. Results revealed that after using different natural mediators, imazalil was removed simultaneously in the appearance of 4-hydroxybenzoic acid. However, in the absence of mediators, no degradation was recorded. Laccase was involved in the oxidation of the mediator, and the oxidized compound enhanced the degradation process.	Maruyama et al., 2007
	Atrazine, prometryn	*Aspergillus* sp.	Laccase was immobilized with microporous starch and combined with mediators to degrade atrazine and prometryn in the water sample collected near the rice field. Both pesticides were removed 61% higher in 7 days than natural environmental conditions.	Chen et al., 2021

Pesticide	Organism	Description	Reference
Methoxychlor	*Trametes versicolor*	Laccase immobilization with chiral mesoporous silica to remediate methoxychlor was investigated. Findings explained that the temperature (45°C) and pH (4.5), the concentration of methoxychlor (30 mg/L), the volume of immobilized enzyme (0.1 g), and the cultivation time (6 h) degraded 85% of methoxychlor, which was 50.75% were higher than free enzymes.	Huang et al., 2019
Chlorpyrifos	*Pichia pastoris*	After the laccase production from strain on day 15 under controlled conditions, the highest activity of laccase was 12,344 U/L. Both synthetic and natural mediators were investigated (ABTS, HBT, VA) and (vanillin, 2,6-DMP guaiacol), respectively. Findings showed that the maximum dosage of vanillin natural mediator led to 98% chlorpyrifos remediation.	Xie et al., 2013a
DDT	White rot fungi	Laccase effectively degrades the residues of DDT in the soil during the initial 15 days. Later on, the degradation level is kept at a stable level during the next 10 days.	Zhao et al., 2010
Chlorpyrifos, profenofos, thiophanate methyl	*Tricholoma giganteum*	It was responsible for removing 29% chlorpyriphos, 7% profenofos, and 72% of thiophanate methyl with incubation of 15 h. Molecular docking examination explained that the maximum binding efficacy of these pesticides was spotted with H83, H320, A95, V384, and P366, which are shown close to the catalytic site.	Rudakiya et al., 2020
Trifluralin	Commercial	Immobilization of laccase carried with bagasse/cellulose nanofibers (CNFs). Their maximum activation was noted in different pH and temperature ranges (6–10 and 30–80°C), respectively. Results revealed that immobilized enzymes effectively degraded trifluralin.	Bansal et al., 2018
Esterase	Carbaryl	*Pantoea ananatis*	Carbohydrate esterase (PaCes7) is a novel carbohydrate esterase purified from microbial specie. Cloned novel PaCes7 presented its maximum catalytic activity (40%) at less than 30°C temperature under mesophilic conditions, rapidly removing carbaryl.	Yao et al., 2020

(Continued)

TABLE 7.1 (Continued)

Enzyme	Type of Pesticide	Source	Degradation Condition	References
	Monocrotophos	*Bacillus subtilis* KPA-1	Monocrotophos was degraded by esterase enzyme, which is purified from bacterial strain. Results revealed that monocrotophos was completely degraded with the initial concentration of 800 ppm at pH 8, temperature 40°C, 3% inoculum size, glucose as carbon source, and ammonium sulfate as nitrogen source within 72 hours.	Acharya et al., 2015
	Parathion-methyl	*Bacillus subtilis* B1	SDS-PAGE and native-PAGE observed a single isozyme with a molecular weight of 86 kDa. Esterase was gained by removing parathion-methyl strain and studying temperature and pH effects on the degradation rate.	Hao et al., 2014
	Lactofen	*Edaphocola flava*	A new gene lane was responsible for lactofen degradation achieved by strain. When the length of the carbon chain increased to (C4-C8), the transformation of *p*-nitrophenyl esters occurred, and their activity was reduced. LanE presented enantioselectivity throughout the removal of lactofen, diclofop-methyl, and quizalofop-ethyl, with a maximum remediation efficiency of (S)-enantiomers than (R)-enantiomers.	Hu et al., 2020b
	Allethrin	*Pseudomonas nitroreducens*	Molecular modeling, docking, and enzyme kinetics were used to explore the binding pocket of the esterase comprising amino acids, such as alanine, arginine, valine, proline, cysteine, glycine, isoleucine, phenylalanine, serine, asparagine, and threonine, which are responsible for allethrin remediation.	Bhatt et al., 2020
	Cadusafos, coumaphos, diazinon, dyfonate, ethoprophos, fenamiphos, methylparathion, and parathion	*Pichia pastoris*	Esterase enzyme was purified from microbial strain under the substrate of p-nitrophenol butyrate, and a maximum expression level of 4 g/L was recorded. The highest activity of the enzyme at temperature 40°C and pH 7 was recorded. Esterase with substrate appearance is responsible for removing cadusafos, coumaphos, diazinon dyfonate, ethoprophos, fenamiphos, methyl parathion, and parathion up to 68, 60, 80, 40, 45, 60, 95, and 100%, respectively.	Kambiranda et al., 2009

Enzyme	Pesticide	Microorganism	Description	Reference
	Lactofen	*Brevundimonas* sp. LY-2	Esterase was involved in cutting down the proper ester bond of the alkanoic side chain of lactofen, and their detoxification was achieved 113.3-fold to homogeneity with a recovery of 6.83%.	Liang et al., 2010
	Lactofen	*Brevundimonas* sp. LY-2	A novel esterase cloned from the bacterium for the removal of lactofen. The esterase revealed enantioselectivity during lactofen remediation, which showed the occurrence of enzyme-mediated enantioselective degradation of chiral herbicides.	Zhang et al., 2017a
	Strobilurin	*Hyphomicrobium* sp. strain DY-1	For the removal of strobilurin fungicides, a novel esterase gene, *strH*, can de-esterify. Esterase catalyzed the de-esterification, generating the corresponding parent acid to gain the detoxification and relieve growth inhibition of chlorella of different strobilurin fungicides.	Jiang et al., 2021
	Fluoroglycofen	*Lysinibacillus* sp. KS-1	The *fluE* gene gained from strain KS-1 contained native esterase involved in the cleavage of carboxyl ester bonds of fluoroglycofen. This gene was further expressed in *Escherichia coli* BL21 and purified by Ni-NTA affinity chromatography. Purified *fluE* could effectively hydrolyze fluoroglycofen and short-chain p-nitrophenol esters.	Huang et al., 2017
Monooxygenase	Beta-cypermethrin	*Streptomyces* sp.	A novel enzyme illustrates a monomeric structure with a molecular mass of 41 kDa and pH of 5.4. The highest activity at pH 7.5 and 30°C and highest stability at pH 6.5–8.5 and temperature less than 10°C were recorded. Their activation rate was considerably accelerated in the presence of Fe^{2+} and remarkably prevented using Ag^+, Al^{3+}, and Cu^{2+}. The enzyme catalyzed the remediation of beta-cypermethrin into five intermediates via hydroxylation and diaryl cleavage.	Chen et al., 2013
	Chlorpyrifos	*Cupriavidus nantongensis* X1^T	Two essential genes (*tcpA* and *fre*) were cloned from strain X1T. Later on, it transferred and expressed in *Escherichia coli* BL21(DE3). Removal of chlorpyriphos by X1^T whole cell was compared with the enzymes 2,4,6- trichlorophenol monooxygenase and NAD(P)H: flavin reductase expressed and purified from *E. coli* BL21(DE3).	Fang et al., 2019

(Continued)

TABLE 7.1 (Continued)

Enzyme Type of Pesticide	Source	Degradation Condition	References
Thiobencarb	*Acidovorax* sp. T1	A native two-component monooxygenase system purified in this strain (TmoAB) was responsible for the initial catabolic reaction. TmoAB was an FMN-dependent monooxygenase and catalyzed the C-S bond, and finally, a breakdown of thiobencarb occurred.	Chu et al., 2017
Chlorimuron-ethyl	*Rhodococcus erythropolis*	The qRT-PCR experiment indicated that carboxylesterase, cytochrome P450, and glycosyltransferase genes were responsible for chlorimuron-ethyl bioremediation.	Cheng et al., 2018b
DDT	*Agrocybe* and *Marasmiellus*	A P450 inhibitor, 1-ABT, inhibited the formation of monohydroxy-DDTs and monohydroxy-DDDs from DDT and DDD, respectively. Findings of this study explained that the oxidative pathway, which was catalyzed by P450 monooxygenase, exist beside reductive dechlorination of DDT.	Suhara et al., 2011
Methyl parathion	*Pseudomonas* sp. strain NyZ402	In the strain of NyZ-402 methyl parathion hydrolase (MPH), *ortho*-nitrophenol 2-monooxygenase (OnpA) and o-benzoquinone reductase (OnpB) were structurally expressed. This strain was free of exogenous antibiotic resistance gene markers, and the introduced genes were genetically stable.	Liu et al., 2010
β-Endosulfan	*Mycobacterium* sp. strain ESD	The *Esd* gene was identified and expressed in a constitutive promoter mycobacterial expression vector and confirmed that degradation of β-endosulfan and activation of *Esd* gene strongly depend on sulfite or sulfate medium. The translation product of this gene had up to 50% sequence resemblance with an unusual family of monooxygenase enzymes that reduced flavins, provided by a separate flavin reductase enzyme, as co-substrates.	Sutherland et al., 2002

Imazalil, a-cypermethrin	*Chironomus riparius*	The inhibition of cytochrome P450 activity was measured in vitro and in vivo. The synergistic potential of a-cypermethrin induced immobilization. Azoles, such as imazalil, have been shown to synergize the effect of pyrethroid insecticides like a-cypermethrin through inhibition of cytochrome P450 monooxygenase responsible for pyrethroid detoxification.	Kuhlmann et al., 2019
Chlorfenapyr	*Rhodopseudomonas capsulate*	Cytochrome P450 monooxygenase (P450) and *cpm* gene activation were identified by HPLC and RT-PCR for mixed wastewater treatment. Molecular analysis showed that chlorfenapyr could elicit *cpm* gene expression to prepare P450 through acting on *MAPKKKs* gene. The elicitation of P450 and *cpm* occurred after one day. This study concluded that this novel molecular mechanism could be applied for mixed wastewater treatment.	Wu et al., 2020
Lindane	*Nicotiana tabacum*	CYP2E1 human monooxygenase was purified by transgenic tobacco plants and examined lindane degradation. Compared to control plants, transgenic tobacco plants with CYP2E1 enzyme showed tremendous resistance against lindane when grown in a hydroponic medium and soil.	Singh et al., 2011
Peroxidase Glyphosate	*Trametes versicolor*	To degrade glyphosate and other xenobiotics, the efficacy of purified manganese peroxidase, laccase, lignin peroxidase, and horseradish peroxidase was investigated in separate in vitro assays with the addition of various mediators. Complete degradation of glyphosate was achieved with MnP, MnSO$_4$, and Tween 80, with or without H$_2$O$_2$.	Pizzul et al., 2009
Lindane	*Ganoderma lucidum*	For the degradation of lindane, in liquid state fermentation, 100.13 U/ml laccases, 50.96 U/ml manganese peroxidase, and 17.43 U/ml lignin peroxidase enzymes were gained with the highest removal of 75.50% after the incubation of twenty-eighth day. While using solid-state fermentation, 156.82 U/g laccases, 80.11 U/g manganese peroxidase, and 18.61 U/g lignin peroxidase enzyme remove 37.50% lindane.	Kaur et al., 2016

(Continued)

TABLE 7.1 (*Continued*)

Enzyme	Type of Pesticide	Source	Degradation Condition	References
	Lindane	*Conidibolus* sp.	For the remediation of lindane, extracellular oxidative enzymes (lignin-modifying enzymes and lignin peroxidase) were purified by a bacterium responsible for the pesticide's degradation.	Nagpal et al., 2008
	Methoxychlor (MC)	*Trametes versicolor*	To remove MC, ligninolytic enzymes, manganese peroxidase (MnP), laccase, and lignin peroxidase (LiP) were excreted by strain. MnP and laccase in the existence of Tween 80 and hydroxybenzotriazole (HBT), respectively, were found to remove MC. MnP-Tween 80 degrade 65% MC after a 24 h treatment.	Hirai et al., 2004
	Lindane	White rot fungus	The production of laccase and manganese was done in both stationary batches and immobilized cultures. Removal of lindane 82±6% was gained in batch cultures at concentrations of 5 and 10 ppm. The maximum removal rate was 81% in the packed bed reactor with a concentration of (1–2 ppm), and the removal rate decreased by 10 ppm.	Tekere et al., 2002
	Atrazine, desethylatrazine (DEA), and desisopropylatrazine (DIA)	*Pleurotus ostreatus* INCQS 40310	Proteomic approaches showed that hydrolases and peroxidases are involved in the degradation of pesticides. Moreover, free cells of the strain were examined to verify intracellular enzymes in the remediation of pesticides. Results revealed that cytochrome P450 is actively involved in the degradation process.	Lopes et al., 2020
	3-phenoxybenzoic acid	*Aspergillus oryzae* M-4	For the degradation of 3-phenoxybenzoic acid cytochrome P450, lignin peroxidase (LiP), laccase, manganese peroxidase (MnP), and dioxygenase were used due to their oxidase's presence in the strain and were responsible for the removal of pesticide with analogous structures. Moreover, $CuSO_4$, NaN_3, $AgNO_3$, EDTA, or piperonyl butoxide (PBO) were selected as the enzyme's inhibitors and inducers.	Zhao et al., 2020
	Diuron	*Phanerochaete chrysosporium*	Manganese peroxidase purified from the strain and removal of diuron was tested during the idiophasic growth. Results showed that this enzyme actively participated in the degradation process of diuron.	Fratila-Apachitei et al., 1999

	Azinphos-methyl, chlorpyrifos, dichlorofenthion, dimethoate, parathion, phosmet, and terbufos	Caldariomyces fumago	The chloroperoxidase enzyme was purified from the microbial strain to remove different organophosphorus pesticides. This enzyme is also involved in the decomposition of various pollutants by substituting a sulfur atom with an oxygen atom and thus oxidizing the phosphorothioate group to an oxon derivative.	Hernandez et al., 1998
	Paraquat	Polyporus tricholoma, Cilindrobasidium laeve, and Deconica citrispora	Three microbial strains were investigated to remove paraquat, and two enzymes (laccase, manganese peroxidase) were purified. The study was carried out with extracellular medium and enzyme extracts. Two strains indicate high efficiency in remediation and manufacturing of enzymes. After one day of cultivation of *D. citrispora*, 49% degradation of paraquat was gained.	Camacho-Morales et al., 2017
Hydrolases	Carbamate	*Pseudomonas* sp. XWY-1	A carbamate hydrolase (mcbA) enzyme was purified from strain responsible for hydrolyzing carbamate pesticides, such as carbaryl, carbofuran, isoprocarb, fenobucarb, propoxur, and aldicarb. The active site of mcbA comprises the histidine that is crucial for catalysis.	Zhu et al., 2018
	Organophosphorus	*Cladosporium cladosporioides* Hu-01	To degrade different organophosphorus pesticides with P-O and P-S bonds, a hydrolase enzyme with a molecular weight of 38.3 KDa was cultivated by the strain. The purified hydrolase enzyme degrades organophosphorus pesticides efficiently.	Gao et al., 2012
	Chlorpyrifos	*Cupriavidus taiwanensis* X1	To remove chlorpyriphos residues, a hydrolase gene was identified from strain X1. This gene extensively showed resemblance to the *opdB* gene encoding parathion hydrolase.	Wang et al., 2015
	Organophosphorus pesticides	*Lactobacillus brevis* WCP902	To remove a variety of organophosphorus pesticides, an organophosphorus hydrolase (OpdB) enzyme was purified. Their highest activity was indicated at pH 6 and 35°C for the removal of chlorpyrifos, coumaphos, diazinon, methyl parathion, and parathion.	Islam et al., 2010

(*Continued*)

TABLE 7.1 (Continued)

Enzyme	Type of Pesticide	Source	Degradation Condition	References
	Organophosphorus pesticides and p-nitrophenol	Moraxella sp.	A hydrolases enzyme was purified from strain and examined their degradation ability against different organophosphorus pesticides. Results showed that the initial hydrolases rate was 0.6 μmol/h/mg dry weight, 1.5 μmol/h/mg dry weight and 9 μmol/h/mg dry weight for methyl parathion, parathion, and paraoxon, respectively.	Shimazu et al., 2001
	Diuron	Arthrobacter globiformis strain D47	Phenylurea hydrolase (puhA gene) was constructed from strain D 47. A smaller subclone of this gene (2.5 kb) coded with the protein expressed in E. coli can remove diuron. This gene and its predicted protein sequence revealed only a low level of protein identity (25% over ca. 440 amino acids) to other database sequences and was named after the enzyme is encoded in (puhA).	Turnbull et al., 2001
	Methyl parathion, parathion, paraoxon, coumaphos, demeton-S, phosmet, and malathion	Penicillium lilacinum BP303	A hydrolase enzyme was cloned from the strain BP303 and examined their degradation ability against various pesticides. At temperature 45°C and pH 7.5, maximum activity was recorded. Moreover, using Hg^{2+}, Fe^{3+}, p-chloromercuribenzoate, iodoacetic acid, and N-ethylmaleimide activity of enzyme reduced, while Cu^{2+}, mercaptoethanol, dithiothreitol, dithioerythritol, glutathione, and detergents slowly enhanced the enzymatic activity. Furthermore, the investigation of catalytic efficiency revealed that paraoxon is the best substrate.	Liu et al., 2004
	Butachlor	Rhodococcus sp. strain B1	The hydrolase designated ChlH, able N-dealkylation of the side chain of butachlor and other herbicides (such as alachlor, acetochlor, butachlor, and pretilachlor) was cloned 185.1-fold to homogeneity with 16.1% recovery. ChlH was highly activated at pH 7–7.5 and temperature 30°C.	Liu et al., 2012
	Pyrethroid pesticides	Aspergillus niger ZD11	Novel pyrethroid hydrolase by the free cells of strain was screened with an overall recovery of 12.6% and 41.5-fold apparent homogeneity. It is a monomeric structure with a molecular mass of 56 kDa and pH of 5.4. At temperature 45°C and pH 6.5, highest enzymatic activity was recorded. In the presence of trans-permethrin substrate purified enzyme hydrolyzed various insecticides with similar carboxyl ester.	Liang et al., 2005

	Pesticide	Organism	Description	Reference
	Coumaphos	*Escherichia coli*	For the biodegradation of coumaphos, a hydrolase was purified by *E. coli* bacteria. It was immobilized on highly porous glass beads and used in a continuous flow packed bed reactor to remove coumaphos pesticide.	Mansee et al., 2000
Carboxylesterases	Malathion	*Alicyclobacillus tengchongensis*	For removing malathion pesticide residues, the carboxylesterases enzyme was purified from strain. At a temperature of 60°C and pH 7 and with b-naphthyl acetate maximum activity and at 25°C and pH 7, the highest stability was recorded. Results showed that this enzyme hydrolyzed 5 mg/L malathion 50% within 25 min and 89% within 100 min.	Xie et al., 2013b
	Malathion	*Escherichia coli* IES-02	Carboxylesterase enzyme with molecular mass of (33, 30, 28 kDa) were secreted (1685.71 U/mg) from bacteria *Escherichia coli* IES-02. Results explained that within 20 min, 81% remediation of malathion was occurred.	Sirajuddin et al., 2020
	Chlorimuron-ethyl	*Rhodococcus erythropolis* D310-1	A carboxylesterase (CarE) enzyme was purified from the strain D310-1 to effectively remove chlorimuron-ethyl. CarE is responsible for catalyzing the de-esterification of chlorimuronethyl. A CarE deletion mutant strain, D310-1 ΔcarE, and wild-type strain D310-1 were cultured with the concentration of 100 mg/L chlorimuron-ethyl, and within 5 days, 86.5% and 58.2% degradation were achieved, respectively.	Zang et al., 2020
	Pyrethroid pesticides	*Bacillus subtilis* BSF01	For the remediation of pyrethroid pesticides, carboxylesterase enzyme was cloned from the strain BSD01. For the purification of enzyme, two techniques were adopted, DNA pull-down and yeast one-hybrid techniques, which indicated that the enzymatic removal of pyrethroids is stimulated through QS signal regulator ComA binding to carboxylesterase gene is, highlighting the synergistic effect of QS regulation and pyrethroid degradation in *B. subtilis* BSF01.	Xiao et al., 2020

(Continued)

TABLE 7.1 (Continued)

Enzyme	Type of Pesticide	Source	Degradation Condition	References
	Aryloxyphenoxypropionate (AOPP)	Brevundimonas sp. QPT-2	To degrade (AOPP) a native family VIII carboxylesterase gene, estwx was purified by the strain QPT-2, and their expression was investigated in E. coli BL21. This novel enzyme hydrolyzed the ester bond cleavage of AOPP herbicides and converted them into corresponding acid and alkyl side chain alcohol.	Xu et al., 2019
	Malathion and parathion	Pseudomonas aeruginosa PA1	For the remediation of malathion and parathion residues, carboxylesterase was purified from the strain PA. Moreover, this enzyme transforms both pesticides into their intermediates (malathion into malathion monocarboxylic acid and dicarboxylic acid).	Singh et al., 2012
	Fenpropathrin, permethrin, cypermethrin, cyhalothrin, deltamethrin, and bifenthrin	Sulfolobus tokodaii	To remove a variety of pyrethroid pesticides, a novel carboxylesterase EstSt7 was purified from the strain. Later, the recombinant enzyme was purified to homogeneity after heat treatment, Ni-NTA affinity, and Superdex200 gel filtration chromatography. Results explained that novel EstSt7 effectively degrade the group of pyrethroid pesticides.	Wei et al., 2013
	Fenpropathrin, cypermethrin, fenvalerate, bifenthrin, and p-nitrophenyl	Pseudomonas synxantha PS1	For the degradation of different pesticides, EstPS1 gene, which encodes a native carboxylesterase of strain PS1 isolated from oil well-produced water, was cloned and sequenced. The maximum recombinant EstPS1 was shown at pH 8 and temperature 60°C, respectively. Moreover, a wide range of temperature (60–100°C) and time interval (14 h to 1 min) EstPS1 indicated strong thermostability and the half-lives (T1/2 thermal inactivation).	Cai et al., 2017
	Organophosphorus pesticide	Micractinium pusillum UUIND2, Chlorella singulari UUIND5, and Chlorella sorokiniana UUIND6	Carboxylesterase was purified from the strains and tested their degradation ability against organophosphorus pesticides. Moreover, carboxylesterase activity was maximum in algal free cells cultured in malathion-containing medium, which verified that malathion transforms into phosphate.	Nanda et al., 2019
	Bioresmethrin, α-cypermethrin, and deltamethrin	Hepatic cells	The carboxylesterase enzyme was purified from the hepatic cells to remove a variety of xenobiotics. This carboxylesterase enzyme metabolize ester-containing xenobiotics and transform them into nontoxic compounds.	Ross et al., 2006

et al., 2021; Unuofin et al., 2019). Commonly deemed that laccase catalytic constituents with the junction of mediator system biologically act together for various industrial use (Bilal et al., 2019). Furthermore, they are also used to decolorize dyes, bleach textiles, treat wastewater, and remove other refractory substances, such as polycyclic aromatic hydrocarbons and hormone-like chemicals (Yang et al., 2015).

For the biodegradation of xenobiotics from the environment, laccase enzymes with and without mediators were investigated. For example, bacterial laccase enzymes without phenolic mediators to remove chlorpyriphos, dichlorophos, monocrotophos, and profenofos were investigated, and findings showed that they were effectively involved in the degradation process (Chauhan and Jha, 2018). Recently, Sarker et al. (2020) studied the remediation of hazardous fungicide metabolite 3,5-dichloroaniline (3,5-DCA) by laccase and MnO_2 mediator with different amendments of phenolic mediators (catechol, syringaldehyde, syringic acid, caffeic acid, and gallic acid). This study explained that the catechol phenolic mediator with laccase significantly degrades the fungicide metabolites.

In another study, Zeng et al. (2017) reported the degradation of isoproturon by laccase enzyme. This study demonstrated that only laccase was inefficient for the degradation of isoproturon herbicide due to a relatively withdrawing solid electron group in the chemical structure of isoproturon. Hence, with the 1 mM 1-hydroxy benzotriazole mediator and 0.3 U/mL laccases, it completely remediates within the duration of 24 h.

Chan-Cheng et al. (2020) combined laccase enzymes with the mixture of (coconut fiber, compost, and soil) with a proportion ratio of 45:13:24 for effective degradation of atrazine, imazalil, pyrimethanil, and metalaxyl from wastewater. After the treatment, the inhibition of laccase enzymes increased without affecting the toxicity of the matrix. Moreover, this study concludes that the handling of accessible bioassays offers a potent tool for monitoring the bioremediation process.

A laccase enzyme from *Trametes versicolor*, responsible for the degradation of agrochemicals, including chlorpyriphos, chlorothalonil, pyrimethanil, atrazine, and isoproturon, was characterized. Results revealed that laccase is highly active at (pH 5–7, temperature 25–30°C), and each pesticide demanded a different laccase enzyme concentration, on average between 4 and 6 mmol/L. Furthermore, this study demonstrated that pyrimethanil and isoproturon removal rates were considerably higher than those of chlorpyrifos, chlorothalonil, and atrazine (Jin et al., 2016).

Das et al. (2020) studied the degradation of chlorpyriphos in soil by laccase enzymes mediated with nanoparticles. The study results indicated that the laccase enzyme with nanoparticles degrades chlorpyriphos three times higher and has a better leaching potential than control. Specific latest remediation findings of different agrochemicals by laccase immobilized enzymes on various support materials are also investigated by multiple researchers. Laccase from *Myceliophthora thermophile* *was purified and immobilized* in monodisperse microspheres (glycidyl methacrylate) to degrade azinphos-methyl. Results indicated that immobilized enzymes completely degraded the azinphos-methyl with the appearance of mediator 2,2'-azino-bis-(3 ethylbenzothiazoline-6-sulfonate) within one hour (Vera et al., 2019).

Vidal-Limon et al. (2018) characterized a laccase enzyme from *Coriollopsis gallica* and, using the physical adsorption method, immobilized it in two carrier materials (mesoporous synthetic silica foam and nanostructured silicon foam). Further, immobilized laccase enzymes reinforced for the degradation of dichlorophen pesticide. The study results indicated that newly synthesized immobilized enzymes oxidized the dichlorophen and reduced acute genotoxicity and apoptotic effects.

Chen et al. (2019) developed biosorbents with the combination of peanut shells and rice straw. The immobilization of *Aspergillus* laccase was investigated to remove nine pesticides (isoproturon, atrazine, prometryn, mefenacet, penoxsulam, nitenpyram, prochloraz, pyrazosulfuron-ethyl, and bensulfuron-methyl) in the agroecosystem. Laccase immobilized enzymes on biomass materials significantly degrade the residues of all the above pesticides. This study concludes that immobilized enzymes are more capable of effectively removing pesticides than free enzymes.

7.2.2 ESTERASE

Many esterase enzymes are reported that play a significant role in remediation processes due to their distinguished biochemical characters, particularly optimal pH, temperature, inhibitors, molecular weight, and enzyme kinetic parameters (Bhatt et al., 2020). Esterases are naturally found in a large number of flora and fauna. The active sites of esterase enzymes suppress serine residues that catalyze reactions by a nucleophilic attack on the substrates (Bhatt et al., 2021).

Esterase enzymes break down the essential ester chemical bond and alter their toxic behavior. This character of these enzymes is compelling for the remediation of different types of hazardous substances such as oil spills, food waste, plastic waste, and agrochemicals (Sharma et al., 2020a). The biochemical properties of multiple esterase enzymes indicated that they exhibit various ideal performing situations, such as pH, temperature, pesticide concentration, microbial strains, and surrounding physicochemical environment. Enlisted, all factors boost the esterase enzyme to degrade pesticides in the surrounding environment. Nearly all of the described hazardous materials were removed by esterases at neutral pH and room temperature (Bhatt et al., 2021; Cycoń and Piotrowska-Seget, 2016).

Recently, a native esterase LanE was purified from novel strain *Edaphocola flava* HME-24. In *E. coli* BL21 (DE3), their expression was carried out to remove lactofen herbicide. This novel type of esterase transformed lactofen into acifluorfen with two major metabolites (diclofopmethyl and quizalofop-ethyl) by breaking the ester bond (Hu et al., 2020b). Esterase was characterized by *Nocardioides* sp. strain SG-4G, and their biochemical characters for the removal of carbendazim were investigated (Gangola et al., 2018). The biodegradation of another carbamate pesticide named carbaryl by esterase enzyme was recently studied. A new carbohydrate esterase, PaCes7, was purified from the lignocellulolytic bacterium *P. ananatis Sd-1*. Free bacterial cells and immobilized enzymes both actively participated in carbaryl remediation (Yao et al., 2020).

Organophosphate pesticides have acute neurotoxicity because they inhibit the acetylcholine esterase (AChE), which stimulates the enzyme of neurotransmitters by decreasing acetylcholine concentration at the synaptic junction (Ahir et al., 2020). Numerous studies reported that organophosphates were removed using different esterases, such as phosphodiesterase, organophosphate hydrolase, and methyl parathion hydrolase based on their distinguished biochemical characters (Bhatt et al., 2021). Chlorpyriphos is a kind of organophosphate extensively used in agriculture to monitor various pests from insect orders Lepidoptera, Diptera, Homoptera, Coleoptera, Orthoptera, Hymenoptera, and Hemiptera (Ahir et al., 2018). To degrade chlorpyriphos in the environment, Wang et al. (2020) purified esterase EstC from *strain Streptomyces lividans* TK24. Results revealed that at the optimal conditions (concentration 5 mg/L, temperature 37°C, 80 min incubation period), enzymes degraded chlorpyriphos residues 79.89%.

Pyrethroids are extensively used to control various kinds of pests in agriculture and households. However, their unwise use spreads different acute diseases and sources of pollution in the environment (Bhatt et al., 2019). The most common pyrethroid pesticides are permethrin, fenpropathrin, cypermethrin, cyhalothrin, fenvalerate, deltamethrin, and bifenthrin (Xu et al., 2020). The biodegradation of their toxic residues characterized a carboxylesterase PytH from the bacterium *Sphingobium faniae* JZ-2. These enzymes efficiently break the ester bond by hydrolyzation pathway and convert their toxic residues into nontoxic substances (Xu et al., 2020).

7.2.3 MONOOXYGENASE

Monooxygenase is an essential group of enzymes that belongs to oxidoreductase and plays a crucial part in the biodegradation, transformation, and cometabolism of xenobiotics (Nair and Jayachandran, 2017). Based on cofactor, they are further characterized in two subgroups (flavin-dependent monooxygenases and P450 monooxygenases) (Singh et al., 2019). P450 plays a significant role in building the biosynthetic pathway of natural products, bioremediation of toxic pollutants, biosynthesis of steroid hormones, and the metabolism of drugs (Zhang et al., 2017c). They are also considered the most versatile catalysts due to their prominent features. They are involved in various chemical oxidation reactions, such as hydroxylation, epoxidation, decarboxylation, N- and O-dealkylation, nitration, and C-C bond coupling or cleavage (Guengerich and Munro, 2013).

Neonicotinoid pesticides are widely used nowadays, and their residues are adversely affecting the environment (Ahmad et al., 2021). To remediate their residues, Wang et al. (2019a) isolated a white rot fungus *Phanerochaete sordida* YK-624, especially against nitenpyram and acetamiprid, and found that cytochrome P450 plays a prominent role in the bioremediation of their residues. Wang et al. (2019b) characterized P450 to degrade acetamiprid and their metabolites (N'-cyano-N-methyl acetamidine and 6-chloro-3-pyridinemethanol).

Carbaryl is a broad-spectrum pesticide group of carbamates widely used in agriculture. Their residues accumulated in soil and water due to their more negligible removal and low solubility in water (Siampiringue et al., 2015). Li et al. (2019) identified the P450 enzyme from *Xylaria* sp. *to degrade carbaryl residues* and proposed their metabolic pathways. In liquid media, the degradation of carbaryl was achieved by the activation of the P450 enzyme. Moreover, the carbaryl esterase strikes on carbaryl and converts it into α-naphthol, which finally degrades into 1,4-naphthoquinone and benzoic acid through the activation of P450 and laccase enzymes.

For the biodegradation of chlorothalonil fungicide, extensively used to prevent foliar diseases of vegetables and other cash crops, Green et al. (2018) characterized two P450 enzymes (CYP561 and CYP65) from *Sclerotinia homoeocarpa*. By the metabolism of P450 enzymes, 4-hydroxy-2,5,6 trichloro-isophthalonitrile metabolite was produced and further detoxified into nontoxic substances. Kammoonah et al. (2018) identified *Pseudomonas putida* strain from soil and purified cytochrome P450cam (camphor hydroxylase) to degrade endosulfan and their metabolites. The strain and enzyme were cultivated in minimal media with 3-chloroindole or the insecticide endosulfan as the source of carbon and nitrogen. Seven P450cam mutants were generated by sequence saturation mutagenesis (SeSaM) and investigated their degradation ability. Results showed that six mutants were accepted 3-chloroindole as a substrate to oxidize 3-chloroindole into isatin.

Hu et al. (2020a) isolated a ligninolytic fungi *T. versicolor*, which is involved in the degradation of diuron and transforms their major metabolite (3,4-dichloroaniline) into more minor toxic compounds. During the degradation process, five transformation products were formed, in which three are proposed; for example, 3-(3,4-chlorophenyl)-1-hydroxymethyl-1-methyl urea (DCPHMU) was supplementarily catalyzed into the N-dealkylated compounds 3-(3,4-chlorophenyl)-1-methyl urea (DCPMU) and 3,4-dichlorophenylurea (DCPU). However, this finding showed that the cytochrome P450 enzymatic system involved intracellularly and catalyzing reaction for diuron removal.

Pentachlorophenol is a type of organochlorine pesticide and a conjugate acid of pentachlorophenolate, a member of pentachlorobenzenes, which contain aromatic fungicides and a chlorophenol (Kim et al., 2019). Recently for the degradation of their residues, Aregbesola et al. (2020) isolated a novel *Bacillus cereus* strain AOA-CPS1 from wastewater. Furthermore, in the same study, P450 enzyme was characterized from the genomic DNA of strain. The gene *pcpABCDE* (cytochrome P450) actively contributed to the removal of pentachlorophenol.

7.2.4 PEROXIDASE

Peroxidases are ubiquitous enzymes that comprise hydrogen peroxide oxidoreductases as donors. Using hydrogen peroxide (H_2O_2), these enzymes catalyze the oxidation of lignin and other phenolic compounds in the occurrence of a mediator (Dave and Das, 2021). It is abundantly present in all living organisms, including bacteria, fungi, algae, plants, and animals. They have an outstanding ability to degrade various types of pollutants accumulated in the environment (Shanmugapriya et al., 2019). The removal of contaminants from the environment by using peroxidase enzymes is somewhat sensitive compared to other enzymes due to less sensitivity to pH, temperature, and concentration of phenols or other toxic substances (Atala et al., 2019). They are classified as heme and nonheme peroxidases. The heme peroxidases comprise a protoporphyrin IX (heme) as a prosthetic group, whereas the nonheme peroxidases do not have that prosthetic group. The latest cataloging phylogenetically splits heme peroxidases into two superfamilies (peroxidase-cyclooxygenase superfamily and peroxidase-catalase superfamily) and three families containing di-heme peroxidases, dyP-type peroxidases (DyPPrx), and haloperoxidases (HalPrx), respectively

(Zámocký and Obinger, 2010). Numerous studies showed the effectiveness of per-oxidases to degrade different kinds of emerging contaminants. The peroxidases extensively used for biodegradation are SBP, MnP, LiP, HRP, and CPO (Alneyadi et al., 2018). Recently, for biodegradation of atrazine herbicide from water, Nejad et al. (2020) investigated hydrogen peroxide enzyme in a sequencing batch reactor (SBR) underneath various functioning circumstances due to in situ generation of H_2O_2-peroxidase. Results explained that the atrazine was completely degraded in the SBR system with a 50 mg/L concentration of atrazine, biomass 328 mg/L, and 10 mM of H_2O_2.

Triclosan is a commercial pesticide extensively used, and its residues cause toxic effects in the agroecosystem (Chen et al., 2018). For the biodegradation of triclosan in the environment, Mallak et al. (2020) isolated two white rot fungal species (*Pleurotus ostreatus*, *Trametes versicolor*) and identified two enzymes (laccase and manganese peroxide). This study revealed that both types of enzymes are significantly involved and correlated in the transformation of triclosan by both isolated white rot fungi. In another study, horseradish peroxidases were purified using a cross-linking immobi-lization method. The material used as a cross-linking agent was ethylene glycol-bis (succinic acid N-hydroxysuccinimide [EG-NHS]). The immobilized enzymes were analyzed in a packed bed reactor system, and results showed that they were effec-tively involved in the bioremediation of dye-based pollutants (Bilal et al., 2017).

Organochlorine pesticides dichlorodiphenyltrichloroethane (DDT) and hexa-chlorocyclohexane (HCH), including alpha-, beta-, gamma-, and delta-HCH isomers, broadly applied for the management of household and agriculture pests globally (Lin et al., 2009). For the biodegradation of that kind of pollutants, Fang et al. (2014) iden-tified 69 degrading microbial genera, which are involved in biodegradation and mostly are belong to Proteobacteria (49.3%) and Actinobacteria (21.7%). All the organic pol-lutants are investigated in freshwater and sediment by metagenomic analysis using six datasets, and these datasets were originated using Illumina high-throughput sequencing. They were based on BLAST against self-established databases of bio-degradation genes of DDT, HCH, and atrazine. The *lip* and *mnp* genes, encoded for peroxidase, and the *carA* gene, encoded for laccase, were the dominant genes for the remediation of hazardous compounds. Peroxidase also plays a vital role to degrade lignin, aromatic pollutants, and other emerging dangerous substances in the presence or absence of mediators. In the oxidation process of phenolic compounds, the byproducts have low solubility and significantly participate in the degradation process (Sharma et al., 2018).

7.2.5 HYDROLASES

Hydrolytic enzymes are extensively applied to biodegrade pesticides and emerging hazardous organic pollutants (Karigar and Rao, 2011). Degraded organic polymers and other hazardous compounds with a molecular weight of 600 Da or less can pass through cell pores. For that purpose, numerous types of extracellular hydrolase enzymes were purified by microbial species (Vasileva-Tonkova and Galabova, 2003). Hydrolases are considered most efficient to remove oil spills, organophosphate, car-bamate, and other types of xenobiotics. Carbamate hydrolase enzymes are able to hydrolyze the carboxyl ester bond by the hydrolysis pathway. This pathway initiates the reaction for the degradation of carbamate compounds. Later, the breakdown of reversible acylation of a serine residue occurs by activating the center protein (Russell et al., 2011).

Jiang et al. (2020a) purified CehA hydrolase enzymes from *Sphingbium* sp. strain CFD-1, which could hydrolyze carbofuran effectively from the ecosystem. The properties of CehA were characterized by site-directed mutagenesis to discover its substrate specificity. Results demonstrated that histidine residues (His 313, His 315, His 453, and His 495) were participating in the hydrolysis of carbofuran via CehA. Fan et al. (2018) isolated a microbial strain, *Paracoccus* sp. TRP for the removal of organophosphate pesticides. Using molecular techniques (BLAST analysis and phylogenetic analysis), a novel organophosphate hydrolase CPD was purified by overexpression in bacteria *E. coli*. Their protein was encoded by the *cpd* gene, which also significantly decreased the residues of organophosphates. For the biodegradation of another type of organophosphate pesticide profenofos, using a hydrolase enzyme purified from bacterium *Rahnella* sp. strain PFF2. Findings of this experiment revealed that the profenofos was rapidly removed by using intracellular hydrolase enzymes (Verma and Chatterjee, 2021).

A microbial species, *Azohydromonas australica*, was studied for the biodegradation of methyl parathion, identified a novel hydrolase enzyme, and investigated their degradation ability against different organophosphorus agrochemicals paraoxon, dichlorvos, and chlorpyrifos. The maximum activation of recombinant hydrolase was at pH 9.5 and temperature 50°C. In the presence of leveraging 1 mM Mn^{2+}, their activation and thermostability were increased by 29.3-fold and 40–50°C, respectively (Zhao et al., 2021). In another study for the biodegradation of methyl parathion, Tiwari et al. (2020) isolated hydrolyze enzyme by bacterial a specie *Cupriavidus* sp. LMGR1. The carbon deficiency instead of substrate significantly induced these enzymes.

They are also involved in the biodegradation of other organic pollutants, such as carbamates. For the degradation of methomyl, a bacterium, *Amintobacter aminovorans* MDW-2, was cultured. Furthermore, by applying molecular approaches, a hydrolase enzyme with encoding genes of *Ameh* was purified. The comparative findings between strain MDW-2 extracted proteins and the mass fingerprints of hydrolase enzymes indicated that enzymes actively participated in detoxification and removal of methomyl. Furthermore, this study also explained that the molecular mass of those enzymes was 34 kDa and contained a functional homodimer. *AmeH* displayed maximal enzymatic activity at 50°C and pH 8.5. The effective catalytic range (k_{cat}/K_m) 3.9 μ/M per second was recorded. This enzyme showed a 27% resemblance with the putative formamidase enzyme purified by *Schizosaccharomyces pombe* ATCC 24843 (Jiang et al., 2020b).

7.2.6 CARBOXYLESTERASES

For the remediation of various kinds of pollutants from the environment, such as synthetic organic compounds, organophosphates, carbamates, and chlorine-containing compounds, carboxylesterases enzymes are widely applied (Cummins et al., 2007). Heidari et al. (2004) screened the active site of a carboxylesterase from *Lucilia cuprina* and *Drosophila melanogaster* for the effective biodegradation of pyrethroid pesticides. Carboxylesterase E3 purified from *Lucilia cuprina* plays a critical role in the removal of organophosphorus pesticides (Scott et al., 2010).

Recently, for the degradation of an extensive range of pyrethroid pesticides from the environment, a carboxylesterase enzyme was purified from the bacterium strain *Sphingobium faniae* JZ-2. That enzyme is responsible for the hydrolyzation of the ester bond of almost all types of pyrethroids. This study also revealed that pyrethroid carboxylesterase shows low sequence identity with reported α/β-hydrolase fold proteins; the typical catalytic center with Ser-His-Asp triad (Ser78, His230, and Asp202) is present and vital for the hydrolase activity (Xu et al., 2020). In another study, Yang et al. (2020) also purified carboxylesterase enzymes from *Geobacillus uzenensisfor to degrade* pyrethroids. They immobilized on epoxy-functionalized supports via a one-pot strategy to gain the immobilized enzyme. Those immobilized enzymes degrade various types of pyrethroids efficiently. After a 40 min reaction, malathion removal was 95.8% with the concentration of 20 mg/L, and bifenthrin was 90.4% with 500 mg/L.

The residues of chlorimuron-ethyl sulfonylurea herbicide are long-lasting and highly persistent in the environment. To control the broad-leaved weeds in various crops, they are extensively used due to their distinguishing characteristics, such as high efficiency, low toxicity, and broad-spectrum activity, especially for soybean (Zhang et al., 2017c). To remove this herbicide, carboxylesterase enzymes are purified from *Escherichia coli* strain BL21. Results showed that carboxylesterase actively participated in catalyzing the ester bond of herbicide. Carboxylesterase deletion mutant strain, D310-1ΔcarE, was cloned, and the herbicide remediation ratio in the existence of 100 mg/L with 120 h was reduced from 86.5% (wild-type strain D310-1) to 58.2% (mutant strain D310-1ΔcarE) (Zang et al., 2020).

7.3 INNOVATIVE ENZYME-BASED TOOLS AND TECHNOLOGIES

The advancement of innovative approaches in genetics and molecular technologies has opened new insights into biodegradation by using microbial species and their associated enzymes (Kumar et al., 2019). For the biodegradation of hazardous substances in the natural environment, microbial species are less effective, and their enzymatic activity decreases. All over the world, many scientists have continuously engaged in the production of novel and most effective enzymes produced by various microbial species for the removal and biotransformation of different types of pollutants, such as dyes, heavy metals, and pesticides, into less-toxic substances (Baweja et al., 2016). Some innovative approaches, such as genetic engineering, immobilization, enzyme engineering, and nano enzymes, could overcome the gaps, as mentioned above in the field of biodegradation (Sharma et al., 2018) (Figure 7.2).

7.3.1 Genetic Engineering

Enzymes' isolation, purification, and optimal activation from their hosts challenge natural environmental conditions. Using genetic engineering techniques, the

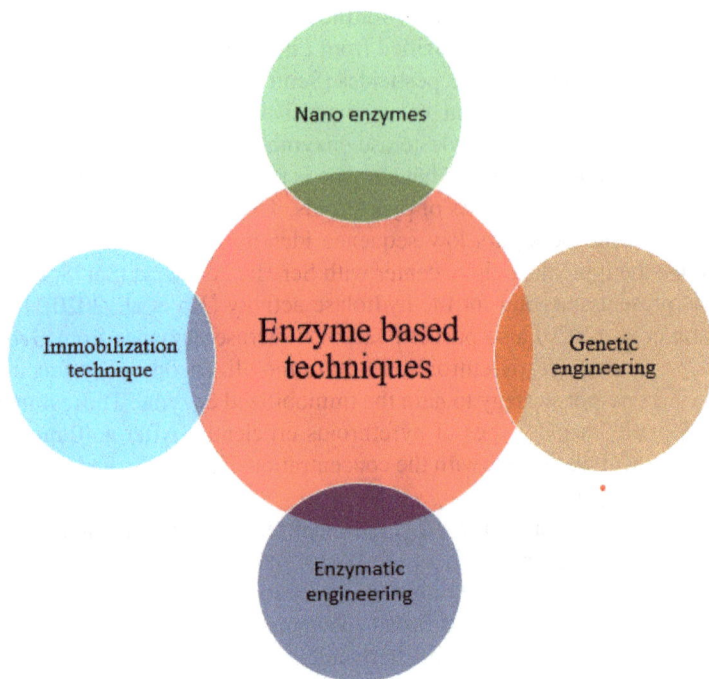

FIGURE 7.2 Innovative enzyme-based tools and technologies.

manufacturing of different enzymes from their indigenous hosts, the transmission of encoding genes, and their expression to another host provide new insights (Gupta et al., 2019). Using genetic engineering tools and techniques in biodegradation has been successfully improved by eradicating the substate transport limitation across the cell membrane, transporting the accumulated hazardous products, or expanding the microbial species substrate spectrum. Furthermore, this technique is also used to boost their shelf life, substrate range, and activation in a wide range of pH and temperature. Recombinant DNA technology is considered a good tool for enhancing the activation of enzymes (Alcalde et al., 2006). Recombinant enzymes exhibit a prominent ability to remove pollutants under different open ecosystems and lab conditions. To give an example, by using microbial species *Thanatephorous cucumeris* strain Dec1, a decolorizing dye peroxidase (DyP) enzyme was purified and expressed in *Aspergillus oryzae* RD 005. Different approaches like DNA shuffling, site-directed mutagenesis, or error-prone PCR improve the substrate range and remediation of pollutants (Dua et al., 2002).

Schofield et al. (2010) took out a study to produce an organophosphate hydrolase enzyme with enhanced hydrolytic strength for rapid remediation of organophosphates using genetic engineering techniques. In another study using genetic engineering tools, carboxylesterase B1 gene was injected into its pET28a plasmid, and its expression in *Escherichia coli* BL-21 was observed. Results revealed that the improved

plasmid, pET28B1, was involved depolluting enzymes. The changed cells were purified in a fermenter to get large numbers of enzymes and maximize their activation
level. Transformed cells and crude enzymes were directly attained from free microbial cells deprived of purification to assess their efficiency. This study demonstrated,
at the range of pH 7 and temperature 37°C and with the incubation time of 90 min,
more than 79% malathion was removed. Apart from malathion, crude enzymes are
considered strong candidates for degrading other pesticides containing ester bonds,
such as parathion (Qiao et al., 2003).

By adopting genetic engineering techniques, the engineered strain was cultured
to remove organophosphates and carbamates. The methyl parathion hydrolase
gene (*mpd*) was injected into the *Sphingomonas* sp. *by molecular approaches*. CDS-1
chromosomes are responsible for degrading carbofuran using the mini-Tn5 transposon system. Both pesticides with the concentration of (100 mg/L) completely
degraded within the incubation period of 16–32 h (Jiang et al., 2007).

7.3.2 ENZYMATIC ENGINEERING

Enzymatic engineering is defined as changing or modifying the primary amino acid
structure to enhance its various characteristics like activation range and harsh environmental tolerance and make it more useful for different industries (Rayu et al.,
2012). Various systematic methods and engineering techniques are responsible for the
purification and screening of novel enzymes from indigenous sources for industrial
applications. Furthermore, genetic engineering techniques also enhance the starting point, manage catalyst restrictions, immobilize enzymes with different carrier
materials, and activate various mediators and substrates (Li et al., 2012). Recently,
enzyme-engineering techniques have been played a vital role in the selective and
high-capacity remediation of heavy metals and other organic and inorganic hazardous substances (Dhanya, 2014).

To degrade numerous pesticide groups, enzymatic engineering-based techniques
and tools are considering more effective options to remove hazardous pollutants in
the environment. A microbial strain, *Pichia pastoris*, was cultured for the biodegradation of malathion, and extracellular expression of PoOPHM$_9$ was investigated
by adopting genetic engineering tools. Subsequently, the improvement findings
explained that the highest spaced tine yield of the enzyme was 640 U/L per hour.
After three days, the highest titer and fermentation yield attained 50.8 kU/L and 4.1
protein g/L, respectively. With the plant-derived detergent volumetric ratio of 0.1%
(w/w), PoOPHM$_9$ demonstrates maximum stability and activity. Within the 20 min,
the 0.15 mM initial concentration of malathion using 0.04 mg/mL was degraded
entirely in an aqueous solution (Bai et al., 2017).

In another study, an engineered microbial strain was actively involved in the
removal of organophosphates and carbofuran. That engineered strain was cultivated and incorporated into the chromosome of carbofuran degrading microbes
named *Sphingomonas* sp. CDS-1 through mini-Tn5 transposon system. Both pesticides (methyl parathion and carbofuran) at the concentration of (100 mg/L) were
rapidly degraded within the time of 32 h and 6 h, respectively (Jiang et al., 2007).

7.3.3 Immobilization Enzyme Technology

Enzyme immobilization strategy is crucial environmental technology to enhance the reusability of high biocatalytic productivity and low cost (Bilal et al., 2018). This technique improves the shelf life of enzymes and provides high biological activity (Zdarta et al., 2018). It also facilitates the binding sites of enzymes, which decreases the leakage of enzymes and boosts the activity of enzymes against hazardous substances (Mateo et al., 2007). The immobilization of enzymes could be carried out using various methods like covalent binding with different carrier materials such as silica gel, cross-linking, encapsulation, and adsorption of microbial enzymes and cells on other support materials, such as sodium alginate (Sirisha et al., 2016). Moreover, the immobilized enzymes can degrade xenobiotics in stressful environmental conditions, enhancing the enzyme stability and permeability (Skoronski et al., 2017). Diao et al. (2013) studied the immobilization of carboxylesterase purified from the insect *Spodoptera litura*. In this study, two different types of mesoporous sieves were used for the supporting enzymes and increasing their effectiveness. This research showed that the immobilized enzyme was more stable than the free enzyme under different environmental conditions. The experiments of pesticides remediation revealed that, relative to removal in a natural environment, the immobilization enzyme had a more remarkable ability to degrade the organic compounds with ester bonds.

For the biodegradation of organophosphates via adopting immobilization enzyme technology organophosphates hydrolase, $PoOPHM_9$ was conjugated and immobilized in a commercial polymer Pluronic F127. This study demonstrated that the immobilized enzymes could effectively degrade organophosphates at high temperatures and pH. Furthermore, the cross-linked enzyme polymer conjugates also increase the presence of various detergents (Cheng et al., 2018a). Laccase enzyme from *Coriollopsis gallica* was purified from the microbial species to degrade dichlorophen. Their immobilization to generate hybrid enzymes was done using the physical adsorption method in mesoporous synthetic silica foam and nanostructured silicon foam. The newly synthesized hybrid enzymes simultaneously degraded the dichlorophen pesticide. Moreover, the other benefit of hybrid nanomaterial composite is to decrease its severe genotoxicity and apoptotic effects (Vidal-Limon et al., 2018).

For the biodegradation of azinphos-methyl, laccase enzymes from *Myceliophthora thermophile* were purified and immobilized in monodisperse microspheres. The immobilized enzymes are much active at variant types of temperature and pH. Moreover, their permeability, optimal activity, and thermal, operational, and storage stability were significantly enhanced compared to the free laccase enzymes. The immobilized enzymes completely removed azinphos-methyl after the reaction of one hour with the appearance of 2,2′-azino-bis-(3-ethylbenzothiazoline-6-sulfonate) mediator (Vera et al., 2019). Yang et al. (2016) investigated the biodegradation of 2,4-dichlorophenol pesticide, adopting another immobilization method called cross-linking. After regulating trials, results indicated that the activation of enzymes was maximum at a temperature of 25°C, pH 5.4, a laccase solution of 0.2 mg/mL and 4% glutaraldehyde, respectively. In the presence of an immobilized biocatalyst, a degradation experiment was carried out at temperature 30°C, pH 5.4, and 0.1 g of immobilized enzyme. Results showed that the remediation ratio of 2-4 dichlorophenol with biocatalyst was 42.28% and in the absence of immobilized bio-catalyst was 15.93%, respectively.

For the biodegradation of organophosphorus pesticides in two crops, such as cucumber and grape, Xue et al. (2019) purified organophospho-hydrolase via supernatant of microbial cell culture, and adsorption of the enzyme was carried out with metal-organic frameworks. To prevent the binding between immobilized enzymes and metal ions, two different carrier materials and a chelating agent, such as unsaturated metal sites and His-tagged proteins, were used to ensure the max-imum activity of enzymes. Results demonstrated that immobilized organophos-phohydrolase, compared with free enzymes under the control conditions, showed five times higher activity (1554 $U/g_{protein}$). Moreover, the findings of this study revealed that immobilized enzymes were reused six times, and this study provides an effective way to degrade organophosphorus pesticides. Recently for the reme-diation of organophosphorus compounds, Santillan et al. (2020) purified microbial phosphotriesterases from different microbial species (*Streptomyces phaeochro-mogenes*, *Streptomyces setonii*, *Nocardia corynebacterioides*, *Nocardia aster-oides,* and *Arthrobacter oxydans*). The isolated enzymes tested as biocatalysts against the degradation of organophosphorus, such as paraoxon, methyl paraoxon, methyl parathion, coroxon, coumaphos, dichlorvos, and chlorpyrifos. This study explained that the immobilized enzymes avoid cellular interruption and centrifu-gation, which can effectively take part in the degradation process, can be reused, and is easy to handle.

The immobilization of enzymes degrades xenobiotics simultaneously and more effectively from the natural environment. Furthermore, this technique on support materials allows its recovery from the proceed solutions, enables reuse, and is low in cost (Gao et al., 2014).

7.3.4 NANO ENZYMES

Nanotechnology is an emerging technique used to detect and remediate toxic sub-stances but plays a vital role in other industries. It involves manipulating atoms and molecules for engineering substances into the nano dimensions (Khatri et al., 2017). Due to various distinguished features, like tiny nanoscale diameter, large surface-to-volume ratio, physicochemical properties, extreme target specificity, and less cost, these materials play a vital role in precisely sensing the xenobiotics (Rawtani et al., 2018). There are different nanomaterials, and their immobilization with enzymes is more capable of removing pollutants. The different types of nano-materials are nanotubes, nanocomposites, and nanoparticles used in the natural envi-ronment and lab conditions (Aragay et al., 2011).

To degrade chlorpyriphos, magnetic iron nanoparticles with the dimensions of 10 and 15 nm and laccase enzymes were immobilized with an extra coating of chitosan and carbodiimide. The chlorpyrifos remediation studies were presented in batch stud-ies under constant shaking for 12 h. Results demonstrated that the immobilization of laccase with magnetic iron nanoparticles in 12 h with the optimal conditions tem-perature 60°C and pH 7, 99% chlorpyrifos was degraded. Magnetic nanoparticles participated in the degradation process 32.3%. In comparison, the immobilized enzymatic nanoparticle was 58.8% to remove chlorpyriphos (Das et al., 2017). Recently, in another study for removing lindane, *Myceliophthora thermophila* laccase enzymes were purified and expressed in *Aspergillus oryzae* (Novozym 51003® laccase) immobilized in nano-silica amino-modified for potential application in the biodegradation process. The adsorption immobilization method completed the binding of enzymes in 40 min, followed by pseudo-first-order kinetics. This study revealed that after seven repetitive uses within one day, 56.8% of lindane was removed, and their original activity was retained about 70% (Bebić et al., 2020). Laccase enzymes were immobilized with silica nanoparticles for the effective remediation of azo dyes. Results explained that after continuous use of 20 times, approximately 80% enzyme activity was gained (Gahlout et al., 2017). Recently, Sharma et al. (2020b) described using dual metallic (Co-Ni) nanocatalyst to remove 2,4-dichlorophenoxyacetic acid (2,4-D) via in situ generations of hydroxyl radical species. The collaborating relation of both metals was investigated and coated with chitosan particles, which recommended tremendous remediation of emerging pollutants even after eight repeated cycles.

Biosensors are nowadays widely used for the welfare of humanity, such as for the assessment of various kinds of diseases and environmental pollutants in the field environment. Mostly paper-based biosensors are enormously applied to an achieved goal. For paper-based bioactive sensors, immobilization of enzymes with nano paper is considered a fundamental step for the transportation of fluid and detection of contaminants biologically in a single-step process. A diversity of models for paper-based biosensors has been established by the involvement of paper microfluidic and dipstick techniques (Martinez et al., 2007). To detect residues of two extensive groups of pesticides (organophosphate and carbamate) in the environment, Badawy and El-Aswad (2018) carried out a study using a bioactive paper-based sensor. The biosensor can detect the residues of methomyl and profenofos with excellent detection limits (methomyl $= 6.16 \times 10{-}4$ mM and profenofos $= 0.27$ mM) and instant response times ($\sim$5 min). The findings indicate that the paper-based biosensor is swift, sensitive, inexpensive, portable, disposable, and easy to use.

To compare with other common techniques of enzymes, the immobilization of enzymes with nanoparticles aided three valuable distinguished qualities. First of all, the nanoparticles and their immobilization with enzymes are effortless, and their manufacturing could be done in hazardous reactants in the absence of any surfactants. Second, the perfectly dense enzyme shells could be gained with similar coreshell nanoparticles. Finally, their particle size can be expediently designed efficiently (Ansari et al., 2012).

7.4 CONCLUSION AND FUTURE PERSPECTIVES

Pesticides are essential for the productivity of crops and pest management in agriculture. Still, despite their beneficial properties due to extensive use, their threats to non-target organisms in the environment cannot be refused. However, an appropriate process is mandatory on a significant basis for the detoxification and remediation of xenobiotics accumulated in the natural ecosystem. The remediation of agrochemicals and other emerging pollutants using enzymes is considered an effective method due to the biology of numerous pesticides, various forms of manipulation and modes of action, resistance against different pathogens, and the targeting of environmental pollutants. The breakdown of hazardous substances by enzymes allows the transformation of electrons from donors to acceptors. The oxidation and reduction pathway of enzymes detoxifies the contaminants more precisely. Using enzymes to degrade hazardous compounds could deliver a long-term advantage. It is ecofriendly and cost-effective compared to the other typical, such as chemical, physical, and incineration methods. The future potential of using microbial enzymes involved in the biodegradation of pesticides can be improved in different ways. One such improvement comprises the detection and selection of microbial strains and their purification of enzymes proficient in surviving various environmental variations. Another prospect is engineered microbes' selection to retain the balance of organisms to improve the remediation of pesticides. New databases, omics, microbial tools, and computational systems approaches can be hired to develop indigenous functional microbial species and their associated enzymes and genes.

REFERENCES

Acharya K, Shilpkar P, Shah M, et al. Biodegradation of insecticide monocrotophos by *Bacillus subtilis* KPA-1, isolated from agriculture soils. *Applied Biochemistry and Biotechnology*, 2015;175(4):1789–1804.

Ahir UN, Vyas TK, Gandhi KD, et al. In vitro efficacy for chlorpyrifos degradation by Novel Isolate *Tistrella sp.* AUC10 isolated from chlorpyrifos contaminated field. *Current Microbiology*, 2020;77(9):2226–2232.

Ahir UN, Vyas TK, Sutaria BP, et al. Screening of bacteria for their chlorpyrifos degrading ability. *International Journal of Current Microbiology and Applied Sciences*, 2018;7:10–17.

Ahmad S, Cui D, Zhong G, Liu J, et al. Microbial technologies employed for biodegradation of neonicotinoids in the agroecosystem. *Frontiers in Microbiology*, 2021;12:759439.

Alcalde M, Ferrer M, Plou FJ, et al. Environmental biocatalysis: From remediation with enzymes to novel green processes. *Trends in Biotechnology*, 2006;24(6):281–287.

Alneyadi AH, Rauf MA, Ashraf SS, et al. Oxidoreductases for the remediation of organic pollutants in water–a critical review. *Critical Reviews in Biotechnology*, 2018;38(7):971–988.

Ansari SA, Husain Q. Potential applications of enzymes immobilized on/in nano materials: A review. *Biotechnology Advances*, 2012;30(3):512–523.

Aragay G, Pons J, Merkoçi A, et al. Recent trends in macro-, micro-, and nanomaterial-based tools and strategies for heavy-metal detection. *Chemical Reviews*, 2011;111(5):3433–3458.

Aregbesola OA, Mokoena MP, Olaniran AO, et al. Biotransformation of pentachlorophenol by an indigenous *Bacillus cereus* AOA-CPS1 isolated from wastewater effluent in Durban, South Africa. *Biodegradation*, 2020;31(4):369–383.

Atala ML, Ghafil JA, Zgair AK, et al. Production of peroxidase from *Providencia spp.* bacteria. *Jouranl of Pharmaceutical Sceinces and Research*, 2019;11(6):2322–2326.

Avila R, Peris A, Eljarrat E, et al. Biodegradation of hydrophobic pesticides by microalgae: Transformation products and impact on algae biochemical methane potential. *Science of the Total Environment*, 2021;754:142114.

Badawy ME, El-Mohammaed, Taktak, Nehad EM. Design and optimization of bioactive paper immobilized with acetylcholinesterase for rapid detection of organophosphorus and carbamate insecticides. *Current Biotechnology*, 2018;2014.

Bai Y-P, Luo X-J, Zhao Y-L, et al. Efficient degradation of malathion in the presence of detergents using an engineered organophosphorus hydrolase highly expressed by *Pichia pastoris* without methanol induction. *Journal of Agricultural and Food Chemistry*, 2017;65(41):9094–9100.

Bansal M, Kumar D, Chauhan GS, et al. Preparation, characterization and trifluralin degradation of laccase-modified cellulose nanofibers. *Materials Science for Energy Technologies*, 2018;1(1):29–37.

Baweja M, Nain L, Kawarabayasi Y, et al. Current technological improvements in enzymes toward their biotechnological applications. *Frontiers in Microbiology*, 2016;7:965.

Bebić J, Banjanac K, Ćorović M, et al. Immobilization of laccase from *Myceliophthora thermophila* on functionalized silica nanoparticles: Optimization and application in lindane degradation. *Chinese Journal of Chemical Engineering*, 2020;28(4):1136–1144.

Bhatt P, Huang Y, Zhan H, et al. Insight into microbial applications for the biodegradation of pyrethroid insecticides. *Frontiers in Microbiology*, 2019;10:1778.

Bhatt P, Rene ER, Kumar AJ, et al. Binding interaction of allethrin with esterase: Bioremediation potential and mechanism. *Bioresource Technology*, 2020;315:123845.

Bhatt P, Zhou X, Huang Y, et al. Characterization of the role of esterases in the biodegradation of organophosphate, carbamate, and pyrethroid pesticides. *Journal of Hazardous Materials*, 2021:125026.

Bilal M, Asgher M, Iqbal HM, et al. Bio-catalytic performance and dye-based industrial pollutants degradation potential of agarose-immobilized MnP using a packed bed reactor system. *International Journal of Biological Macromolecules*, 2017;102:582–590.

Bilal M, Iqbal HM, Barceló DJSoTTE. Persistence of pesticides-based contaminants in the environment and their effective degradation using laccase-assisted biocatalytic systems. *Science of The Total Environment*, 2019;695:133896.

Bilal M, Iqbal HM, Guo S, et al. State-of-the-art protein engineering approaches using biological macromolecules: A review from immobilization to implementation view point. *International Journal of Biological Macromolecules*, 2018;108:893–901.

Brouwer M, Huss A, van der Mark M, et al. Environmental exposure to pesticides and the risk of Parkinson's disease in the Netherlands. *Environment International*, 2017;107:100–110.

Cai X, Wang W, Lin L, et al. Autotransporter domain-dependent enzymatic analysis of a novel extremely thermostable carboxylesterase with high biodegradability towards pyrethroid pesticides. *Scientific Reports*, 2017;7(1):1–11.

Camacho-Morales RL, Gerardo-Gerardo JL, Sánchez JE, et al. Ligninolytic enzyme production by white rot fungi during paraquat (herbicide) degradation. *Revista Argentina de Microbiologia*, 2017;49(2):189–196.

Chan-Cheng M, Cambronero-Heinrichs JC, Masís-Mora M, et al. Ecotoxicological test based on inhibition of fungal laccase activity: Application to agrochemicals and the monitoring of pesticide degradation processes. *Ecotoxicology and Environmental Safety*, 2020;195:110419.

Chauhan PS, Jha B. Pilot scale production of extracellular thermo-alkali stable laccase from *Pseudomonas sp.* S2 using agro waste and its application in organophosphorous pesticides degradation. *Journal of Chemical Technology and Biotechnology*, 2018;93(4):1022–1030.

Chen J, Hartmann EM, Kline J, et al. Assessment of human exposure to triclocarban, triclosan and five parabens in US indoor dust using dispersive solid phase extraction followed by liquid chromatography tandem mass spectrometry. *Journal of Hazardous Materials*, 2018;360:623–630.

Chen S, Lin Q, Xiao Y, et al. Monooxygenase, a novel beta-cypermethrin degrading enzyme from *Streptomyces sp. PloS One*, 2013;8(9):e75450.

Chen X, Zhou Q, Liu F, et al. Removal of nine pesticide residues from water and soil by biosorption coupled with degradation on biosorbent immobilized laccase. *Chemosphere*, 2019;233:49–56.

Chen X, Zhou Q, Liu F, et al. Performance and kinetic of pesticide residues removal by microporous starch immobilized laccase in a combined adsorption and biotransformation process. *Environmental Technology & Innovation*, 2021;21:101235.

Cheng H, Zhao Y-L, Luo X-J, et al. Cross-linked enzyme-polymer conjugates with excellent stability and detergent-enhanced activity for efficient organophosphate degradation. *Bioresources and Bioprocessing*, 2018a;5(1):1–9.

Cheng Y, Zang H, Wang H, et al. Global transcriptomic analysis of *Rhodococcus erythropolis* D310-1 in responding to chlorimuron-ethyl. *Ecotoxicology and Environmental Safety*, 2018b;157:111–120.

Chu C-W, Liu B, Li N, et al. A novel aerobic degradation pathway for thiobencarb is initiated by the TmoAB two-component flavin mononucleotide-dependent monooxygenase system in *Acidovorax sp.* strain T1. *Applied and Environmental Microbiology*, 2017;83(23):e01490–17.

Cimino AM, Boyles AL, Thayer KA, et al. Effects of neonicotinoid pesticide exposure on human health: A systematic review. *Environmental Health Perspectives*, 2017;125(2):155–162.

Cummins I, Landrum M, Steel PG, et al. Structure activity studies with xenobiotic substrates using carboxylesterases isolated from *Arabidopsis thaliana. Phytochemistry*, 2007;68(6):811–818.

Cycoń M, Piotrowska-Seget Z. Pyrethroid-degrading microorganisms and their potential for the bioremediation of contaminated soils: A review. *Frontiers in Microbiology*, 2016;7:1463.

Daam MA, Chelinho S, Niemeyer JC, et al. Environmental risk assessment of pesticides in tropical terrestrial ecosystems: Test procedures, current status and future perspectives. *Ecotoxicology and Environmental Safety*, 2019;181:534–547.

Das A, Jaswal V, Yogalakshmi KJC. Degradation of chlorpyrifos in soil using laccase immobilized iron oxide nanoparticles and their competent role in deterring the mobility of chlorpyrifos. *Chemosphere*, 2020;246:125676.

Das A, Singh J, Yogalakshmi K. Laccase immobilized magnetic iron nanoparticles: Fabrication and its performance evaluation in chlorpyrifos degradation. *International Biodeterioration and Biodegradation*, 2017;117:183–189.

Dave S, Das J. Role of microbial enzymes for biodegradation and bioremediation of environmental pollutants: Challenges and future prospects. In *Bioremediation for Environmental Sustainability*. Elsevier, USA, 2021, pp. 325–346.

Dhanya MS. Advances in microbial biodegradation of chlorpyrifos. *Journal of Environmental Research and Development*, 2014;9(1):232–240.

Diao J, Zhao G, Li Y, et al. Carboxylesterase from *Spodoptera Litura*: Immobilization and use for the degradation of pesticides. *Procedia Environmental Sciences*, 2013;18:610–619.

Dua M, Singh A, Sethunathan N, et al. Biotechnology and bioremediation: Successes and limitations. *Applied Microbiology and Biotechnology*, 2002;59(2):143–152.

Fan S, Li K, Yan Y, et al. A novel chlorpyrifos hydrolase CPD from *Paracoccus sp.* TRP: Molecular cloning, characterization and catalytic mechanism. *Electronic Journal of Biotechnology*, 2018;31:10–16.

Fang H, Cai L, Yang Y, et al. Metagenomic analysis reveals potential biodegradation pathways of persistent pesticides in freshwater and marine sediments. *Science of the Total Environment*, 2014;470:983–992.

Fang L, Shi T, Chen Y, et al. Kinetics and catabolic pathways of the insecticide chlorpyrifos, annotation of the degradation genes, and characterization of enzymes TcpA and Fre in *Cupriavidus nantongensis* X1T. *Journal of Agricultural and Food Chemistry*, 2019;67(8):2245–2254.

Fratila-Apachitei LE, Hirst JA, Siebel MA, et al. Diuron degradation by *Phanerochaete chrysosporium* BKM-F-1767 in synthetic and natural media. *Biotechnology Letters*, 1999;21(2):147–154.

Gahlout M, Rudakiya DM, Gupte S, et al. Laccase-conjugated amino-functionalized nanosilica for efficient degradation of Reactive Violet 1 dye. *International Nano Letters*, 2017;7(3):195–208.

Gangola S, Sharma A, Bhatt P, et al. Presence of esterase and laccase in *Bacillus subtilis* facilitates biodegradation and detoxification of cypermethrin. *Scientific Reports*, 2018;8(1):1–11

Gao Y, Chen S, Hu M, et al. Purification and characterization of a novel chlorpyrifos hydrolase from *Cladosporium cladosporioides* Hu-01. *PLoS One*, 2012;7(6):e38137.

Gao Y, Truong YB, Cacioli P, et al. Bioremediation of pesticide contaminated water using an organophosphate degrading enzyme immobilized on nonwoven polyester textiles. *Enzyme and Microbial Technology*, 2014;54:38–44.

Gonçalves AM, Rocha CP, Marques JC, et al. Enzymes as useful biomarkers to assess the response of freshwater communities to pesticide exposure–A review. *Ecological Indicators*, 2021;122:107303.

Green R, Sang H, Im J, et al. Chlorothalonil biotransformation by cytochrome P450 monooxygenases in *Sclerotinia homoeocarpa*. *FEMS Microbiology Letters*, 2018;365(19):fny214.

Guengerich FP, Munro AW. Unusual cytochrome P450 enzymes and reactions. *Journal of Biological Chemistry*, 2013;288(24):17065–17073.

Gupta J, Rathour R, Singh R, et al. Production and characterization of extracellular polymeric substances (EPS) generated by a carbofuran degrading strain *Cupriavidus sp.* ISTL7. *Bioresource Technology*, 2019;282:417–424.

Hao J, Liu J, Sun M. Identification of a marine *Bacillus* strain C5 and parathion-methyl degradation characteristics of the extracellular esterase B1. *BioMed Research International*, 2014;2014.

Heidari R, Devonshire A, Campbell B, et al. Hydrolysis of organophosphorus insecticides by in vitro modified carboxylesterase E3 from *Lucilia cuprina*. *Insect Biochemistry and Molecular Biology*, 2004;34(4):353–363.

Hernandez J, Robledo NR, Velasco L, et al. Chloroperoxidase-mediated oxidation of organophosphorus pesticides. *Pesticide Biochemistry and Physiology*, 1998;61(2):87–94.

Hirai H, Nakanishi S, Nishida T. Oxidative dechlorination of methoxychlor by ligninolytic enzymes from white-rot fungi. *Chemosphere*, 2004;55(4):641–645.

Hu K, Torán J, López-García E, et al. Fungal bioremediation of diuron-contaminated waters: Evaluation of its degradation and the effect of amendable factors on its removal in a trickle-bed reactor under non-sterile conditions. *Science of the Total Environment*, 2020a;743:140628.

Hu T, Xiang Y, Chen Q, et al. A novel esterase LanE from *Edaphocola flava* HME-24 and the enantioselective degradation mechanism of herbicide lactofen. *Ecotoxicology and Environmental Safety*, 2020b;205:111141.

Huang X, Chen F, Sun B, et al. Isolation of a *fluoroglycofen*-degrading KS-1 strain and cloning of a novel esterase gene *fluE*. *FEMS Microbiology Letters*, 2017;364(16).

Huang Y, Li J, Yang Y, et al. Characterization of enzyme-immobilized catalytic support and its exploitation for the degradation of methoxychlor in simulated polluted soils. *Environmental Science and Pollution Research*, 2019;26(27):28328–28340.

Islam SMA, Math RK, Cho KM, et al. Organophosphorus hydrolase (OpdB) of *Lactobacillus brevis* WCP902 from kimchi is able to degrade organophosphorus pesticides. *Journal of Agricultural and Food Chemistry*, 2010;58(9):5380–5386.

Jamwal S, Kumar R, Sharma A, et al. Response surface methodology (RSM) approach for improved extracellular RNase production by a *Bacillus sp. Journal of Advance Microbiology*, 2017;3:131–144.

Jia J, Zhang S, Wang P, et al. Degradation of high concentration 2, 4-dichlorophenol by simultaneous photocatalytic–enzymatic process using TiO_2/UV and laccase. *Journal of Hazardous Materials*, 2012;205:150–155.

Jiang J, Zhang R, Li R, et al. Simultaneous biodegradation of methyl parathion and carbofuran by a genetically engineered microorganism constructed by mini-Tn5 transposon. *Biodegradation*, 2007;18(4):403.

Jiang W, Gao Q, Zhang L, et al. Identification of the key amino acid sites of the carbofuran hydrolase CehA from a newly isolated carbofuran-degrading strain *Sphingbium sp.* CFD-1. *Ecotoxicology and Environmental Safety*, 2020;189:109938.

Jiang W, Gao Q, Zhang L, et al. Detoxification esterase StrH initiates strobilurin fungicide degradation in *Hyphomicrobium sp.* Strain DY-1. *Applied and Environmental Microbiology*, 2021;87(11):e00103–21.

Jiang W, Zhang C, Gao Q, et al. Carbamate CN hydrolase gene ameH responsible for the detoxification step of methomyl degradation in *Aminobacter aminovorans* strain MDW-2. *Applied and Environmental Microbiology*, 2020;87(1):e02005–20.

Jin X, Yu X, Zhu G, et al. Conditions optimizing and application of laccase-mediator system (LMS) for the laccase-catalyzed pesticide degradation. *Scientific Reports*, 2016;6(1):1–7.

Kambiranda DM, Asraful-Islam SM, Cho KM, et al. Expression of esterase gene in yeast for organophosphates biodegradation. *Pesticide Biochemistry and Physiology*, 2009;94(1):15–20.

Kammoonah S, Prasad B, Balaraman P, et al. Selecting of a cytochrome P450cam SeSaM library with 3-chloroindole and endosulfan–identification of mutants that dehalogenate 3-chloroindole. *Biochimica et Biophysica Acta (BBA)-Proteins and Proteomics*, 2018;1866(1):68–79.

Karigar C, Rao S. Role of microbial enzymes in the bioremediation of pollutants: A review. *Enzyme Research*, 2011;1.

Kaur H, Kapoor S, Kaur G, et al. Application of ligninolytic potentials of a white-rot fungus *Ganoderma lucidum* for degradation of lindane. *Environmental Monitoring and Assessment*, 2016;188(10):1–10.

Khatri N, Tyagi S, Rawtani D. Recent strategies for the removal of iron from water: A review. *Journal of Water Process Engineering*, 2017;19:291–304.

Kim S, Chen J, Cheng T, et al. PubChem 2019 update: Improved access to chemical data. *Nucleic Acids Research*, 2019;47(D1):D1102–D1109.

Kuhlmann J, Kretschmann AC, Bester K, et al. Enantioselective mixture toxicity of the azole fungicide imazalil with the insecticide α-cypermethrin in *Chironomus riparius*: Investigating the importance of toxicokinetics and enzyme interactions. *Chemosphere*, 2019;225:166–173.

Kumar M, Prasad R, Goyal P, Teotia P, Tuteja N. et al. Environmental biodegradation of xenobiotics: Role of potential microflora. In *Xenobiotics in the Soil Environment*. Springer, Cham, 2017, pp. 319–334.

Kumar SS, Ghosh P, Malyan SK, et al. A comprehensive review on enzymatic degradation of the organophosphate pesticide malathion in the environment. *Journal of Environmental Science and Health, Part C*, 2019;37(4):288–329.

Li F, Di L, Liu Y, et al. Carbaryl biodegradation by *Xylaria sp.* BNL1 and its metabolic pathway. *Ecotoxicology and Environmental Safety*, 2019;167:331–337.

Li N, Yao L, He Q, et al. 3, 6-Dichlorosalicylate catabolism is initiated by the DsmABC cytochrome P450 monooxygenase system in *Rhizorhabdus dicambivorans* Ndbn-20. *Applied and Environmental Microbiology*, 2018;84(4):e02133–17.

Li S, Yang X, Yang S, et al. Technology prospecting on enzymes: Application, marketing and engineering. *Computational and Structural Biotechnology Journal*, 2012;2(3):e201209017.

Liang B, Zhao Y-K, Lu P, et al. Biotransformation of the diphenyl ether herbicide lactofen and purification of a lactofen esterase from *Brevundimonas sp.* LY-2. *Journal of Agricultural and Food Chemistry*, 2010;58(17):9711–9715.

Liang WQ, Wang ZY, Li H, et al. Purification and characterization of a novel pyrethroid hydrolase from *Aspergillus niger* ZD11. *Journal of Agricultural and Food Chemistry*, 2005;53(19):7415–7420.

Lin T, Hu Z, Zhang G, et al. Levels and mass burden of DDTs in sediments from fishing harbors: The importance of DDT-containing antifouling paint to the coastal environment of China. *Environmental Science & Technology*, 2009;43(21):8033–8038.

Liu H-M, Cao L, Lu P, et al. Biodegradation of butachlor by *Rhodococcus sp.* strain B1 and purification of its hydrolase (ChlH) responsible for N-dealkylation of chloroacetamide herbicides. *Journal of Agricultural and Food Chemistry*, 2012;60(50):12238–12244.

Liu Y, Wei Q, Wang S-J, et al. Construction of an engineered strain free of antibiotic resistance gene markers for simultaneous mineralization of methyl parathion and ortho-nitrophenol. *Applied Microbiology and Biotechnology*, 2010;87(1):281–287.

Liu Y-H, Liu Y, Chen Z-S, et al. Purification and characterization of a novel organophosphorus pesticide hydrolase from *Penicillium lilacinum* BP303. *Enzyme and Microbial Technology*, 2004;34(3–4):297–303.

Lopes RdO, Pereira PM, Pereira ARB, et al. Atrazine, desethylatrazine (DEA) and desisopropylatrazine (DIA) degradation by *Pleurotus ostreatus* INCQS 40310. *Biocatalysis and Biotransformation*, 2020;38(6):415–430.

Mallak AM, Lakzian A, Khodaverdi E, et al. Effect of *Pleurotus ostreatus* and *Trametes versicolor* on triclosan biodegradation and activity of laccase and manganese peroxidase enzymes. *Microbial Pathogenesis*, 2020;149:104473.

Mansee AH, Chen W, Mulchandani AJB, et al. Biodetoxification of coumaphos insecticide using immobilized *Escherichia coli* expressing organophosphorus hydrolase enzyme on cell surface. *Biotechnology and Bioprocess Engineering*, 2000;5(6):436–440.

Martinez AW, Phillips ST, Butte MJ, et al. Patterned paper as a platform for inexpensive, low-volume, portable bioassays. *Angewandte Chemie*, 2007;119(8):1340–1342.

Maruyama T, Komatsu C, Michizoe J, et al. Laccase-mediated degradation and reduction of toxicity of the postharvest fungicide imazalil. *Process Biochemistry*, 2007;42(3):459–461.

Mateo C, Grazu V, Palomo JM, et al. Immobilization of enzymes on heterofunctional epoxy supports. *Nature Protocols*, 2007;2(5):1022–1033.

Materu SF, Heise S, Urban B. Seasonal and spatial detection of pesticide residues under various weather conditions of agricultural areas of the Kilombero Valley Ramsar Site, Tanzania. *Frontiers in Environmental Science*, 2021;9:30.

Mishra S, Zhang W, Lin Z, et al. Carbofuran toxicity and its microbial degradation in contaminated environments. *Chemosphere*, 2020;259:127419.

Mustapha MU, Halimoon N, Johar WLW, et al. An overview on biodegradation of carbamate pesticides by Soil Bacteria. *Pertanika Journal of Science & Technology*, 2019;27(2).

Nagpal V, Srinivasan M, Paknikar KM. Biodegradation of γ-hexachlorocyclohexane (Lindane) by a non-white rot fungus conidiobolus 03-1-56 isolated from litter. *Indian Journal of Microbiology*, 2008;48(1):134–141.

Nair IC, Jayachandran K. Enzymes for bioremediation and biocontrol. In *Bioresources and Bioprocess in Biotechnology*. Springer, Singapore, 2017, pp. 75–97.

Nanda M, Kumar V, Fatima N, et al. Detoxification mechanism of organophosphorus pesticide via carboxylestrase pathway that triggers de novo TAG biosynthesis in oleaginous microalgae. *Aquatic Toxicology*, 2019;209:49–55.

Nejad HM, Moussavi G. Advanced biodegradation process of atrazine in the peroxidase-mediated sequencing batch reactor (SBR) and moving-bed SBR (MSBR): Mineralization and detoxification. *Journal of Environmental Health Science and Engineering*, 2020;18(2):433–439.

Pizzul L, del Pilar Castillo M, Stenström J. Degradation of glyphosate and other pesticides by ligninolytic enzymes. *Biodegradation*, 2009;20(6):751–759.

Qiao C-L, Yan Y-C, Shang H, et al. Biodegradation of pesticides by immobilized recombinant *Escherichia coli*. *Bulletin of Environmental Contamination and Toxicology*, 2003;71(2):0370–0374.

Rashtbari S, Dehghan G. Biodegradation of malachite green by a novel laccase-mimicking multicopper BSA-Cu complex: Performance optimization, intermediates identification and artificial neural network modeling. *Journal of Hazardous Materials*, 2021;406:124340.

Rawtani D, Khatri N, Tyagi S, et al. Nanotechnology-based recent approaches for sensing and remediation of pesticides. *Journal of Environmental Management*, 2018;206:749–762.

Rayu S, Karpouzas DG, Singh BK. Emerging technologies in bioremediation: Constraints and opportunities. *Biodegradation*, 2012;23(6):917–926.

Ross MK, Borazjani A, Edwards CC, et al. Hydrolytic metabolism of pyrethroids by human and other mammalian carboxylesterases. *Biochemical Pharmacology*, 2006;71(5):657–669.

Rudakiya DM, Patel DH, Gupte A. Exploiting the potential of metal and solvent tolerant laccase from *Tricholoma giganteum* AGDR1 for the removal of pesticides. *International Journal of Biological Macromolecules*, 2020;144:586–595.

Russell RJ, Scott C, Jackson CJ, et al. The evolution of new enzyme function: Lessons from xenobiotic metabolizing bacteria versus insecticide-resistant insects. *Evolutionary Applications*, 2011;4(2):225–248.

Sabarwal A, Kumar K, Singh RP, et al. Hazardous effects of chemical pesticides on human health–Cancer and other associated disorders. *Environmental Toxicology and Pharmacology*, 2018;63:103–114.

Santillan J, Muzlera A, Molina M, et al. Microbial degradation of organophosphorus pesticides using whole cells and enzyme extracts. *Biodegradation*, 2020;31(4):423–433.

Sarker A, Lee S-H, Kwak S-Y, et al. Comparative catalytic degradation of a metabolite 3, 5-dichloroaniline derived from dicarboximide fungicide by laccase and MnO_2 mediators. *Ecotoxicology and Environmental Safety*, 2020;196:110561.

Schofield D, DiNovo AA. Generation of a mutagenized organophosphorus hydrolase for the biodegradation of the organophosphate pesticides malathion and demeton-S. *Journal of Applied Microbiology*, 2010;109(2):548–557.

Scott C, Lewis SE, Milla R, et al. A free-enzyme catalyst for the bioremediation of environmental atrazine contamination. *Journal of Applied Microbiology*, 2010;91(10):2075–2078.

Shanmugapriya S, Manivannan G, Selvakumar G, et al. Extracellular fungal peroxidases and Laccases for waste treatment: Recent improvement. In *Recent Advancement in White Biotechnology Through Fungi*. Springer, Cham, 2019, pp. 153–187.

Sharma B, Dangi AK, Shukla P. Contemporary enzyme based technologies for bioremediation: A review. *Journal of Environmental Management*, 2018;210:10–22.

Sharma A, Sharma T, Sharma T, Sharma S, et al. Role of microbial hydrolases in bioremediation. In *Microbes and Enzymes in Soil Health and Bioremediation*. Springer, Singapore, 2019, pp. 149–164.

Sharma B, Shukla P. Futuristic avenues of metabolic engineering techniques in bioremediation. *Biotechnology and Applied Biochemistry*, 2020a;69:51–60.

Sharma RK, Arora B, Sharma S, et al. In situ hydroxyl radical generation using the synergism of the Co–Ni bimetallic centres of a developed nanocatalyst with potent efficiency for degrading toxic water pollutants. *Materials Chemistry Frontiers*, 2020b;4(2):605–620.

Shimazu M, Mulchandani A, Chen W, et al. Simultaneous degradation of organophosphorus pesticides and p-nitrophenol by a genetically engineered *Moraxella sp.* with surface-expressed organophosphorus hydrolase. *Biotechnology and Bioengineering*, 2001;76(4):318–324.

Siampiringue M, Wong-Wah-Chung P, Sarakha M, et al. Impact of the soil structure and organic matter contents on the photodegradation of the insecticide carbaryl. *Journal of Soils and Sediments*, 2015;15(2):401–409.

Singh B, Kaur J, Singh K, et al. Biodegradation of malathion by *Brevibacillus sp.* strain KB2 and *Bacillus cereus* strain PU. *World Journal of Microbiology and Biotechnology*, 2012;28(3):1133–1141.

Singh N, Kumar A, Sharma B. Role of fungal enzymes for bioremediation of hazardous chemicals. In *Recent Advancement in White Biotechnology Through Fungi*. Springer, Cham, 2019, pp. 237–256.

Singh S, Sherkhane PD, Kale SP, et al. Expression of a human cytochrome P4502E1 in *Nicotiana tabacum* enhances tolerance and remediation of γ-hexachlorocyclohexane. *New Biotechnology*, 2011;28(4):423–429.

Sirajuddin S, Khan MA, Qader SAU, et al. A comparative study on degradation of complex malathion organophosphate using of *Escherichia coli* IES-02 and a novel carboxylesterase. *International Journal of Biological Macromolecules*, 2020;145:445–455.

Sirisha VL, Jain A, Jain AJ Aif, et al. Enzyme immobilization: An overview on methods, support material, and applications of immobilized enzymes. *Advances in Food and Nutrition Research*, 2016;79:179–211.

Skoronski E, Souza DH, Ely C, et al. Immobilization of laccase from *Aspergillus oryzae* on graphene nanosheets. *International Journal of Biological Macromolecules*, 2017;99:121–127.

Srinivasan P, Selvankumar T, Kamala-Kannan S, et al. Production and purification of laccase by *Bacillus sp.* using millet husks and its pesticide degradation application. *3 Biotech*, 2019;9(11):1–10.

Suhara H, Adachi A, Kamei I, et al. Degradation of chlorinated pesticide DDT by litter-decomposing basidiomycetes. *Biodegradation*, 2011;22(6):1075–1086.

Sutherland TD, Horne I, Russell RJ, et al. Gene cloning and molecular characterization of a two-enzyme system catalyzing the oxidative detoxification of β-endosulfan. *Applied and Environmental Microbiology*, 2002;68(12):6237–6245.

Tekere M, Ncube I, Read J, et al. Biodegradation of the organochlorine pesticide, lindane by a sub-tropical white rot fungus in batch and packed bed bioreactor systems. *Environmental Technology*, 2002;23(2):199–206.

Tiwari B, Sindhu V, Mishra AK, et al. Carbon catabolite repression of methyl parathion degradation in a bacterial isolate characterized as a *Cupriavidus sp.* LMGR1. *Water, Air, & Soil Pollution*, 2020;231(7):1–14.

Turnbull GA, Ousley M, Walker A, et al. Degradation of substituted phenylurea herbicides by *Arthrobacter globiformis* strain D47 and characterization of a plasmid-associated hydrolase gene, *puhA*. *Applied and Environmental Microbiology*, 2001;67(5):2270–2275.

Unuofin JO, Okoh AI, Nwodo UU. Aptitude of oxidative enzymes for treatment of wastewater pollutants: A laccase perspective. *Molecules*, 2019;24(11):2064.

Vasileva-Tonkova E, Galabova D. Hydrolytic enzymes and surfactants of bacterial isolates from lubricant-contaminated wastewater. *Zeitschrift für Naturforschung C*, 2003;58(1–2):87–92.

Vera M, Nyanhongo GS, Pellis A, et al. Immobilization of *Myceliophthora thermophila* laccase on poly (glycidyl methacrylate) microspheres enhances the degradation of azinphos-methyl. *Journal of Applied Polymer Science*, 2019;136(16):47417.

Verma S, Chatterjee S. Biodegradation of profenofos, an acetylcholine esterase inhibitor by a psychrotolerant strain *Rahnella sp.* PFF2 and degradation pathway analysis. *International Biodeterioration & Biodegradation*, 2021;158:105169.

Vidal-Limon A, García Suárez PCn, Arellano-García E, et al. Enhanced degradation of pesticide dichlorophen by laccase immobilized on nanoporous materials: A cytotoxic and molecular simulation investigation. *Bioconjugate Chemistry*, 2018;29(4):1073–1080.

Wang B, Wu S, Chang X, et al. Characterization of a novel hyper-thermostable and chlorpyrifos-hydrolyzing carboxylesterase EstC: A representative of the new esterase family XIX. *Pesticide Biochemistry and Physiology*, 2020;170:104704.

Wang D, Xue Q, Zhou X, et al. Isolation and characterization of a highly efficient chlorpyrifos degrading strain of *Cupriavidus taiwanensis* from sludge. *Journal of Basic Microbiology*, 2015;55(2):229–235.

Wang J, Ohno H, Ide Y, et al. Identification of the cytochrome P450 involved in the degradation of neonicotinoid insecticide acetamiprid in *Phanerochaete chrysosporium*. *Journal of Hazardous Materials*, 2019a;371:494–498.

Wang J, Tanaka Y, Ohno H, et al. Biotransformation and detoxification of the neonicotinoid insecticides nitenpyram and dinotefuran by *Phanerochaete sordida* YK-624. *Environmental Pollution*, 2019b;252:856–862.

Wei T, Feng S, Shen Y, et al. Characterization of a novel thermophilic pyrethroid-hydrolyzing carboxylesterase from *Sulfolobus tokodaii* into a new family. *Journal of Molecular Catalysis B: Enzymatic*, 2013;97:225–232.

Wu P, Zhao R, Zhang X, Niu T, et al. *Rhodopseudomonas capsulata* enhances cleaning of chlorfenapyr from environment. *Journal of Cleaner Production*, 2020;259, 120271.

Xiao Y, Lu Q, Yi X, et al. Synergistic degradation of pyrethroids by the quorum sensing-regulated carboxylesterase of *Bacillus subtilis* BSF01. *Frontiers in Bioengineering and Biotechnology*, 2020;8:889.

Xie H, Li Q, Wang M, et al. Production of a recombinant laccase from *Pichia pastoris* and biodegradation of chlorpyrifos in a laccase/vanillin system. *Journal of Microbiology and Biotechnology*, 2013a;23(6):864–871.

Xie Z, Xu B, Ding J, et al. Heterologous expression and characterization of a malathion-hydrolyzing carboxylesterase from a thermophilic bacterium, *Alicyclobacillus tengchongensis*. *Biotechnology Letters*, 2013b;35(8):1283–1289.

Xu D, Gao Y, Sun B, et al. Pyrethroid carboxylesterase PytH from *Sphingobium faniae* JZ-2: Structure and catalytic mechanism. *Applied and Environmental Microbiology*, 2020;86(12):e02971-19.

Xu X, Wang J, Yu T, et al. Characterization of a novel aryloxyphenoxypropionate herbicide-hydrolyzing carboxylesterase with R-enantiomer preference from *Brevundimonas sp.* QPT-2. *Process Biochemistry*, 2019;82:102–109.

Xue S, Li J, Zhou L, et al. Simple purification and immobilization of His-tagged organophosphohydrolase from cell culture supernatant by metal organic frameworks for degradation of organophosphorus pesticides. *Journal of Agricultural and Food Chemistry*, 2019;67(49):13518–13525.

Yang J, Yang X, Lin Y, et al. Laccase-catalyzed decolorization of malachite green: Performance optimization and degradation mechanism. *PloS One*, 2015;10(5):e0127714.

Yang X, Tang X, Dong F, et al. Facile one-pot immobilization of a novel thermostable carboxylesterase from *Geobacillus uzenensis* for continuous pesticide degradation in a packed-bed column reactor. *Catalysts*, 2020;10(5):518.

Yang Y-X, Pi N, Zhang J-B, et al. USPIO assisting degradation of MXC by host/guest-type immobilized laccase in AOT reverse micelle system. *Environmental Science and Pollution Research*, 2016;23(13):13342–13354.

Yao Q, Huang M, Bu Z, et al. Identification and characterization of a novel bacterial carbohydrate esterase from the bacterium *Pantoea ananatis* Sd-1 with potential for degradation of lignocellulose and pesticides. *Biotechnology Letters*, 2020;42(8):1479–1488.

Zámocký M, Obinger C. Molecular phylogeny of heme peroxidases. In *Biocatalysis Based on Heme Peroxidases*. Springer, Berlin and Heidelberg, 2010, pp. 7–35.

Zang H, Wang H, Miao L, et al. Carboxylesterase, a de-esterification enzyme, catalyzes the degradation of chlorimuron-ethyl in *Rhodococcus erythropolis* D310-1. *Journal of Hazardous Materials*, 2020;387:121684.

Zdarta J, Meyer AS, Jesionowski T, et al. Developments in support materials for immobilization of oxidoreductases: A comprehensive review. *Advances in Colloid and Interface Science* 2018;258:1–20.

Zeng S, Qin X, Xia LJBEJ. Degradation of the herbicide isoproturon by laccase-mediator systems. *Biochemical Engineering Journal*, 2017;119:92–100.

Zhan H, Huang Y, Lin Z, et al. New insights into the microbial degradation and catalytic mechanism of synthetic pyrethroids. *Environmental Research*, 2020;182:109138.

Zhang J, Zhao M, Yu D, et al. Biochemical characterization of an enantioselective esterase from *Brevundimonas sp.* LY-2. *Microbial Cell Factories*, 2017a;16(1):1–9.

Zhang W, Lu Z, Yang K, et al. Impacts of conversion from secondary forests to larch plantations on the structure and function of microbial communities. *Applied Soil Ecology*, 2017b;111:73–83.

Zhang X, Li S. Expansion of chemical space for natural products by uncommon P450 reactions. *Natural Product Reports*, 2017c;34(9):1061–1089.

Zhao J, Chen X, Jia D, et al. Identification of fungal enzymes involving 3-phenoxybenzoic acid degradation by using enzymes inhibitors and inducers. *MethodsX*, 2020;7:100772.

Zhao S, Xu W, Zhang W, et al. In-depth biochemical identification of a novel methyl parathion hydrolase from *Azohydromonas australica* and its high effectiveness in the degradation of various organophosphorus pesticides. *Bioresource Technology*, 2021;323:124641.

Zhao Y, Yi X, Zhang M, et al. Fundamental study of degradation of dichlorodiphenyltrichloroethane in soil by laccase from white rot fungi. *International Journal of Environmental Science & Technology*, 2010;7(2):359–366.

Zhu M, Feng X, Qiu G, et al. Synchronous response in methanogenesis and anaerobic degradation of pentachlorophenol in flooded soil. *Journal of Hazardous Materials*, 2019;374:258–266.

Zhu S, Qiu J, Wang H, et al. Cloning and expression of the carbaryl hydrolase gene *mcbA* and the identification of a key amino acid necessary for carbaryl hydrolysis. *Journal of Hazardous Materials*, 2018;344:1126–1135.

8 Esterases and Their Industrial Applications

Hamza Rafeeq, Asim Hussain, Ayesha Safdar,
Sumaira Shabbir, Muhammad Bilal, Farooq Sher,
Marcelo Franco, and Hafiz M. N. Iqbal

ABSTRACT

Esterases are hydrolase-class enzymes involved in cleavage catalysis and ester-bond formation. Esterases are part of inter-esterification, intra-esterification, and responses to transesterification. To isolate esterases, several techniques are employed to screen and identify esterases. For microorganisms that produce esterases, several screening and identification procedures are used. The greatest purity of esterase is achieved via several purifying procedures. Esterases' industrial applications contribute significantly to environmentally friendly nature initiatives, as well as the food and textile industries. The current review focuses on different aspects, namely screening and purification processes for esterase enzyme analyses. The current evaluation also highlights its potential uses in several domains. To examine it in-depth, we strive to provide a vast array of esterase enzyme knowledge. It is critical to note that esterases are less explored enzymes with a low literature survey than lipases, implying that the future scope of the study will be quite beneficial.

Keywords: esterase, biocatalysis, purification, thermophilic esterases, biomass degradation, biosensor development

CONTENTS

8.1   Introduction.. 170
8.2   Classification of Esterases 171
8.3   Aldridge 1953 171
8.4   Enzyme Commission 1978 172
8.5   Enzyme Committee 1989 173
8.6   Sources of Esterase 173
        8.6.1   Screening of Esterases 176
8.8   Plate Assay Technique 176
8.9   UV Fluorescence Technique 177
8.10 Synthesis of Cell Fractions 177
8.11 Fermentation 177
8.12 Purification of Enzymes.. 177

DOI: 10.1201/9781003202998-8

8.13 Ammonium Sulfate Precipitation 177
8.14 Dialysis 178
8.15 Acid Hydrolysis 178
8.16 Chromatography 178
8.17 Extremophilic Esterases... 178
     8.17.1 Thermophilic Esterases/Lipases 178
8.18 Psychrophilic Esterases 178
8.19 Halophiles... 179
8.20 Alkalophiles.. 179
8.21 Applications of Esterases.. 179
     8.21.1 Enzymatic Degradation of Plastic... 179
8.22 Synthesis of Ester, Chiral Drugs, and Optically Active Compounds........... 179
8.23 Deinking 180
8.24 Neuropathy 180
8.25 Anti-Tumor and Anti-Cancer.. 181
8.26 Biosensor Fabrication 181
8.27 Treatment of Hypercholesterolemia and Diagnostics.............................. 182
8.28 Biomass Depolymerization... 182
8.29 Food and Feed Industry 183
8.30 Flavorings and Alcoholic Beverage Industry... 184
8.31 Conclusion 184
8.32 Acknowledgments 184
8.33 Conflict of Interests.. 185
References... 185

8.1 INTRODUCTION

Esterases are enzymes that catalyze an ester group hydrolysis from a range of substrates to produce the esterified acid (Gopalan and Nampoothiri, 2016). They comprise both lipolytic enzymes, such as enzymes that use lipids as substrate, also called lipases, and non-lipolytic esterases, which are active on the substrates of the water-soluble ester. Regrettably, in scientific journals, the words "esterases" and "lipases" are sometimes used indiscriminately (Thierry et al., 2017). The three major criteria that differentiate between true lipases and nonlipolytic esterases are length, the physicochemical composition of the substrate, and enzyme kinetics of the hydro-lyzed acyl ester chain (Wilkinson, 2004).

Esterase influences the rates of reversible reactions—that is, the organic phase supports ester synthesis that can be broken down by the same enzyme in the aqueous phase (Khodami et al., 2001). On the other hand, esterases vary from lipases primarily in substrate specificity and interfacial activation (Panda and Gowrishankar, 2005). Lipases, which have a hydrophobic domain surrounding the active site, prefer long-chain fatty acid triglycerides, whereas esterases have an acyl-binding pocket (Shukla, 2012).

Lipase is the main category of esterases utilized for industrial applications. Lipase is a catalytic enzyme for de-esterification or transesterification, a process that may be adapted to produce free acid or to produce a second ester, usually an alkyl ester. Lipases are also used in the degreasing of leather as part of the arsenal enzyme used in detergents. The biotechnological production of enzymes utilize agro-industrial waste (Gopalan and Nampoothiri, 2016). The use of lipases and proteases in lens-cleaning solutions for the cleansing of contact lenses is a new application (Binod et al., 2013). Immobilized lipases are also employed to produce biodiesel from vegetable oils or unedible oils, such as jatropha, by transesterification of methanol triglycerides to yield long-chain fatty acid methyl esters (Carvalho et al., 2013). Other widely utilized esterases include acetyl xylanesterases, feruloyl esterases that are components of enzyme mixture used to saccharify lignocellulosic substances and bleaching enzymes from pulp and paper.

Because of its good qualities and its wide variety of nonnatural substratum specialties, high stability, and strong enantioselectivity, esterases are useful for many industrial processes. Because flavors account for more than a quarter of the global food additive market, customers prefer food items labeled "natural," as well as biological flavor esters produced by enzymes. Esterases have several potential uses in food industries in flavor-esters manufacturing processes (Ahmed et al., 2010). Various flavored esters have so far been produced using microbial lipolytic enzymes (Dandavate et al., 2009).

Most esterases in α/β hydrolase folded proteins (Pfam PF00561 domain) belong to the superfamily (Punta et al., 2012) of carboxylesterase gene families (Hotelier et al., 2010). A variety of functionally different enzymes can hydrolyze a broad range of substrates in the α/β hydrolase folding region. This superfamily, for example, comprises proteases, lipases, esterases, dehalogenases, peroxidases, and epoxy hydrolases and is one of the most popular protein folds (Hotelier et al., 2004). Each carboxylesterase enzyme core is an α/β sheet, not a barrel, with eight strands linked by helixes. The proteins in this intimate have very different substrate specialties, and their main DNA sequences bear little resemblance. However, esterases are believed to come from a shared ancestor because of their structural similarities and the preserved residue arrangement in the catalytic location (Oakeshott et al., 2010). The recognized structure of the α/β hydrolase fold comprises of six parallel α helices and eight β sheets, with the exemption of the β-2 sheet that would in few circumstances not have a parallel orientation.

8.2 CLASSIFICATION OF ESTERASES

Esterase classification is complicated, and various nomenclatures are commonly used by a specific species or community of generally similar species to classify these enzymes (Montella et al., 2012).

8.3 ALDRIDGE 1953

The classification scheme focused on association with organophosphate (OP) molecules was introduced by Aldridge (1953). This classification scheme enables esterases A (Est-A) (catalyzes the hydrolysis of OPs), esterases B (Est-B), and esterases C (Est-C) (do not react with OPs). It was assumed that Est A and B had a similar association process with OPs, with a somewhat slow speed of phosphorylated Est-B enzyme related to Est-A being the main difference (Walker and Mackness, 1983). The serine remaining at the catalytic location of Est-B was then indicated to be vulnerable to phosphorylation that did not seem to exist with corresponding Est A residue (Walker and Mackness, 1983). Augustinsson also suggested that Est-A had a cysteine moiety in an active site rather than the serine nucleophile (Satoh et al., 2002; Li et al., 2015). Instead of having cysteine in an active site, which can hydrolyze, OPs may fall into a separate domain (Yu et al., 2009; Li et al., 2010).

8.4   ENZYME COMMISSION 1978

Enzymes are classified into four numbers before EC letters. The Classification Scheme Enzymes Committee is sponsored by the Nomenclature Committee of the International Union of Biochemistry and Molecular Biology (NC-IUBMB). The first number represents reaction type, the second number represents the chemical bond, and the third number shows the substrate nature. The last figure applies to an enzyme-cognized particular substratum (Testa and Kraemer, 2007). For esterases, EC3 defines the classes of hydrolases (enzymes to encourage hydrolysis), and EC 3.1 hydrolases influence ester bonds (e.g., esterase, lipases, and exonucleases). The esterases referred here apply to two separate groups in terms of the structure of the substratum. The first part is classified as EC 3.1.1, which contains a wide variety of carboxylic substrate esters and enzymes such as acetylcholinesterases and other recognized and unknown esterases. The second class is known to include two specific enzymes called EC 3.1.8 (Pfam domain of PF0126) and diisopropyl-fluorophosphatase (EC 3.1.8.2; domain of Pfam PF08450) (Punta et al., 2012).

In 1978, a new classification scheme was introduced by the NC-IUBMB (Walker and Mackness, 1983). Carboxylesterases (EC 3.1.1.1) or carboxylic ester hydrolases, also known as aliesterases, are most commonly found on open chains of organic molecules (i.e., aliphatic molecules), in comparison with arylesterases (EC 3.1.1.2), also referred to as paraoxonases, Est-A hydrolyases, or aryl ester hydrolyses that are primarily active in treating aromatic compounds and capable of hydrolysis. However, EC 3.1.1.1 enzymes may also hydrolyze compounds of aryl ester, while EC 3.1.1.2 may hydrolyze open-chain esters of carboxylic esters but are not their preferred substrate.

EC 3.1.1.2 is able to hydrolyze phenylacetate and other phenolic esters according to the 1978 classification scheme (Aldridge, 1953). Later evidence showed that the situation was more difficult. Esterases that were capable of hydrolyzing paraoxon but not phenylacetate and enzymes capable of hydrolyzing phenylacetate or other phenolic esters have been identified but not paraoxon (Walker, 1993). A corresponding cataloging scheme has been developed to differentiate from Est-A (Aldridge classification), esterases of phenolic hydrolysis esters such as phenyl acetate, the hydrolysis trial composites of OP. Different peaks of an activation without correlation have been identified in human serum fractions, which means that each reaction involves different enzymes (Mackness et al., 1987).

8.5 ENZYME COMMITTEE 1989

More research was required for the 1989 revision of the NC-IUBMB esterase classification scheme (Walker, 1993). Subsequently, formerly, esterases that especially have hydrolytic OP molecules (comprising phosphorus derivatives-H3PO4, most of which include OP and H_3PO_3 esters) have been listed as EC 3.1.8. EC 3.1.8 enzymes are hydrolases of phosphorus triesters, which function in phosphoric bonds of the triester. These enzymes are called Est-A, OP hydrolyses, or aryldialkyphosphates (the suggested term) by certain enzymes generating dialkyl and aryl alcohols (Wheelock et al., 2005). The inhibition of these enzymes requires divalent ions. EC 3.1.8.1 enzymes were known formerly as EC 3.1.1.2, partly explaining the literature misunderstanding concerning esterase classification and, particularly, Est-A.

Est-A and Est-B are currently described as two classes of enzymes: the ones with hydrolysis OPs (Est-A) and the ones which are gradually subdued by operating agents (Est-B). Familiarized by Aldridge (1953), this nomenclature applies not to specific classes of enzymes. One of the main features of the carboxylesterase family is that a great deal has occurred, and the high resistance toward changes in their primary structure will lead to significant changes in the specificity of the substrates, thereby promoting functional diversification of the family (Hotelier et al., 2010). Although the definition of esterases diverges from the classification, the opinion is that Aldridge's enzymes known as Est-B (i.e., those with OPs inhibited) belong to the serum esterases or carboxylesterases families. The serine moiety of the active enzyme shows a key function in the hydrolyze reaction in carboxylesterases (Satoh et al., 2002; Satoh and Hosokawa, 2006). Some researchers indicated, for example, that the Aldridge-distinct enzymes Est-A (those which hydrolyze OPs) have particular distinctiveness (e.g., the prerequisite for divalent ions and the need for cysteine moiety, a substitute for the serine at the active site). The NC-IUBMB categorizes both EC 3.1.1.2 and EC 3.1.8.1 with Est-A, and Wheelock et al. (2005) observed both groups of OP enzymes. Table 8.1 lists a classification of esterases on the basis of EC number.

8.6 SOURCES OF ESTERASE

Plants, animals, and microorganisms have been recognized as a source for esterases production (Figure 8.1). The expenses of cultivation and maintenance are inexpensive and simple to adjust; thus, microorganism enzymes are preferred. Either intrinsically or through induction, all species of microbes produce esterases. Diverse bacterial species are usually purchased or isolated from cultural collections for esterase biosynthesis. Microbes are collected from the cheese area, oil-polluted municipal waste, or sea squid to be used in the production of esterase (Ranjitha et al., 2009; Sayali et al., 2013). A new technique for esterase metagenome testing was applied, and the source of esterase originated through metagenomic libraries in many circumstances (Fan et al., 2012). The sequence automatic processes and shotgun cloning have been responsible for the launching of several genome projects that include a great deal of genetic data. To date, the Genome Atlas Database has included 1,078 bacterial genomes and 82 archaea (Hallin and Ussery, 2004). Selected results reveal a lot of enzymes, which were subsequently cloned, overexpressed, and purified for

TABLE 8.1

Classification of Esterases on the Basis of EC Number

Enzyme	EC No.	Function	References
Acetylesterase	3.1.1.6	Splits off acetyl groups	Ding et al. (2019)
Cholinesterase		Lyses choline-based esters	Sharma (2019)
• Acetylcholinesterase	3.1.1.7	Inactivates the neurotransmitter acetylcholine	Colovic et al. (2013)
• Butyrylcholinesterase	3.1.1.7		Colovic et al. (2013)
Pectinesterase	3.1.1.11	Clarifies fruit juices	Toushik et al., 2017
Thiolester hydrolases	3.1.2	Catalyzes the hydrolysis of thiolester into a carboxylic acid and a thiol	Sibon and Strauss, 2016
Thioesterases		Splits off ester groups, specifically thiol groups	Wang et al. (2018)
• Ubiquitin carboxy-terminal hydrolase L1	3.1.2.15	Cleaves ubiquitin from proteins	Matuszczak et al. (2020)
Phosphatase	3.1.3.x	Hydrolyses phosphoric acid monoesters into a phosphate ion and an alcohol	Krysiak et al. (2018)
Alkaline phosphatase	3.1.3.1		Whyte (2020)
• Phosphodiesterase (cGMP-specific phosphodiesterase type 5)	3.1.4.17	Removes phosphate groups from many types of molecules, including nucleotides, proteins, and alkaloids. Inactivates the second messenger cAMP	Qureshi et al. (2018)
Fructose bisphosphatase	3.1.3.11	Converts fructose-1,6-bisphosphate to fructose-6-phosphate in gluconeogenesis	Park et al. (2020)
Phosphoric diester hydrolases	3.1.4	Catalyzes the hydrolysis of a phosphodiester to give a phosphomonoester and a free hydroxyl group	Qureshi et al. (2018)
Triphosphoric monoester hydrolases	3.1.5	Hydrolyzes a triphosphoester to give a triphosphate group and a free hydroxyl group	Wu et al. (2020)
Sulfuric ester hydrolases (sulfatases)	3.1.6	Catalyzes the hydrolysis of sulfate esters	Korban et al. (2017)

Enzyme	EC number	Description	Reference
Diphosphoric monoester hydrolases	3.1.7	Hydrolyzes a diphosphoester to give a diphosphate group and a free hydroxyl group	Lange and Srividya (2019)
Phosphoric triester hydrolases	3.1.8	Catalyzes the hydrolysis of a phosphoric triester	Rajai et al. (2019)
Exonucleases (deoxyribonucleases and ribonucleases)	3.1.11	Cleaves nucleotides from the end (exo) of a polynucleotide chain	Mukherjee et al. (2004); Rather et al. (2020)
• Exodeoxyribonucleases producing 5'phosphomonoesters	3.1.13	Catalyzes digestion of the ends of linear DNA	
• Exoribonucleases producing 5'-phosphomonoesters	3.1.14	Degrades RNA by removing terminal nucleotides from either 5' end of the RNA molecule	
• Exoribonucleases producing 3'-phosphomonoesters	3.1.15	Degrades RNA by removing terminal nucleotides from either 3' end	
• Exonucleases active with either ribo- or deoxy-		Cleaves nucleotides one at a time from the end (exo) of a polynucleotide chain	
Endonucleases (deoxyribonucleases and ribonucleases)	3.1.21–	Cleaves the phosphodiester bond within a polynucleotide chain	Maruyama et al. (2019)
• Endodeoxyribonuclease	3.1.25	Catalyzes cleavage of the phosphodiester bonds in DNA	Al-Rashedi et al. (2020)
• Endoribonuclease		Cleaves either single-stranded or double-stranded RNA	

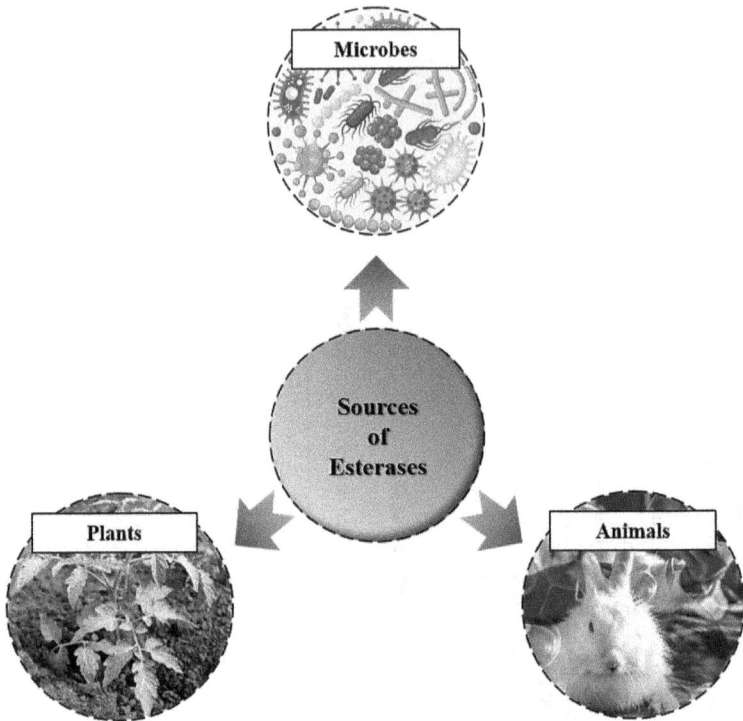

FIGURE 8.1 Sources of esterases.

biochemical characterization by genome mining for new genes through homology with identified lipase and esterases. Thus, a few lipolytic enzymes were cloned and expressed in mesophilic hosts from *Thermus thermophilus* HB27, whose genome is fully sequenced and accessible publicly (Henne et al., 2004). Extremely thermal stability and a very high behavior at mesophilic temperatures were obtained with a significant proprietary reciprocal esterase, a significant fact of its thermophilic nature (López-López et al., 2010).

8.6.1 SCREENING OF ESTERASES

Several methods of screening are available and are used for the selection and detection of esterase-producing organisms.

8.8 PLATE ASSAY TECHNIQUE

This is the most frequently used assay, in which several substances may be utilized. After 24 to 48 hours of incubation at 37°C, esterase-producing microorganisms produce clear areas around the colony. Ethyl acetate, sodium lactate, tween-20, tween-80, rhodium olive oil, tributyrin, and naphthyl acetate are several substrates. Certain additional substances may be used to designate areas for cleaning.

Fluorescent substances such as rhodamine B detectable in UV light or Fast Blue R R chemicals may be used to generate a brown product. Such Lugol's iodine solution, which helps to expand clear areas, may be exposed to plates even after incubation (Faiz et al., 2007).

8.9   UV FLUORESCENCE TECHNIQUE

The method for the management of esterase-producing microbiological colonies was reported in 1971 in mixed culture. The process comprises the cultivation and growing at room temperature for 3 minutes with the sterile fiber filter saturated with 8 x10–5M of 4-MUB (ester of 7-hydroxy-4-methyl coumarin). After culturing, a second Petri plate with the same alignment is fitted onto the glass fiber filter. The image is then taken using UV light. Colonies producing extracellular esterase are responsible for the highly light patches of the 4-MU (7-hydroxy-4-umbelliferone) (Pancholy and Lynd, 1971).

8.10   SYNTHESIS OF CELL FRACTIONS

Three cell divisions may be created to determine the location of an enzyme (i.e., whether the enzyme is external, internal, or membranous). The external fraction is supernatant after centrifugation of the culture broth. The resulting pellet is agitated and dissolved using mechanical or chemical procedures in the grown medium. The supernatant is then centrifuged, and the internal fraction is retained. The produced pellet is subsequently washed and considered as a membranous fraction in the growth media. Each fraction has an estimated enzyme activity. The location of the enzyme is shown with the highest activity (Sayali et al., 2013).

8.11   FERMENTATION

Submerged fermentation, solid-state fermentation, and slurry-state fermentation are all three fermentations that may lead to enzyme production (Jacob and Prema, 2006). All fermentation process parameters are regulated, and enzymes are manufactured.

8.12   PURIFICATION OF ENZYMES

The crude enzyme may be processed in numerous ways to get the highest enzyme purity.

8.13   AMMONIUM SULFATE PRECIPITATION

Precipitation with ammonium sulfate is a technique for enzyme isolation based on the salting technique. Because of its great susceptibility to water, ammonium sulfate is used commonly; it has no harmful and cost-effective influence on enzyme function. The technique involves the insertion of the purifying enzyme of an increased ammonium sulfate concentration. After 24 hours of chilling, precipitation is accomplished. If the magnetic stirring is supplied, a higher yield is achieved.

8.14   DIALYSIS

This is a process in which semipermeable membranes are used for the extraction of enzymes. The perforations of these membranes may easily pass the small molecules, such as salt ions, while larger molecules, such as enzymes, remain in the membrane. A crucial factor for the isolation of the target protein is the molecular cut off the dialysis membrane. Cellulose or cellophane are the different membranes that may be used for dialysis.

8.15   ACID HYDROLYSIS

It usually includes the treatment of enzymes with acid at high temperatures. In the most common acid hydrolysis procedures, 6N HCl is utilized at 110°C for 20–24 hours. This is an indispensable component in enzyme purification and is not applicable to many investigators.

8.16   CHROMATOGRAPHY

Instead of inspection, preparatory chromatography is often used to purify a compound either in sufficient amounts or for eventual application. Complicated blends may be collected separately at two stationary and mobile phases in this physical process. Column chromatography, ion-exchange chromatography, hydrophobic chromatography interaction, reversed chromatography, and affinity chromatogram are a few chromatographic methodologies used for enzyme purification (Sayali et al., 2013).

8.17   EXTREMOPHILIC ESTERASES

8.17.1   THERMOPHILIC ESTERASES/LIPASES

The species adapted to high temperatures live between 45°C and 122°C and exhibit several molecular variations in comparison to mesophilic organisms and are mostly archaea and eubacteria. In reality, the GC content of the coding regions is correlated with the weather, and at higher temperatures, the GC content is higher, which is the main difference between mesophilic species (Zheng and Wu, 2010). In relation to the mesophilic equivalent, membranes have a distinct structure. Many reports have shown that the membranes of thermophilic species produce higher levels of lipids that are stable at temperatures, in particular ether lipids and esters, which have long acyl chains (Koga, 2012). Probably, several lipases and esterases have now been identified, and their possible uses from food to the pharmaceutical industry are high thermal stability, greater half-life, and organic solvent constancy (Mandrich et al., 2012).

8.18   PSYCHROPHILIC ESTERASES

The most common feature of psychrophilic species are microbes, archaea, yeasts, and algae in low-temperature regions, such as high mountain regions and perennial glaciers. In this respect, psychrophilic species are undergoing a set of adaptations, which are designed to flourish and to live best at low temperatures, sometimes at high temperatures. In this era, a number of psychrophilic esterases have been isolated and primarily studied in order to explain the cold tolerance molecular determinants in areas where the substrate is temperature-sensitive, such as fruit, organic synthesis, animal feed, textiles, detergents, and beverages (Ramnath et al., 2017).

8.19   HALOPHILES

The species are mostly archaea and bacteria, which are suited to living at elevated salt levels (maximum 5 M NaCl). They are capable of maintaining the osmotic equilibrium of salt accumulated at isotonic levels. Halophilic proteins mostly adjust to protein surface levels to avoid their precipitation by growing the amount of negative waste, but this adaptation often provides consistency with low water content. Numerous esterases have been isolated and categorized from halophilic species (De Luca and Mandrich, 2020).

8.20   ALKALOPHILES

These species may survive in high- or low-pH environments. It is their capacity to preserve internal pH close to neutrality by proton pumps, and thus, besides those in periplasm space, the proteins do not require particular adaptation. Few lipases can be separated for high-pH adjustment and used in detergent preparation for fat hydrolysis, where high pH values are normally used (De Luca and Mandrich, 2020).

8.21   APPLICATIONS OF ESTERASES

8.21.1   Enzymatic Degradation of Plastic

The genera *Pseudomonas*, *Comamonas*, and *Bacillus* are plastic-degrading microorganisms. A three-step process is the mechanism of plastic enzyme breakdown. First, the microorganism released enzyme binds to the plastic surface and performs hydrolytic polyethylene cleavage. The second phase is depolymerization, when complex polymers become little monomers, dimers. The mineralization follows, which is a process of deterioration where CO_2, H_2O, and CH_4 are the final products. Some esterases for plastic breakdown have been studied. Clamping the ester link between polymers results in a low molecular weight of polyurethane esterase from *Comamonas acidovorans* to PLA and ES-PU from a polyurethylene adipate (Bhardwaj et al., 2012).

8.22   SYNTHESIS OF ESTER, CHIRAL DRUGS, AND OPTICALLY ACTIVE COMPOUNDS

The production of esterase linkages involves esterases. Esterases catalyze the enzyme production of esters. Researchers accomplished the ester synthesis using non-starter bacteria isolated from the cheese surface (Gandolfi et al., 2000). Esterases are used principally in the manufacture of optically pure substances and of pharmaceutical products, such as antibiotics and anti-inflammatory medicinal products. Esterase developed chiral medicines, including anti-inflammatory medicines used in the pharmaceutical industry as an agent to destroy pain. An esterase from *Trichosporon brassicae* has been widely used to produce optically pure (S)-and/or I-ketoprofen (2-(3 benzoylphenyl) propionic acid), which is very useful in reducing inflammation and pain caused by asthma, sunburn, menstruation, and fever (Kohli and Gupta, 2016). Stereospecific transformations in taxol-semi-syntheses (e.g., throumboxane-A2-antagonist, acetylcholine esterase inhibitors, anticholesterol drugs) have been identified in the synthesis of pharmaceutical intermediates. An esterase of *Pseudomonas stutzeri* A1501, with uniquely stereospecific characteristics, has been described for its use in industrial synthesis (Lehmann et al., 2014). The chemicals that are optically active are chemicals that allow plane-polarized light to spin. The high chemical yield optically pure 2S-6 acid was manufactured. The tetramethylsaline standard was used to measure the optical rotation on a polarimeter. HPLC analysis determined the enantiomeric surplus (Tombo et al., 1987).

8.23 DEINKING

Deinking is the process of removing and separating the ink from the substance produced. Due to its great efficiency and poor-quality influence on the final paper, the enzymatic diminution is becoming more and more important. For the enzyme-deinking process, several enzymes, such as cellulases, xylanases, esterases, lipases, and ligninolytic enzymes, have been utilized (Bolanča and Bolanča, 2004).

8.24 NEUROPATHY

Neuropathy is a nerve condition that may adversely influence feeling, reaction, or activity of the gland or organ. NTE (neuropathy target esterase) is a membrane-bound protein present in vertebrate neurons that plays a key role in chemically induced and naturally occurring neurological disorders. During the testing of possible organophosphorus neurotoxicants (paraoxon, malaoxon, chlorpyrifos-oxon, dichlorovos, and trichlorfon) on neuroblastoma cell lines (human SHSY5Y and murine NB41A3), it was discovered that organophosphorus compounds inhibit target esterases acetylcholinesterase (AChE) and NTE, resulting in acute and delayed neurotoxicity. The function of NTE in neurodegeneration was demonstrated experimentally in NTE knockout mice produced by cre-loxP site-specific recombination, which revealed that NTE deficiency resulted in neuronal vacuolation and extensive membrane deformities in hippocampal and thalamic neurons (Akassoglou et al., 2004). The loss of NTE phospholipase activity and accumulation of phosphatidylcholine attributable to organophosphorus-mediated delayed neuropathy (OPIDN) resulted in endoplasmic reticulum dysfunction and axonal transport hindrance, according to studies in mammalian cell lines and yeasts (Glynn, 2007). This is illustrated by the pathway in which NTE deacylates phosphatidylcholine (PtdCho) at the cytoplasmic face of the endoplasmic reticulum membrane to shape soluble products such as free fatty acids (FFA) and glycerophosphocholine (GroPCho), but organophosphate inhibition results in OPIDN (Zaccheo et al., 2004). The function of NTE in metabolism and pathophysiology was recently reviewed. NTE-mediated synthesis of glycerophos-phocholine, an abundant organic osmolyte in renal medullary cells, protects renal medullary cells from elevated interstitial concentrations of NaCl and urea. The function of NTE in controlling the cytotoxic aggregation of lysophospholipid in mammalian membranes and maintaining lipid bilayer fluidity has been established. The influence of the hydrolysis of 1-palmitoyl-2-hydroxy-sn-glycero-3-phosphocholine (p-lysoPC) by the catalytic domain of NTE on different assisted bilayer membranes (sBLMs) formulations was studied using the fluorescence recovery after pattern photobleaching (FRAPP), and it was concluded that the fluidity of sBLMsm reconstituted on silica decreased significantly (Greiner et al., 2010).

8.25   ANTI-TUMOR AND ANTI-CANCER

Different lethal cancers have been identified, and one of them is lung cancer. There is a lack of precision and efficacy in recent chemotherapeutic methods for lung cancer. The nanotherapeutic medicinal product β-lapachone (β-lap) has been transformed into porcine liver esterase by means of bioconsistent and biodegradable poly(ethyleneglycol)-b-poly (D, L-26 lactic acid) (PEG-b-PLA) micellulose β-lap-dC3 and by 28-dC6 (PLE). Antitumor efficacy and long-term survival with cytotoxicity assays in A549 and H596 lung cancer cells were demonstrated in the β-lapachone product (Ma et al., 2015). The anti-tumor effect of carboxylesterase (CE) expressive NSCs has been shown to treat primary lung cancer or metastatic lung cancer in the brain in a neural stem cell (NSC) dependent enzyme/prodrug therapy (NDEPT) (Yi et al., 2014). The development of A549 human non-small-cell lung adenocarcinomas in vitro and in vivo, thereby supplying therapeutic genes to brain tumors has been used as an important therapy for brain metastases from lung cancer (NSC) expressing rabbit carboxylesterase (F3. CE) (Hong et al., 2013).

8.26   BIOSENSOR FABRICATION

While manufacturing cholesterol-dependent biosensors, cholesterol esterase was immobilized on polyaniline films in combination with cholesterol oxidase and peroxidase and was used as sensing agents for cholesterol estimations and enhanced biosensor electrodes shelf durability (Singh et al., 2006). In the clinical diagnosis and prevention of a variety of clinical disorders, such as hypertension, arteriosclerosis disorders, cerebral thrombosis, and coronary heart diseases, the estimation of metabolites, such as glucose, urea, and cholesterol in the blood sample is essential (Kohli and Gupta, 2016).

Recently, extremely responsive fluorogenic esterase probes, obtained from the far-red fluorophore 7-hydroxy-9H-(1,3 dichloro-9,9-dimethylacridin-2-one) (DDAO), were used for the detection of low PIC levels at various stages of tuberculosis infection in mycobacterial lysates (Tallman and Beatty, 2015). It has been challenging to classify *Mycobacterium tuberculosis* esterases in the disease since most of the inclusion bodies develop in heterologous hosts. Esterase with ferrocene capped gold nanoparticles was used in blood samples for the analysis of cholesterol in a recent study (Davis-Lorton, 2015).

8.27   TREATMENT OF HYPERCHOLESTEROLEMIA AND DIAGNOSTICS

Hypercholesterol is distinguished by extremely elevated serum cholesterol levels and is a known risk factor for atherosclerosis and CHD in humans (Heidrich et al., 2004). In a study, it was noticed that targeting cholesterol esterase inhibitors may be helpful therapies for limiting the absorption of cholesterol (Ellidag et al., 2014). Multiple myeloma is a cancer of plasma cells, representing 1% of neoplastic and 13% of hematological disorders in the USA. In patients with multiple myeloma, arylesterase functions in the controls and patients with elevated oxidative stress was found to be substantially lower (Howell et al., 2014). The degradation of cocaine in rats and defense against convulsive and fatal effects of cocaine were reported to avoid harmful cocaine effects on the central nervous system (Aïzoun et al., 2013). The function of esterases in the implementation of malaria control strategies has also been established, which assists insecticide resistance to bendiocarb in *Anopheles gambiae* Tanguieta. Leukocyte esterase has recently been suggested in the synovial fluid as a proxy for periprothesis joint infection (Tischler et al., 2014).

8.28   BIOMASS DEPOLYMERIZATION

Biomass degradation includes the synergistic effect of a number of cellulolytic, xylanolytic, and/or pectinolytic esterases. Because of its capacity to hydrolyze ester bonds between cellulose residuals and phenolic compounds, cinnamoyl esterases are active in a disorganization network as "helper" enzymes, thereby making it easy for hydrolases to reach the mainstay of the cellular wall polymers. This preparation is beneficial for a range of uses as described underneath (Benoit et al., 2008). The main challenge in obtaining fibers of good quality is the removal of lignin, which is anatomically rooted in the pulp network and which is accountable for the black color of the pulp. In order to bleach and eliminate lignin, the kraft method utilizes chemical therapies of chlorine compounds, which contain high dioxin and chlorolignins, which are contaminating complexes. The increased biologic conversion characteristics provide an intriguing substitute for organic bleaching. Hemicellulases and oxidreductases, such as xylanases and laccases, are used in the bleaching of pulp to reduce intake of chlorine and to improve the final luminosity of pulp (Mayer and Staples, 2002). *A. niger* FaeA was seen to lead to effective delignification of the pulp of wheat straw in conjunction with xylanase and laccase therapy. *A. niger* FaeA has also been used with oilseed flax straw in a completely free chlorine phase, resulting in a very small amount of kappas (directly commensurate with lignin content), a good influence on pulp lightness and phenolic compounds of interest (Tapin et al., 2006).

The development of fuel ethanol from sustainable lignocellulosic materials is another non-food use of feruloyl esterases. *A. niger* FaeA was also used to convert lignocellulosic biomass to fermentable sugar in conjunction with xylanases and laccases to produce bioethanol. The effectiveness of the enzyme therapy was assessed in the saccharification step by calculating sugar yield with the best results with a FaeA and xylanase combination (Tabka et al., 2006). Phenolics, such as ferulic, p-coumaric, caffeic, and sinapic acids are released from the wall of the plant through feruloyl esterases. These phenolic compounds in the kingdom of plants are commonly dispersed and are increasingly being looked upon in the fields of fruit, hygiene, cosmetics, and drug applications. Ferulic acid can serve various biological roles, including UV absorbing, antioxidant, and anti-inflammatory functions. It is one of beer's main antioxidant components, although its production during storage is triggered by orange juice. The antioxidant function of phenolic acids is mostly attributed to their chemical composition and the inclusion of the aromatic ring of hydroxy classes. There is also an increase in antioxidant efficiency of two hydroxy groups on caffeic acid relative to one on ferulic acid (Benoit et al., 2008).

8.29   FOOD AND FEED INDUSTRY

Esterases may promote the division of esters into acid and alcohol in the aqueous solution. Moreover, esterases hydrolyze short-chain acylglycerols instead of long-chain ones and are also distinct from lipases. Esterases play a leading part in the food and alcohol industry, where they are often used to modify oil and fat in different fruit juices and create fragrances and flavors (Raveendran et al., 2018). The esteric bond between ferolic acid and various polysaccharides in plant cell walls is breached by feruloyl esterases, an essential category of enzymes from the esterase family. As feruloyl esterases, lignocellulosic biomass hydrolyses are unavoidable in the management of waste (Faulds, 2010).

In a metagenomic library from the Cow Rumen Microbial Community, Cheng et al. (2012) examined the behavior of feruloyl esterase and determined that feruloyl esterase, which may release acid from wheat straw, could be protease resistance. Due to its strong pH, thermal stability, and protease tolerance, this specific esterase has great commercial applications. Diverse methyl or ethyl esters of short-chain fatty acids provide fruity flavor in cheese manufacture. Ethyl esters and thioesters are known to be generated by bacteria. The new thermostable esterase from the highly thermotolerant *Bacillus licheniformis* heterologously expressed in *E. coli* was generated for the development of short-chain flavor esters by Alvarez-Macarie and Baratti (2000). Alvarez-Macarie and Baratti recorded feruloyl esterase, the precursor to vanillin, the flavor compound found in food and drink, is one of the key enzymes of ferulic acid biosynthesis. Microbial synthesis of ferulylesterase has been confirmed by several researchers (Raveendran et al., 2018).

The basic criterion for animal feed is fiber digestibility. Failed ingestion can impede animal development and trigger immunological stress, leading to feed conversion in animals and thus limiting farmers' profitability. Ferulic and hydroxycinnamic acids may promote animal health by themselves. However, feruloylation is a major inhibitor of the ruminant digestive system on plant cell walls, chiefly with an increased drilling diet (Dilokpimol et al., 2016). The addition of FAEs or FAE enzyme cocktails may enhance the access of major enzymes that degrade the chain, leading to increased fiber digestion and bioavailability of phytonutrients, accelerating animal development, and reducing immune stress (Jayaraman et al., 2015).

8.30 FLAVORINGS AND ALCOHOLIC BEVERAGE INDUSTRY

FAEs have surprisingly been utilized both to remove odors and to improve the fragrance of many seasonings. FAEs have often been used to enhance the aroma. Flavor and smell are essential to performance in the luxurious fermented seasonings, in particular, in Japanese rice wines (sake and mirin) and in the alcoholic beverage industry. Ferulic acid and its by-products, including 4-vinyl guaiacarol, vanillic acid, and vanillin are the main aroma components of these drugs. FAEs may be used as a koji generating FAE or as an intermediate along with xylanases and cellulases in the saccharification phase to enhance the discharge of fertilizers from the cell wall of rice and other cereal grain and turn them in fermentation and aging to aromatic derived products (Kanauchi, 2012). Figure 8.2 illustrates the potential application of esterases in various fields.

8.31 CONCLUSION

Esterases are multiple, widely used enzymes, as detailed above. In high-quality applications in industrial domains and fine chemicals, esterases are becoming more significant. Syntheses of key fatty acid esters, which are used in many chemical, medicinal, cosmetic, and food products, are rapidly replacing older harsh chemical processes. Interesting characteristics of pure esterase may be used for commercial use. There is a need to conduct novel research work on esterases to take advantage of their potential applications in various fields.

8.32 ACKNOWLEDGMENTS

Consejo Nacional de Ciencia y Tecnología (MX) is thankfully acknowledged for partially supporting this work under Sistema Nacional de Investigadores (SNI) program awarded to Hafiz M. N. Iqbal (CVU: 735340).

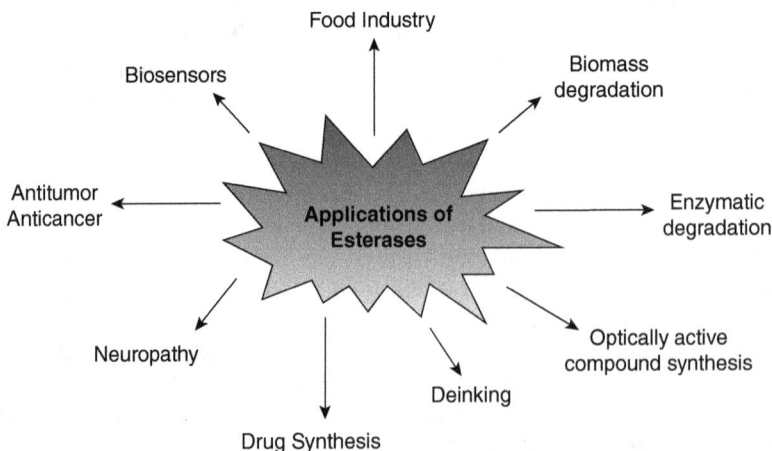

FIGURE 8.2 Application of esterases.

8.33   CONFLICT OF INTERESTS

The authors declare no conflicting interests.

REFERENCES

Ahmed, E. H., Raghavendra, T., & Madamwar, D. (2010). An alkaline lipase from organic solvent tolerant Acinetobacter sp. EH28: Application for ethyl caprylate synthesis. *Bioresource Technology*, *101*(10), 3628–3634.

Aïzoun, N., Aïkpon, R., Padonou, G. G., Oussou, O., Oké-Agbo, F., Gnanguenon, V., . . . Akogbéto, M. (2013). Mixed-function oxidases and esterases associated with permethrin, deltamethrin and bendiocarb resistance in Anopheles gambiae sl in the south-north transect Benin, West Africa. *Parasites & Vectors*, *6*(1), 1–11.

Akassoglou, K., Malester, B., Xu, J., Tessarollo, L., Rosenbluth, J., & Chao, M. V. (2004). Brain-specific deletion of neuropathy target esterase/swisscheese results in neurodegeneration. *Proceedings of the National Academy of Sciences*, *101*(14), 5075–5080.

Aldridge, W. N. (1953). Serum esterases. 1. Two types of esterase (A and B) hydrolysing p-nitrophenyl acetate, propionate and butyrate, and a method for their determination. *Biochemical Journal*, *53*(1), 110–117.

Al-Rashedi, N. A., Munahi, M. G., & AH ALObaidi, L. (2020). Prediction of potential inhibitors against SARS-CoV-2 endoribonuclease: RNA immunity sensing. *Journal of Biomolecular Structure and Dynamics*, 1–14.

Alvarez-Macarie, E., & Baratti, J. (2000). Short chain flavour ester synthesis by a new esterase from Bacillus licheniformis. *Journal of Molecular Catalysis B: Enzymatic*, *10*(4), 377–383.

Benoit, I., Danchin, E. G., Bleichrodt, R. J., & de Vries, R. P. (2008). Biotechnological applications and potential of fungal feruloyl esterases based on prevalence, classification and biochemical diversity. *Biotechnology Letters*, *30*(3), 387–396.

Bhardwaj, H., Gupta, R., & Tiwari, A. (2012). Microbial population associated with plastic degradation. *Scientific Reports*, *1*, 1–4.

Binod, P., Palkhiwala, P., Gaikaiwari, R., Nampoothiri, K. M., Duggal, A., Dey, K., & Pandey, A. (2013). Industrial enzymes-present status and future perspectives for India. *Journal of Scientific & Industrial Research*, *72*.

Bolanča, I., & Bolanča, Z. (2004). *Chemical and Enzymatic Deinking Flotation of Digital Prints*. 4th International DAAAM Conference "Industrial Engineering—Innovation as Competitive Edge for SME". Tallinn, Estonia (pp. 173–176).

Carvalho, Ana K. F., Da Rós, Patricia C. M., Teixeira, Larissa F., Andrade, Grazielle S. S., Zanin, Gisella M., & de Castro, Heizir F. (2013). Assessing the potential of non-edible oils and residual fat to be used as a feedstock source in the enzymatic ethanolysis reaction. *Industrial Crops and Products*, *50*: 485–493.

Cheng, F., Sheng, J., Cai, T., Jin, J., Liu, W., Lin, Y., . . . Shen, L. (2012). A protease-insensitive feruloyl esterase from China Holstein cow rumen metagenomic library: Expression, characterization, and utilization in ferulic acid release from wheat straw. *Journal of Agricultural and Food Chemistry*, *60*(10), 2546–2553.

Colovic, M. B., Krstic, D. Z., Lazarevic-Pasti, T. D., Bondzic, A. M., & Vasic, V. M. (2013). Acetylcholinesterase inhibitors: Pharmacology and toxicology. *Current Neuropharmacology*, *11*(3), 315–335.

Dandavate, V., Jinjala, J., Keharia, H., & Madamwar, D. (2009). Production, partial purification and characterization of organic solvent tolerant lipase from Burkholderia multivorans V2 and its application for ester synthesis. *Bioresource Technology*, *100*(13), 3374–3381.

Davis-Lorton, M. (2015). An update on the diagnosis and management of hereditary angioedema with abnormal C1 inhibitor. *Journal of Drugs in Dermatology: JDD*, *14*(2), 151–157.

De Luca, V., & Mandrich, L. (2020). Lipases/esterases from extremophiles: Main features and potential biotechnological applications. In *Physiological and Biotechnological Aspects of Extremophiles* (pp. 169–181). San Diego, CA 92101, United States, Academic Press.

Dilokpimol, A., Mäkelä, M. R., Aguilar-Pontes, M. V., Benoit-Gelber, I., Hildén, K. S., & de Vries, R. P. (2016). Diversity of fungal feruloyl esterases: Updated phylogenetic classification, properties, and industrial applications. *Biotechnology for Biofuels*, *9*(1), 1–18.

Ding, J., Zhou, Y., Zhu, H., Deng, M., Long, L., Yang, Y., . . . Huang, Z. (2019). Identification and characterization of an acetyl esterase from Paenibacillus sp. XW-6–66 and its novel function in 7-aminocephalosporanic acid deacetylation. *Biotechnology Letters*, *41*(8), 1059–1065.

Ellidag, H. Y., Eren, E., Aydin, O., Yıldırım, M., Sezer, C., & Yilmaz, N. (2014). Multiple myeloma: Relationship to antioxidant esterases. *Medical Principles and Practice*, *23*(1), 18–23.

Faiz, O., Colak, A., Saglam, N., Çanakçi, S., & Belduz, A. O. (2007). Determination and characterization of thermostable esterolytic activity from a novel thermophilic bacterium Anoxybacillus gonensis A4. *BMB Reports*, *40*(4), 588–594.

Fan, X., Liu, X., Huang, R., & Liu, Y. (2012). Identification and characterization of a novel thermostable pyrethroid-hydrolyzing enzyme isolated through metagenomic approach. *Microbial Cell Factories*, *11*(1), 1–11.

Faulds, C. B. (2010). What can feruloyl esterases do for us? *Phytochemistry Reviews*, *9*(1), 121–132.

Gandolfi, R., Gaspari, F., Franzetti, L., & Molinari, F. (2000). Hydrolytic and synthetic activities of esterases and lipases of non-starter bacteria isolated from cheese surface. *Annals of Microbiology*, *50*(2), 183–190.

Glynn, P. (2007). Axonal degeneration and neuropathy target esterase. *Arhiv za higijenu rala i toksikologiju*, *58*(3), 355–358.

Gopalan, N., & Nampoothiri, K. M. (2016). Biotechnological production of enzymes using agro-industrial wastes: Economic considerations, commercialization potential, and future prospects. In *Agro-Industrial Wastes as Feedstock for Enzyme Production* (pp. 313–330). San Diego, CA 92101, United States, Academic Press.

Greiner, A. J., Richardson, R. J., Worden, R. M., & Ofoli, R. Y. (2010). Influence of lysophospholipid hydrolysis by the catalytic domain of neuropathy target esterase on the fluidity of bilayer lipid membranes. *Biochimica et Biophysica Acta (BBA)-Biomembranes*, *1798*(8), 1533–1539.

Hallin, P. F., & Ussery, D. W. (2004). CBS genome atlas database: A dynamic storage for bioinformatic results and sequence data. *Bioinformatics*, *20*(18), 3682–3686.

Heidrich, J. E., Contos, L. M., Hunsaker, L. A., Deck, L. M., & Vander Jagt, D. L. (2004). Inhibition of pancreatic cholesterol esterase reduces cholesterol absorption in the hamster. *BMC Pharmacology*, *4*(1), 1–9.

Henne, A., Brüggemann, H., Raasch, C., Wiezer, A., Hartsch, T., Liesegang, H., . . . Fritz, H. J. (2004). The genome sequence of the extreme thermophile Thermus thermophilus. *Nature Biotechnology*, *22*(5), 547–553.

Hong, S. H., Lee, H. J., An, J., Lim, I., Borlongan, C., Aboody, K. S., & Kim, S. U. (2013). Human neural stem cells expressing carboxyl esterase target and inhibit tumor growth of lung cancer brain metastases. *Cancer Gene Therapy*, *20*(12), 678–682.

Hotelier, T., Nègre, V., Marchot, P., & Chatonnet, A. (2010). Insecticide resistance through mutations in cholinesterases or carboxylesterases: Data mining in the ESTHER database. *Journal of Pesticide Science*, 1006190139–1006190139.

Hotelier, T., Renault, L., Cousin, X., Negre, V., Marchot, P., & Chatonnet, A. (2004). ESTHER, the database of the α/β-hydrolase fold superfamily of proteins. *Nucleic Acids Research*, *32*(suppl_1), D145–D147.

Howell, L. L., Nye, J. A., Stehouwer, J. S., Voll, R. J., Mun, J., Narasimhan, D., . . . Woods, J. H. (2014). A thermostable bacterial cocaine esterase rapidly eliminates cocaine from brain in nonhuman primates. *Translational Psychiatry*, *4*(7), e407.

Jacob, N., & Prema, P. (2006). Influence of mode of fermentation on production of polygalacturonase by a novel strain of streptomyces lydicus. *Food Technology & Biotechnology*, *44*(2).

Jayaraman, S., Mukkalil, R., & Chirakkal, H. (2015). Use of ferulic acid esterase to improve performance in monogastric animals. *U.S. Patent Application No. 14/522,968*.

Kanauchi, M. (2012). Characteristics and role of feruloyl esterase from Aspergillus awamori in Japanese spirits, 'Awamori' production. *Scientific, Health and Social Aspects of the Food Industry. Rijeka: InTech*, 145–162.

Khodami, A., Morshed, M., & Tavanaei, H. (2001). Effects of enzymatic hydrolysis on drawn polyester filament yarns. *Iranian Polymer Journal*, *10*(6).

Kim, Y. J., Choi, G. S., Kim, S. B., Yoon, G. S., Kim, Y. S., & Ryu, Y. W. (2006). Screening and characterization of a novel esterase from a metagenomic library. *Protein Expression and Purification*, *45*(2), 315–323.

Koga, Y. (2012). Thermal adaptation of the archaeal and bacterial lipid membranes. *Archaea*, *2012*.

Kohli, P. O. O. J. A., & Gupta, R. E. E. N. A. (2016). Medical aspects of esterases: A mini review. *International Journal of Pharmacology and Pharmaceutical Sciences*, *8*, 21–26.

Korban, S. A., Bobrov, K. S., Maynskova, M. A., Naryzhny, S. N., Vlasova, O. L., Eneyskaya, E. V., & Kulminskaya, A. A. (2017). Heterologous expression in Pichia pastoris and biochemical characterization of the unmodified sulfatase from Fusarium proliferatum LE1. *Protein Engineering, Design and Selection*, *30*(7), 477–488.

Krysiak, J., Unger, A., Beckendorf, L., Hamdani, N., von Frieling-Salewsky, M., Redfield, M. M., . . . Linke, W. A. (2018). Protein phosphatase 5 regulates titin phosphorylation and function at a sarcomere-associated mechanosensor complex in cardiomyocytes. *Nature Communications*, *9*(1), 1–14.

Lange, B. M., & Srividya, N. (2019). Enzymology of monoterpene functionalization in glandular trichomes. *Journal of Experimental Botany*, *70*(4), 1095–1108.

Lehmann, S. C., Maraite, A., Steinhagen, M., & Ansorge-Schumacher, M. B. (2014). Characterization of a novel Pseudomonas stutzeri lipase/esterase with potential application in the production of chiral secondary alcohols. *Advances in Bioscience and Biotechnology*, *5*(13), 1009.

Li, B., Wang, Y. H., Liu, H. T., Xu, Y. X., Wei, Z. G., Chen, Y. H., & Shen, W. D. (2010). Genotyping of acetylcholinesterase in insects. *Pesticide Biochemistry and Physiology*, *98*(1), 19–25.

Li, P. Y., Chen, X. L., Ji, P., Li, C. Y., Wang, P., Zhang, Y., . . . Zhang, X. Y. (2015). Interdomain hydrophobic interactions modulate the thermostability of microbial esterases from the hormone-sensitive lipase family. *Journal of Biological Chemistry*, *290*(17), 11188–11198.

López-López, O., Fuciños, P., Pastrana, L., Rúa, M. L., Cerdán, M. E., & González-Siso, M. I. (2010). Heterologous expression of an esterase from Thermus thermophilus HB27 in Saccharomyces cerevisiae. *Journal of Biotechnology*, *145*(3), 226–232.

Ma, X., Huang, X., Moore, Z., Huang, G., Kilgore, J. A., Wang, Y., . . . Gao, J. (2015). Esterase-activatable β-lapachone prodrug micelles for NQO1-targeted lung cancer therapy. *Journal of Controlled Release*, *200*, 201–211.

Mackness, M. I., Thompson, H. M., Hardy, A. R., & Walker, C. H. (1987). Distinction between 'A'-esterases and arylesterases. Implications for esterase classification. *Biochemical Journal*, *245*(1), 293–296.

Mandrich, L., De Santi, C., de Pascale, D., & Manco, G. (2012). Effect of low organic solvents concentration on the stability and catalytic activity of HSL-like carboxylesterases: Analysis from psychrophiles to (hyper) thermophiles. *Journal of Molecular Catalysis B: Enzymatic*, *82*, 46–52.

Maruyama, K., Nakagawa, N., & Hasebe, N. (2019). SUN-156 Apurinic/apyrimidinic endode-oxyribonuclease 1 (APE1), an antioxidant and DNA-repair enzyme, has a renoprotective effect during kidney injury. *Kidney International Reports*, *4*(7), S222.

Matuszczak, E., Tylicka, M., Komarowska, M. D., Debek, W., & Hermanowicz, A. (2020). Ubiquitin carboxy-terminal hydrolase L1—physiology and pathology. *Cell Biochemistry and Function*, *38*(5), 533–540.

Mayer, A. M., & Staples, R. C. (2002). Laccase: New functions for an old enzyme. *Phytochemistry*, *60*(6), 551–565.

Montella, I. R., Schama, R., & Valle, D. (2012). The classification of esterases: An important gene family involved in insecticide resistance-A review. *Memorias do Instituto Oswaldo Cruz*, *107*(4), 437–449.

Mukherjee, D., Fritz, D. T., Kilpatrick, W. J., Gao, M., & Wilusz, J. (2004). Analysis of RNA exonucleolytic activities in cellular extracts. In *mRNA Processing and Metabolism* (pp. 193–211). Totowa, New Jersey, USA, Humana Press.

Oakeshott, J., Claudianos, C., Campbell, P. M., Newcomb, R. D., & Russell, R. (2010). Biochemical genetics and genomics of insect esterases. *Comprehensive Molecular Insect Science*, 5.

Pancholy, S. K., & Lynd, J. Q. (1971). Microbial esterase detection with ultraviolet fluorescence. *Applied Microbiology*, *22*(5), 939–941.

Panda, T., & Gowrishankar, B. S. (2005). Production and applications of esterases. *Applied Microbiology and Biotechnology*, *67*(2), 160–169.

Park, H. J., Jang, H. R., Park, S. Y., Kim, Y. B., Lee, H. Y., & Choi, C. S. (2020). The essential role of fructose-1, 6-bisphosphatase 2 enzyme in thermal homeostasis upon cold stress. *Experimental & Molecular Medicine*, *52*(3), 485–496.

Punta, M., Coggill, P. C., Eberhardt, R. Y., Mistry, J., Tate, J., Boursnell, C., . . . Finn, R. D. (2012). The Pfam protein families database. *Nucleic acids research*, *40*(D1), D290–D301.

Qureshi, B. M., Behrmann, E., Schöneberg, J., Loerke, J., Bürger, J., Mielke, T., . . . Heck, M. (2018). It takes two transducins to activate the cGMP-phosphodiesterase 6 in retinal rods. *Open Biology*, *8*(8), 180075.

Rajai, M., Goudarzian, A. H., Robatjazi, M., & Akbari, H. (2019). Formulation and preparation of oil in water cream sample based on OPH enzyme and evaluate enzyme performance in the cream. *Tabari Biomedical Student Research Journal*, *1*(2), 27–30.

Ramnath, L., Sithole, B., & Govinden, R. (2017). Classification of lipolytic enzymes and their biotechnological applications in the pulping industry. *Canadian Journal of Microbiology*, *63*(3), 179–192.

Ramos Tombo, G. M., Schär, H. P., & Ghisalba, O. (1987). Application of Microbes and microbial esterases to the preparation of optically active N-acetylindoline-2-carboxylic acid. *Agricultural and Biological Chemistry*, *51*(7), 1833–1838.

Ranjitha, P., Karthy, E. S., & Mohankumar, A. (2009). Purification and partial characterization of esterase from marine vibrio fischeri. *Modern Applied Science*, *3*(6).

Rather, S. A., Masoodi, F. A., Rather, J. A., Ganaie, T. A., Akhter, R., & Wani, S. M. (2020). Proteins as Enzymes. *Food Biopolymers: Structural, Functional and Nutraceutical Properties*, 299.

Raveendran, S., Parameswaran, B., Beevi Ummalyma, S., Abraham, A., Kuruvilla Mathew, A., Madhavan, A., . . . Pandey, A. (2018). Applications of microbial enzymes in food industry. *Food Technology and Biotechnology*, *56*(1), 16–30.

Satoh, T., & Hosokawa, M. (2006). Structure, function and regulation of carboxylesterases. *Chemico-Biological Interactions*, *162*(3), 195–211.

Satoh, T., Taylor, P., Bosron, W. F., Sanghani, S. P., Hosokawa, M., & La Du, B. N. (2002). Current progress on esterases: From molecular structure to functi on. https://doi.org/10.1124/dmd.30.5.488.

Sayali, K., Sadichha, P., & Surekha, S. (2013). Microbial esterases: An overview. *International Journal of Current Microbiology and Applied Sciences*, *2*(7), 135–146.

Sharma, K. (2019). Cholinesterase inhibitors as Alzheimer's therapeutics. *Molecular Medicine Reports*, *20*(2), 1479–1487.

Shukla, A. (2012). Characterization of mycobacterial estrases/lipases using combined biochemical and computational enzymology. Bachelor of Technology, Bioinformatics, SRM University, India.

Sibon, O. C., & Strauss, E. (2016). Coenzyme A: To make it or uptake it? *Nature Reviews Molecular Cell Biology*, *17*(10), 605–606.

Singh, S., Solanki, P. R., Pandey, M. K., & Malhotra, B. D. (2006). Cholesterol biosensor based on cholesterol esterase, cholesterol oxidase and peroxidase immobilized onto conducting polyaniline films. *Sensors and Actuators B: Chemical*, *115*(1), 534–541.

Tabka, M. G., Herpoël-Gimbert, I., Monod, F., Asther, M., & Sigoillot, J. C. (2006). Enzymatic saccharification of wheat straw for bioethanol production by a combined cellulase xylanase and feruloyl esterase treatment. *Enzyme and Microbial Technology*, *39*(4), 897–902.

Tallman, K. R., & Beatty, K. E. (2015). Far-red fluorogenic probes for esterase and lipase detection. *ChemBioChem*, *16*(1), 70–75.

Tapin, S., Sigoillot, J. C., Asther, M., & Petit-Conil, M. (2006). Feruloyl esterase utilization for simultaneous processing of nonwood plants into phenolic compounds and pulp fibers. *Journal of Agricultural and Food Chemistry*, *54*(10), 3697–3703.

Testa, B., & Kraemer, S. D. (2007). The biochemistry of drug metabolism—an introduction: Part 3. Reactions of hydrolysis and their enzymes. *Chemistry & Biodiversity*, *4*(9), 2031–2122.

Thierry, A., Collins, Y. F., Mukdsi, M. A., McSweeney, P. L., Wilkinson, M. G., & Spinnler, H. E. (2017). Lipolysis and metabolism of fatty acids in cheese. In *Cheese* (pp. 423–444). San Diego, CA 92101, United States, Academic Press.

Tischler, E. H., Cavanaugh, P. K., & Parvizi, J. (2014). Leukocyte esterase strip test: Matched for musculoskeletal infection society criteria. *JBJS*, *96*(22), 1917–1920.

Toushik, S. H., Lee, K. T., Lee, J. S., & Kim, K. S. (2017). Functional applications of lignocellulolytic enzymes in the fruit and vegetable processing industries. *Journal of Food Science*, *82*(3), 585–593.

Walker, C. H. (1993). The classification of esterases which hydrolyse organophosphates: Recent developments. *Chemico-Biological Interactions*, *87*(1–3), 17–24.

Walker, C. H., & Mackness, M. I. (1983). Esterases: Problems of identification and classification. *Biochemical Pharmacology*, *32*(22), 3265–3269.

Wang, T. P., Su, Y. C., Chen, Y., Severance, S., Hwang, C. C., Liou, Y. M., . . . Wang, E. C. (2018). Corroboration of Zn (II)—Mg (II)-tertiary structure interplays essential for the optimal catalysis of a phosphorothiolate thiolesterase ribozyme. *RSC Advances*, *8*(57), 32775–32793.

Wheelock, C. E., Shan, G., & Ottea, J. (2005). Overview of carboxylesterases and their role in the metabolism of insecticides. *Journal of Pesticide Science*, *30*(2), 75–83.

Whyte, M. P. (2020). Hypophosphatasia: Nature's window on alkaline phosphatase function in humans. In *Principles of Bone Biology* (pp. 1569–1599). San Diego, CA 92101, United States, Academic Press.

Wilkinson, M. G. (2004). Lipolysis and catabolism of fatty acids in cheese. *Cheese: Chemistry, Physics and Microbiology, Volume 1: General Aspects*, *1*, 373.

Wu, L., Li, X., Zhao, M., & Bai, Y. (2020). Grazing regulation of phosphorus cycling in grassland ecosystems: Advances and prospects. *Chinese Science Bulletin*, *65*(23), 2469–2482.

Yi, B. R., Kim, S. U., & Choi, K. C. (2014). Co-treatment with therapeutic neural stem cells expressing carboxyl esterase and CPT-11 inhibit growth of primary and metastatic lung cancers in mice. *Oncotarget*, *5*(24), 12835.

Yu, Q. Y., Lu, C., Li, W. L., Xiang, Z. H., & Zhang, Z. (2009). Annotation and expression of carboxylesterases in the silkworm, Bombyx mori. *BMC Genomics*, *10*(1), 1–14.

Zaccheo, O., Dinsdale, D., Meacock, P. A., & Glynn, P. (2004). Neuropathy target esterase and its yeast homologue degrade phosphatidylcholine to glycerophosphocholine in living cells. *Journal of Biological Chemistry*, *279*(23), 24024–24033.

Zheng, H., & Wu, H. (2010). Gene-centric association analysis for the correlation between the guanine-cytosine content levels and temperature range conditions of prokaryotic species. *BMC Bioinformatics*, *11*(11), 1–10.

9 Soil Microbial Enzymes and Their Importance, Significance, and Industrial Applications

Hemant Dasila, Sarita Joshi, and Sudipta Ramola

ABSTRACT

Soil enzymes are an important class of enzymes. They regulate several vital biochemical and physiological reactions of soil microflora that help in maintaining soil health. One of the important functions of soil enzymes is the regulation of nutrients, including mobilization and decay of organic matter in the soil. Soil enzymes are also industrially important for their wide-scale use. Soil enzymes with desired properties can be produced in a large amount by optimizing fermentation conditions. Soil enzymes are a diverse range of enzymes that are industrially important for the development of varied products, including vitamins, antibiotics, and cosmetics. These include enzymes such as laccase, esterase, dehydrogenase, peroxidase, and esterase. These soil enzymes are also potential candidates for the degradation of toxic and xenobiotic compounds from the environment. The present chapter focuses on the application of soil enzymes for industrial purposes and factors affecting their production.

Keywords: soil enzymes, laccase, dehydrogenase, esterase, monooxygenase, dehalogenase, peroxidase

CONTENTS

9.1   Introduction.. 192
9.2   Industrially Important Soil Enzymes 192
    9.2.1   Microbial Laccase.. 192
        9.2.1.1   Industrial Application of Laccase 194
    9.2.2   Microbial Dehydrogenase.. 194
        9.2.2.1   Industrial Application of Dehydrogenase 195
    9.2.3   Microbial Esterase 195
        9.2.3.1   Industrial Application of Esterase..................................... 196
    9.2.4   Microbial Dehalogenase 196
        9.2.4.1   Industrial Application of Dehalogenase............................ 197
    9.2.5   Oxygenase.. 197

DOI: 10.1201/9781003202998-9

9.2.5.1 Industrial Application of Monooxygenase........................ 198
9.2.6 Microbial Peroxidase.. 198
9.2.6.1 Industrial Application of Peroxidase Enzyme.................. 199
9.3 Factors Affecting Industrial Enzyme Production... 199
9.4 Conclusion and Future Prospect 200
References.. 201

9.1 INTRODUCTION

Soil supports a wide range of life forms. The quality of soil is affected by various biological activities that are very sensitive to different environmental factors. Soil microbiota and enzymes are potential indicators of soil health (Pajares et al., 2011). These soil enzymes play a very important role in maintaining soil fertility and microbiological activities (Benitez et al., 2000). Rhizospheric soil is biologically very active and has diverse soil enzymes (Gianfreda, 2015). The presence of soil enzymes in the rhizosphere region depends on several nutrient sources, such as plant roots, microbes, and animals. Various biochemical processes are carried out by soil enzymes by releasing different enzymes, such as β-xylosidase, β-glucosidase, α-glucosidase, amylase, chitinase, dehydrogenase, urease, protease, glucosaminidase, phosphatase, phenoloxidase, aminopeptidase, N-acetyl-arylsulphatases, and L-leucine (Herold et al., 2014). The industrial use of soil enzymes is important for various purposes, e.g. for degradation of xenobiotic compounds and production of industrial products, including drugs, cosmetics products, dye-decolorizing agents, and fermented products. Optimization of different reaction conditions is also very important in order to increase the production of soil enzymes at the industrial level. The basic outlay of enzyme recovery in a fermenter is shown in Figure 9.1, depicting different steps of cell disruption or cell exclusion, product separation via chromatographic technique, and finally formulation to yield product. Monitoring of soil enzymes is important to know how soil microbiota influence soil properties. Precise quantitative and qualitative estimation of soil enzymes remains a challenge (Baldrian, 2009). Extraction of soil enzyme depends upon the nature of the soil; for example, extraction of soil enzyme from clay soil is difficult as compared to forests soil, which is rich in organic matter (Vepsäläine n et al., 2001). Some commonly used assays to measure soil enzymes include spectrophotometric and fluorometric assays to measure soil hydrolases, such as glucosidase, galactosidase, and glucosaminidase (Dick et al., 2013). In this chapter, some of the important soil enzymes, such as laccase, dehydrogenase, esterase, dehalogenase, monooxygenase, and peroxidase, as well as their sources and industrial application are discussed.

9.2 INDUSTRIALLY IMPORTANT SOIL ENZYMES

9.2.1 MICROBIAL LACCASE

Laccases are the most studied enzymatic system (Williamson, 1994). These are extracellular enzymes secreted by several fungi (Agematu et al., 1993) during some secondary metabolic reactions. Some of the exceptions of fungi that do not secret laccase

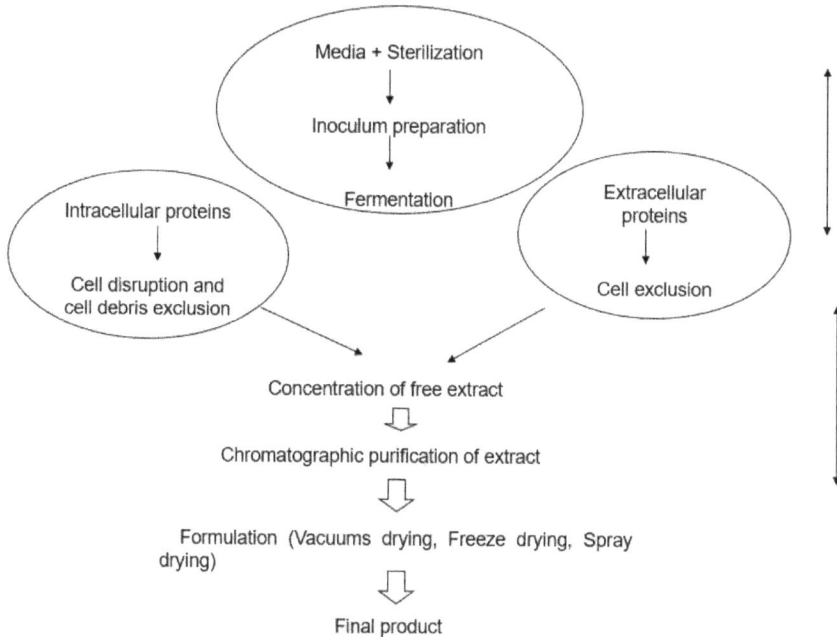

FIGURE 9.1 Upstream and downstream processes regulating enzyme production at the substrate level.

are Chytridiomycetes and Zygomycetes. Laccase production is also reported by some freshwater Ascomycetes. Its production has also been reported in *Melanocarpus albomyces, Gaeumannomyces graminis, Magnaporthe grisea, Neurospora crassa, Ophiostoma novo-ulmi, Monocillium indicum, Podospora anserina*, and *Marginella* (Thakker et al., 1992). Laccase contains 15–30% carbohydrates and has a molecular mass of 60–90 kDa. It is a copper-containing 1,4-benzenediol: oxygen oxidoreductase (EC 1.10.3.2) found in microorganisms and some higher plants. It is glycosylated poly-phenol oxidases having four copper ions per molecule (Couto and Herrera, 2006). It generally performs oxidation reactions by producing oxygen-free radicals that are responsible for polymerization, hydration, and disproportionation. For large-scale production of laccase, two processes are very important i.e. solid-state fermentation and submerged fermentation.

i) **Submerged fermentation:** It involves the use of microorganisms in the presence of high oxygen liquid nutrient medium. Bioreactor operations are run in a continuous manner to get maximum efficiency (Blánquez et al., 2007).

ii) **Solid State fermentation:** It involves the production of enzymes by using some natural substrate like agricultural residue in which fungi can be grown naturally (Brijwani et al., 2010). The use of lignin, cellulose and hemicelluloses in bioreactors allows feasible growth to fungi as they are very good sources of sugar, but the major problem is their limited supply of heat and mass transfer. However different fermenter design has been developed to rectify this. These include tray, inert (nylon) noninert support, expanded bed, and immersion configuration (barley bran). Among these methods, the tray method gives the best results (Couto et al., 2003).

After the cultivation of laccase, next step is its purification. The most common method of purification of the laccase is using ammonium sulfate via precipitation of protein. Some other methods are anion exchange, gel filtration chromatography, and buffer exchange protein. Single-step purification of laccase in *Neurospora crassa* has been reported by using celite chromatography with a 54-fold increase in the enzyme-specific activity (e.g., 333 U mg^{-1}) (Grotewold et al., 1998).

9.2.1.1   Industrial Application of Laccase

  i) **Dye decolorization**: The textile industry requires a high amount of chemicals and water for wet processing. The chemical structure of these dyes offers great resistance to fading when exposed to water, light, and other chemicals. Laccase enzyme is very efficient in dye decolorization and hence can be used in a variety of industries (Domínguez et al., 2005). Laccase produced by *T. versicolor* pellets has been successfully used in treating black liquors discharge for reducing and detoxifying the aromatic compounds, chemical oxygen demand (COD), and color. Reduction in aromatic compounds up to 70–80% and COD up to 60% has been observed (Blánquez et al., 2004). Laccase produced by *T. versicolor* has also been shown to completely discolorize the Tropaeolin O, amaranth, Congo red, Reactive Black 5, and Reactive Blue 15 with no dye sorption (Ramsay and Nguyen, 2002). Hair dyes that contain laccase are less irritant and much easier to handle than conventional hair dye as laccase replaces hydrogen peroxide in the dye formulation (Roriz et al., 2009).

 ii) **Food industry:** Laccase is also used in the food industry for the removal of undesirable phenolic compounds in juice processing, bioremediation of wastewater, and wine stabilization (Couto and Herrera, 2006). Laccase improves the sensory properties and also its functionality (Madhavi and Lele, 2009). In the beer industry, it not only increases the shelf life of beer but also provides stability. Laccase has been used in polyphenol oxidation to remove excess oxygen, as a result of which, the shelf life of beer increases (Minussi et al., 2002).

iii) **Paper and pulp industry:** Laccase and chlorine are used as oxidants for the degradation and separation of lignin. Toxic components released by the paper and pulp industry can be reduced by the laccase enzyme. Laccase can also be used as an alternative substrate to replace ClO_2 in the pulp mill industry (Kunamneni et al., 2007).

9.2.2   Microbial Dehydrogenase

Among all the microbial enzymes present in soil microbiota, microbial dehydrogenases are most vital, as these are the only microbial enzyme class, which occur in all living microorganisms (Moeskops et al., 2010). Microbial dehydrogenase enzymes are involved in most of the redox reactions (Gu et al., 2009). Redox reactions are the most active reaction that occurs continuously in soil, and thus, they are considered to be the general indicators of soil microbiological activity (Moeskops et al., 2010). In the environment, different types of dehydrogenase are present, which differ in terms of their coenzyme type (e.g., flavin adenine dinucleotide or nicotinamide adenine dinucleotide phosphate) and nicotinamide adenine dinucleotide. Dehydrogenases can carry out various dehydrogenation and hydrogenation reactions (Subhani et al., 2001). Dehydrogenase uses oxygen and various other inorganic electron acceptors present in microbial cells (Brzezińska et al., 2001).

9.2.2.1 Industrial Application of Dehydrogenase

There are various classes of dehydrogenase enzymes, but here we have mentioned some important classes of dehydrogenase enzymes that possess industrial applications. Some of them are discussed as follows:

i) **Alcohol dehydrogenase:** Alcohol dehydrogenases (ADHs) (E.C. 1.1.1.x; x = 1 or 2, ADHs) are members of the oxidoreductase family, which catalyze the reversible conversion between aldehydes, alcohols, or ketones by transferring carbon 4 or C-4 hydride from NAD(P)H to the carbonyl carbon of a ketone or aldehyde substrate (Musa and Phillips, 2011). Alcohol dehydrogenase possesses various industrial applications, such as its use in the beverage industry and drug industry. From the last decade, ADHs have been widely used in the asymmetric synthesis of chiral alcohols (Musa and Phillips, 2011; Zhang et al., 2015).

ii) **Drug industry:** Dehydrogenases are widely used in the drug industry for making different types of drugs, including antidepressant drugs. Buspirone, a strong serotonin drug, is produced by the activity of the dehydrogenase enzyme released by the microorganism *Rhizopus stolonifer* SC 13898 (Patel et al., 2005).

9.2.3 MICROBIAL ESTERASE

Esterases (EC 3.1.1.x) are a diverse group of hydrolases that catalyze the formation and cleavage of ester bonds. They are widely distributed in plants, animals, and microorganisms. Esterases possess high stereospecificity, which makes them one of the food biocatalyst agents for the production of pure compounds. Another very interesting fact about esterase is that it does not require any cofactors for reaction to complete. Two major classes of hydrolases that are of utmost importance include "true" esterases (EC 3.1.1.3, carboxyl ester hydrolases) and lipases (EC 3.1.1.1, triacylglycerol hydrolases). Some important properties of esterase are given below (Table 9.1).

Sources of esterase are diverse in nature. It has been isolated from animals, microorganisms, and plants. Microorganisms that produce esterase include fungi, bacteria, and actinomycetes. They express esterase either in an inducible manner or in a constitutive manner. In addition to that, esterase also offers an additional advantage of low cost to grow. Esterase-producing microorganisms can be isolated from

TABLE 9.1

Physical and Solvent Solubility Property of Esterase Enzyme

S. No.	Property	Esterase
1.	Preferred substrates	Simple esters, triglycerides
2.	Interfacial activation/lid	No
3.	Substrate hydrophobicity	High to low
4.	Enantioselectivity	High to low
5.	Solvent stability	High to low

cheese surfaces, oil contamination areas, areas with high garbage load, and marine squid (Ranjitha et al., 2009). Some important genera of microorganisms that are reported to produce esterase include *Bacillus* sp. (Kim et al., 2004), *Streptomyces* sp. (Nishimura and Inouye, 2000), *Thermoanaerobacterium* sp. (Shao and Wiegel, 1995), *Micrococcus* sp. (Fernández et al., 2004), and *Penicillium* sp. (Horne et al., 2002).

9.2.3.1 Industrial Application of Esterase

i) **Chemical industry:** Esterase is used in paper and pulp, leather, baking, and textile industries. Sterol esterase is generally used in paper manufacturing as it is very efficient in hydrolyzing both sterol ester and triglycerides. Cholesteryl esterase and steryl esterase isolated from *Candida rugosa* and *Chromobacterium viscosum* play a very important role in reducing pitch issues that pose a threat to paper machines (Kontkanen et al., 2004).

ii) **Food industry:** Aryl esterase isolated from *S. cerevisae* has been widely used for developing flavor in food and alcoholic beverages (Lomolino et al., 2003).

iii) **Drug development:** Erythromycin esterase isolated from *Pseudomonas* spp. has been used in clinical studies (Kim et al., 2002).

iv) **Cosmetic industry:** Ferulolyl esterase isolated from *A. niger* is frequently used in the cosmetic industry (Giuliani et al., 2001).

9.2.4 MICROBIAL DEHALOGENASE

Dehalogenase enzymes catalyze the removal of a halogen atom from a substrate. These enzymes were first reported in the chloroacetate and fluoroacetate mechanism (Davies and Evans, 1962). Dehalogenase enzymes are classified on the basis of their functions. For example, some dehalogenase are hydrolytic in nature, some play a crucial role in redox reaction, and some are coenzyme dependent in nature. Iodotyrosine dehalogenase (IYD) is very efficient in catalyzing halide removal from halotryosine by the process of reductive dehalogenation (Citterio et al., 2019; Sun et al., 2017). IYD undergoes reductive half reaction, which results in the formation of $IYD\text{-}FMN_{red}$ catalyzed by unidentified reductase by two-electron transfers. For IYD reduction, NADPH act as an electron donor. However, the reduction ability of NADPH was lost after the extraction and purification of IYD from its original cells

(Matthews et al., 2020). Some hydrolytic enzymes have been reported which possess properties, such as thermal stability, catalytic efficacy, and tolerance to extreme saline conditions (Zhang and Kim, 2010).

There are various types of hydrolytic dehalogenases, which can be categorized into haloacid dehalogenases, aliphatic dehalogenases, fluoroacetate dehalogenases, and halohydrin dehalogenases. Aliphatic dehalogenases include haloalkane dehalogenases, which convert haloalkanes to their corresponding proton, halides, and alcohol as a result of catalysis of hydrolytic cleavage of carbon halogen bonds. Haloacid-dehalogenase-like enzymes cover phosphohydrolases found in the bacterial population of marine and other environments. Dehalogenation of fluoroacetate performed by fluoroacetate dehalogenases is capable of hydrolyzing the strongest bond (i.e., carbon-fluorine bond). During the process of degradation, the displacement of chlorine by nucleophilic attack of aromatic halogens is catalyzed by haloaromatic dehalogenase (Oyewusi et al., 2020).

9.2.4.1 Industrial Application of Dehalogenase

i) **Environmental remediation:** The most common persistent chlorinated pollutants are polychlorinated biphenyls, DDT, chlorofluorocarbons, and dioxins. These compounds can concentrate and biomagnify in the living organisms, causing disruption of the endocrine system, and have carcinogenic effects (Hileman, 1993). The most important enzyme for their degradation is the dehalogenase enzyme, which has a catalytic effect on the cleavage of carbon halogen bonds and uses the halogenated organic compounds as electron donors, electron acceptors, and potential carbon sources.

ii) **Production of useful products:** Different dehalogenases have been used to produce useful compounds, such as chiral compounds in herbicides, amino acids, optically active hydroxy acids, and chiral reagents useful in the synthesis of herbicides and pharmaceuticals.

iii) **Applications in drinking water treatment:** Halogenated compounds present in drinking water can be biodegraded by microorganisms producing dehalogenase. Some of the successful examples include the reduction of haloacetic acids by *Afipia* spp. and the removal of 1,2-dichloroethane using *Xanthobacter autotrophicus* GJ 10.

9.2.5 OXYGENASE

Oxygenases are enzymes that insert oxygen atoms into the substrates. Depending on the number of oxygen atoms used in the oxidation, oxygenases can be either monooxygenase (incorporate one oxygen atom into the substrate) or dioxygenase (incorporate two atoms of oxygen into the substrate). Monooxygenase first activates molecular oxygen by donating electrons to it, and then the oxygenation of the substrate occurs. During this process of activation, the formation of reactive oxygen intermediate takes place. The type of intermediate depends on the type of cofactor present in monooxygenase. Some monooxygenases obtain these electrons from the substrate, whereas others from external electron donors. Some of the monooxygenases that do not require any cofactors for their activity are

ActVAorf6 monooxygenase from *Streptomyces coelicolor* A3(2) (Fetzner, 2002), tetracenomycin F1 monooxygenase (TcmH) from *Streptomyces glauscens* (Shen and Hutchinson, 1993), Rv0793 monooxygenase from *Mycobacterium tuberculosis* (Lemieux et al., 2005), and quinol monooxygenase (YgiN) from *E. coli* (Adams and Jia, 2005), whereas others require cofactors, such as heme, flavin, copper, and pterin, for their function.

9.2.5.1 Industrial Application of Monooxygenase

i) **Environmental remediation:** Monooxygenase has high regioselectivity and stereoselectivity, so these act as excellent biocatalysts and perform biotransformation and biodegradation of various pollutants (e.g., heme-containing bacterial oxygenase, such as CYP102 from *Bacillus megaterium* BM3, hydroxylate alkanes, fatty acids, and aromatic compounds) (Urlacher et al., 2004). Two flavin-dependent monooxygenase present in *Rhodococcus erythropolis* IGTS8; dibenzothiophene monooxygenase (DszC) and dibenzothiophene sulfone monooxygenases (DszA) help in biodesulfurization of dibenzothiophene (Kilbane and Jackowski, 1992). Dibenzothiophene monooxygenase (EC 1.14.13.-) helps in the conversion of dibenzothiophene to dibenzothiophene sulphoxide and further to dibenzothiophene sulphone. Dibenzothiophene sulfone is further converted to 2'-hydroxybiphenyl 2-sulfinate (HBPS) by dibenzothiophene sulfone monooxygenase (EC 1.14.13.-). One more enzyme, HBPS desulfinase (DszB), is also involved in the final step of bio-desulfurization and converts HBPS to hydroxybiphenyl (HBP) and sulfate. Similarly, bioremediation of toxic nitrophenols is also initiated by aerobic degradation with the removal of the nitro group by oxygenase activity (Zeyer and Kocher, 1988) of 4-nitrophenol 4-monooxygenase (EC 1.14.13.-) from *Pseudomonas* sp. strain WBC-3 (Zhang et al., 2009). This enzyme converts p-nitrophenol to p-benzoquinone and nitrocatechol to 1,2,4-benzenetriol (Wei et al., 2010) in the presence of NADPH.

ii) **Pharmaceutical industry:** Monooxygenases such as styrene monooxygenase (EC 1.14.13.-) from *Pseudomonas* sp. VLB 120 are also used for the synthesis of pharmaceutical compounds. It is a two-component enzyme—Sty A and Sty B, which act as oxidase and reductase, respectively, and convert styrene to s-styrene oxide (Panke et al., 1998). Other examples of monooxygenases of pharmaceutical interest are 2-hydroxybiphenyl 3-monooxygenase (HbpA, EC 1.14.13.44) from *Pseudomonas azelaica* HBP1 and HbpA from recombinant *E. coli* JM101 that catalyzes *ortho*-hydroxylation of 2-hydroxybiphenyl to 2,3-dihydroxybiphenyl, conversion of different 2-substituted phenols to 3-substituted catechols and the production of 3-tert-butylcatechol, respectively (Held et al., 1998, 1999; Meyer et al., 2003).

9.2.6 Microbial Peroxidase

Microbial peroxidases (EC 1.11.1.7) are oxidoreductases that catalyze the reduction of hydrogen peroxide (H_2O_2) and oxidation of various inorganic and organic compounds (Hamid and Rehman, 2009). Peroxidases are widely distributed throughout nature. These enzymes are produced by different microbes and animals. The majority of microorganisms that produce peroxidase include *Bacillus subtiis, Bacillus sphaericus, Pseudomonas* sp., *Citrobacter* sp., *Cyanobacteria, Candida krusei, Coprinosis cinerea,* and actinomycetes (*Thermobifida fusca, Streptomyces* sp.).

9.2.6.1   Industrial Application of Peroxidase Enzyme

i) **Decolorization of dyes:** The textile industry often releases rhodamine dyes, which are resistant against many decolorizing agents, but microbial peroxidases can decolorize these dyes (Huber and Carré, 2012). Different bacterial peroxidase has been used for decolorization of azo dye and Cr (VI) by peroxidase enzyme secreted by *Brevibacterium casei* under nutrient limiting conditions. *Phanerochaete chrysosporium* RP 78 has also been reported to produce peroxidase enzyme during dye decolorization. *Bacillus* sp. VUS isolated from soil treated with textile effluent also showed degradation capability of different types of dyes (Dawkar et al., 2008).

ii) **Polychlorinated biphenyl degradation:** Pesticides are broad-range substances that are commonly used to control weed, insects, and fungi. Long pesticide exposure to humans is associated with chronic health problems, such as respiratory problems, memory loss, dermatologic conditions, birth defects, and neurologic deficits (McCauley et al., 2006). Biological decomposition of these biphenyl derivatives is the most effective way to eliminate them from the environment. Peroxidases isolated from some fungal species have been reported to transform several pesticides into harmless compounds. Organophosphorus transformation by white rot fungi (Jauregui et al., 2003) and organophosphorus pesticides by chloroperoxidase isolated from *Caldariomyces fumago* have also been reported.

9.3   FACTORS AFFECTING INDUSTRIAL ENZYME PRODUCTION

The activity of soil enzymes is very sensitive to the external environment, including both climatic factors and induced anthropogenic disturbances (Vepsäläinen et al., 2001). Several biological and physicochemical factors also affect soil enzymes. Soil enzymes work at optimum pH and temperature. Some important factors that control industrial enzyme production are described as follows:

i) **Initial pH:** Initial pH of the fermentation medium is one major factor that regulates industrial microbial enzyme production. pH affects the availability of the nutrient in the medium and the transport of various nutrient components across the microbial membrane that directly affect the industrial enzyme production (Niyonzima and More, 2015). Low pH (i.e., 5 to 5.5) was reported in the case of phytase by *Bacillus lehensis* MLB2 (More et al., 2015) and amylase production by *Chryseobacterium* sp. (Hasan et al., 2017). Alkaline pH of about 10.5 has been reported during industrial production of CG-Tase by *Bacillus halodurans* (More et al., 2012). In the case of fungal industrial production, pH ranges from acidic to basic (5 to 9) (Mini et al., 2012). The lower pH range of about 4 to 4.5 has been reported in the production of phytase, pectinase, and tannase by *Aspergillus species* (Sandhya et al., 2015). Any deviation in the optimum initial pH may lead to a decrease in enzyme products due to the disruption of transport mechanisms in the bacterial and fungal membrane.

ii) **Effect of incubation temperature:** Incubation temperature is also a very important environmental parameter in industrial microbial enzyme production. The optimum temperature for incubation in the case of bacteria lies between 30 and 50°C, with the optimum temperature being 37°C in most cases (More et al., 2012). Low incubation temperature of about 20 to 28°C was reported in industrial amylase production by *Bacillus* sp. (Lailaja and Chandrasekaran, 2013). In context to fungi, the optimum incubation temperature range is about 25 to 47°C (Dhital et al., 2013).

iii) **Effect of inoculum:** Inoculum concentration is also one of the important parameters for microbial growth and thus directly affects microbial enzyme production. The optimum inoculum concentration in the case of bacterial industrial ranges from 0.6 to 4% (Beena et al., 2012). However, in the case of lipase production by *Staphylococcus arlettae* JPBW-1, the inoculum size was around 10% (Chauhan et al., 2013). Different inoculum levels were found to be effective with different fungi. Two percent of inoculum was found to be optimum for protease and tannase production by *Aspergillus terreus* gr. (Niyonzima et al., 2013) and *Mucor circinelloides* (El-Refai et al., 2017), respectively. However, 9% of inoculum size was found to be optimum for protease production by *Scopulariopsis* sp. (Niyonzima et al., 2013). Generally, it is found that industrial microbial enzyme production is directly correlated with inoculum concentration till optimum concentration is reached (Niyonzima et al., 2013), accompanied by rapid substrate degradation.

iv) **Effect of carbon source:** The primary energy source for bacterial and fungal growth is carbon, and thus, it plays an important role in the industrial production of enzymes. Different types of carbon sources, such as starch, malt extract, soluble pullulan, carboxymethylcellulose (CMC), sucrose, and glucose, have been used for producing important industrial bacterial enzymes. In some reports, a mixture of bean flour and corn flour, skim milk with glucose, and sesame oil with glucose (Lailaja and Chandrasekaran, 2013) were also used for bacterial enzyme production. For industrial fungal enzymes production, sucrose, lactose, pectin, CMC, lactose, tannate, glucose, and maltose are used. Generally, it is observed that industrial enzymes produced by microorganisms with low carbon content are most effective, and this makes production cost-effective. Sometimes inexpensive substances like green tea leaves, orange peel moistened with molasses, and jamun leaves (Lincoln and More, 2017) are also used.

9.4   CONCLUSION AND FUTURE PROSPECT

In this chapter, an overview is given about the importance of soil enzymes, which are particularly used in the development of industrial products. The growing demand for industrially important soil enzymes may result in modification of optimizing conditions in the fermentation medium. However, with so much advancement in technology, there is room for improvement to quantify maximum soil enzymes. New, integrated approaches include discovery and quantification of soil enzymes, and the application part simultaneously has to be developed, which may result in the full utilization of soil enzymes for both human welfare and environmental restoration.

REFERENCES

Adams, M. A., & Jia, Z. (2005). Structural and biochemical evidence for an enzymatic quinone redox cycle in Escherichia coli: Identification of a novel quinol monooxygenase*[boxs]. *Journal of Biological Chemistry*, *280*(9), 8358–8363.

Agematu, H., Tsuchida, T., Kominato, K., Shibamoto, N., Yoshioka, T., Nishida, H., & Murao, S. (1993). Enzymatic dimerization of penicillin X. *The Journal of Antibiotics*, *46*(1), 141–148.

Baldrian, P. (2009). Ectomycorrhizal fungi and their enzymes in soils: Is there enough evidence for their role as facultative soil saprotrophs? *Oecologia*, *161*(4), 657–660.

Beena, A. K., Geevarghese, P. I., & Jayavardanan, K. K. (2012). Detergent potential of a spoilage protease enzyme liberated by a psychrotrophic spore former isolated from sterilized skim milk. *American Journal of Food Technology*, *7*(2), 89–95.

Benitez, E., Melgar, R., Sainz, H., Gomez, M., & Nogales, R. (2000). Enzyme activities in the rhizosphere of pepper (Capsicum annuum, L.) grown with olive cake mulches. *Soil Biology and Biochemistry*, *32*(13), 1829–1835.

Blánquez, P., Caminal, G., Sarrà, M., & Vicent, T. (2007). The effect of HRT on the decolourisation of the Grey Lanaset G textile dye by Trametes versicolor. *Chemical Engineering Journal*, *126*(2–3), 163–169.

Blánquez, P., Casas, N., Font, X., Gabarrell, X., Sarrà, M., Caminal, G., & Vicent, T. (2004). Mechanism of textile metal dye biotransformation by Trametes versicolor. *Water Research*, *38*(8), 2166–2172.

Brijwani, K., Oberoi, H. S., & Vadlani, P. V. (2010). Production of a cellulolytic enzyme system in mixed-culture solid-state fermentation of soybean hulls supplemented with wheat bran. *Process Biochemistry*, *45*(1), 120–128.

Brzezińska, M., Stępniewska, Z., & Stępniewski, W. (2001). Dehydrogenase and catalase activity of soil irrigated with municipal wastewater. *Polish Journal of Environmental Studies*, *10*(5), 307–311.

Chauhan, M., Chauhan, R. S., & Garlapati, V. K. (2013). Evaluation of a new lipase from Staphylococcus sp. for detergent additive capability. *BioMed Research International*, 374967.

Citterio, C. E., Targovnik, H. M., & Arvan, P. (2019). The role of thyroglobulin in thyroid hormonogenesis. *Nature Reviews Endocrinology*, *15*(6), 323–338.

Couto, S. R., & Herrera, J. L. T. (2006). Industrial and biotechnological applications of laccases: A review. *Biotechnology Advances*, *24*(5), 500–513.

Couto, S. R., Moldes, D., Liébanas, A., & Sanromán, A. (2003). Investigation of several bioreactor configurations for laccase production by Trametes versicolor operating in solid-state conditions. *Biochemical Engineering Journal*, *15*(1), 21–26.

Davies, J. I., & Evans, W. C. (1962). Elimination of halide ions from aliphatic halogen-substituted organic acids by an enzyme preparation from *Pseudomonas dehalogenans*. *Biochemical Journal*, *82*(3).

Dawkar, V. V., Jadhav, U. U., Jadhav, S. U., & Govindwar, S. P. (2008). Biodegradation of disperse textile dye Brown 3REL by newly isolated Bacillus sp. VUS. *Journal of applied microbiology*, *105*(1), 14–24.

Dhital, R., Panta, O. P., & Karki, T. B. (2013). Optimization of cultural conditions for the production of pectinase from selected fungal strain. *Journal of Food Science and Technology Nepal, 8,* 65–70.

Dick, W. A., Thavamani, B., Conley, S., Blaisdell, R., & Sengupta, A. (2013). Prediction of β-glucosidase and β-glucosaminidase activities, soil organic C, and amino sugar N in a diverse population of soils using near infrared reflectance spectroscopy. *Soil Biology and Biochemistry, 56,* 99–104.

Domínguez, A., Couto, S. R. & Sanromán, M. (2005). Dye decolorization by Trametes hirsuta immobilized into alginate beads. *World Journal of Microbiology and Biotechnology, 21*(4), 405–409.

El-Refai, H. A., Abdel-Naby, M. A., Mostafa, H., Amin, M. A., & Salem, H. A. A. (2017). Statistical optimization for tannase production by Mucor circinelloides isolate F6–3–12 under submerged and solid state fermentation. *Current Trends in Biotechnology and Pharmacy, 11*(2), 167–180.

Fetzner, S. (2002). Oxygenases without requirement for cofactors or metal ions. *Applied Microbiology and Biotechnology, 60,* 243–257.

Fernández, J., Mohedano, A. F., Fernández-García, E., Medina, M., & Nuñez, M. (2004). Purification and characterization of an extracellular tributyrin esterase produced by a cheese isolate, Micrococcus sp. INIA 528. *International Dairy Journal, 14*(2), 135–142.

Gianfreda, L. (2015). Enzymes of importance to rhizosphere processes. *Journal of Soil Science and Plant Nutrition, 15*(2), 283–306.

Giuliani, S., Piana, C., Setti, L., Hochkoeppler, A., Pifferi, P. G., Williamson, G., & Faulds, C. B. (2001). Synthesis of pentylferulate by a feruloyl esterase from Aspergillus niger using water-in-oil microemulsions. *Biotechnology Letters, 23*(4), 325–330.

Grotewold, E., Taccioli, G. E., Aisemberg, G. O., & Judewicz, N. D. (1998). A single-step purification of an extracellular fungal laccase. *MIRCEN Journal of Applied Microbiology and Biotechnology, 4*(3), 357–363.

Gu, Y., Wang, P., & Kong, C. H. (2009). Urease, invertase, dehydrogenase and polyphenoloxidase activities in paddy soil influenced by allelopathic rice variety. *European Journal of Soil Biology, 45*(5–6), 436–441.

Hamid, H., & Rehman, K. U. (2009). Potential applications of peroxidases. *Food Chemistry, 115*(4), 1177–1186.

Hasan, M. M., Marzan, L. W., Hosna, A., Hakim, A., & Azad, A. K. (2017). Optimization of some fermentation conditions for the production of extracellular amylases by using Chryseobacterium and Bacillus isolates from organic kitchen wastes. *Journal of Genetic Engineering and Biotechnology, 15*(1), 59–68.

Held, M., Schmid, A., Kohler, H.P.E., Suske, W., Witholt, B., & Wubbolts, M. G. (1999). An integrated process for the production of toxic catechols from toxic phenols based on a designer biocatalyst. *Biotechnology and Bioengineering, 62*(6), 641–648.

Held, M., Suske, W., Schmid, A., Engesser, K.H., Kohler, H.P.E., Witholt, B., & Wubbolts, M.G. (1998). Preparative scale production of 3-substituted catechols using a novel monooxygenase from Pseudomonas azelaica HBP 1. *Journal of Molecular Catalysis B: Enzymatic, 5*(1–4), 87–93.

Herold, N., Schöning, I., Gutknecht, J., Alt, F., Boch, S., Müller, J., & Schrumpf, M. (2014). Soil property and management effects on grassland microbial communities across a latitudinal gradient in Germany. *Applied Soil Ecology, 73,* 41–50.

Hileman, B. (1993). Concerns broaden over chlorine and chlorinated hydrocarbons. *Chemical & Engineering News, 93,* 11–20.

Horne, I., Sutherland, T. D., Oakeshott, J. G., & Russell, R. J. (2002). Cloning and expression of the phosphotriesterase gene hocA from Pseudomonas monteilii C11. *Microbiology, 148*(9), 2687–2695.

Huber, P., & Carré, B. (2012). Decolorization of process waters in deinking mills and similar applications: A review. *BioResources, 7*(1), 1366–1382.

Jauregui, J., Valderrama, B., Albores, A., & Vazquez-Duhalt, R. (2003). Microsomal transformation of organophosphorus pesticides by white rot fungi. *Biodegradation, 14*(6), 397–406.

Kilbane, J. J., & Jackowski, K. (1992). Biodesulfurization of water-soluble coal-derived material by Rhodococcus rhodochrous IGTS8. *Biotechnology and Bioengineering, 40*(9), 1107–1114.

Kim, G. J., Choi, G. S., Kim, J. Y., Lee, J. B., Jo, D. H., & Ryu, Y. W. (2002). Screening, production and properties of a stereospecific esterase from Pseudomonas sp. S34 with high selectivity to (S)-ketoprofen ethyl ester. *Journal of Molecular Catalysis B: Enzymatic, 17*(1), 29–38.

Kim, H. K., Na, H. S., Park, M. S., Oh, T. K., & Lee, T. S. (2004). Occurrence of ofloxacin ester-hydrolyzing esterase from Bacillus niacini EM001. *Journal of Molecular Catalysis B: Enzymatic, 27*(4–6), 237–241.

Kontkanen, H., Tenkanen, M., Fagerström, R., & Reinikainen, T. (2004). Characterisation of steryl esterase activities in commercial lipase preparations. *Journal of Biotechnology, 108*(1), 51–59.

Kunamneni, A., Ballesteros, A., Plou, F. J., & Alcalde, M. (2007). Fungal laccase-a versatile enzyme for biotechnological applications. *Communicating Current Research and Educational Topics and Trends in Applied Microbiology, 1*, 233–245.

Lailaja, V. P., & Chandrasekaran, M. (2013). Detergent compatible alkaline lipase produced by marine Bacillus smithii BTMS 11. *World Journal of Microbiology and Biotechnology, 29*(8), 1349–1360.

Lemieux, M. J., Ference, C., Cherney, M. M., Wang, M., Garen, C., & James, M. N. (2005). The crystal structure of Rv0793, a hypothetical monooxygenase from M. tuberculosis. *Journal of Structural and Functional Genomics, 6*(4), 245–257.

Lincoln, L., & More, S. S. (2017). Screening and Enhanced Production of neutral invertase from Aspergillus sp. by utilization of Molasses-A by-product of Sugarcane industry. *Advances in Bioresearch, 8*(4).

Lomolino, G., Rizzi, C., Spettoli, P., Curioni, A., & Lante, A. (2003). Cell vitality and esterase activity of Saccharomyces cerevisiae is affected by increasing calcium concentration. *Agro Food Industry Hi-Tech, 14*(6), 32–35.

Madhavi, V., & Lele, S. S. (2009). Laccase: Properties and applications. *Bio Resources, 4*(4), 1694–1717.

Matthews, A., Saleem-Batcha, R., Sanders, J. N., Stull, F., Houk, K. N., & Teufel, R. (2020). Aminoperoxide adducts expand the catalytic repertoire of flavin monooxygenases. *Nature Chemical Biology, 16*(5), 556–563.

McCauley, L. A., Anger, W. K., Keifer, M., Langley, R., Robson, M. G., & Rohlman, D. (2006). Studying health outcomes in farmworker populations exposed to pesticides. *Environmental Health Perspectives, 114*(6), 953–960.

Meyer, A., Held, M., Schmid, A., Kohler, H. P. E., & Witholt, B. (2003). Synthesis of 3-tert-butylcatechol by an engineered monooxygenase. *Biotechnology and Bioengineering, 81*(5), 518–524.

Mini, K. D., Mini, K. P., & Mathew, J. (2012): Screening of fungi isolated from poultry farm soil for keratinolytic activity. *Advances in Applied Science Research* (3), 2073–2077.

Minussi, R. C., Pastore, G. M., & Durán, N. (2002). Potential applications of laccase in the food industry. *Trends in Food Science & Technology, 13*(6–7), 205–216.

Moeskops, B., Sukristiyonubowo, Buchan, D., Sleutel, S., Herawaty, L., Husen, E., Saraswati, R., Setyorini, D., & Neve, S. D. (2010). Soil microbial communities and activities under intensive organic and conventional vegetable farming in West Java, Indonesia. *Applied Soil Ecology, 45*(2), 112–120.

More, S. S., Niraja, R., Evelyn, C., Byadgi, A., Shwetha, V., & Das Mangaraj, S. (2012). Isolation, purification and biochemical characterization of CGTase from Bacillus halodurans. *Hrvatski časopis za prehrambenu tehnologiju, biotehnologiju i nutricionizam, 7*(1–2), 90–97.

More, S. S., Shrinivas, S., Agarwal, A., Chikkanna, A., Janardhan, B., & Niyonzima, F. N. (2015). Purification and characterization of phytase from Bacillus lehensis MLB2. *Biologia, 70*(3), 294–304.

Musa, M. M., & Phillips, R. S. (2011). Recent advances in alcohol dehydrogenase-catalyzed asymmetric production of hydrophobic alcohols. *Catalysis Science & Technology, 1*(8), 1311–1323.

Nishimura, M., & Inouye, S. (2000). Inhibitory effects of carbohydrates on cholesterol esterase biosynthesis in Streptomyces lavendulae H646-SY2. *Journal of Bioscience and Bioengineering, 90*(5), 564–566.

Niyonzima, F. N., & More, S. S. (2015). Detergent-compatible proteases: Microbial production, properties, and stain removal analysis. *Preparative Biochemistry and Biotechnology, 45*(3), 233–258.

Niyonzima, F. N., More, S. S., & Muddapur, U. (2013). Optimization of fermentation culture conditions for alkaline lipase production by Bacillus flexus XJU-1. *Current Trends in Biotechnology and Pharmacy, 7*(3), 793–803.

Oyewusi, H. A., Wahab, R. A., & Huyop, F. (2020). Dehalogenase-producing halophiles and their potential role in bioremediation. *Marine Pollution Bulletin, 160*, 111603.

Pajares, S., Gallardo, J. F., Masciandaro, G., Ceccanti, B., & Etchevers, J. D. (2011). Enzyme activity as an indicator of soil quality changes in degraded cultivated Acrisols in the Mexican Trans-volcanic Belt. *Land Degradation & Development, 22*(3), 373–381.

Panke, S., Witholt, B., Schmid, A., & Wubbolts, M. G. (1998). Towards a biocatalyst for (S)-styrene oxide production: characterization of the styrene degradation pathway of Pseudomonas sp. strain VLB120. *Applied and Environmental Microbiology, 64*(6), 2032–2043.

Patel, R., Chu, L., Nanduri, V., Li, J., Kotnis, A., Parker, W., Liu, M., & Mueller, R. (2005). Enantioselective microbial reduction of 6-oxo-8-[4-[4-(2-pyrimidinyl)-1-piperazinyl] butyl]-8-azaspiro [4.5] decane-7, 9-dione. *Tetrahedron: Asymmetry, 16*(16), 2778–2783.

Ramsay, J. A., & Nguyen, T. (2002). Decoloration of textile dyes by Trametes versicolor and its effect on dye toxicity. *Biotechnology Letters, 24*(21), 1757–1761.

Ranjitha, P., Karthy, E. S., & Mohankumar, A. (2009). Purification and characterization of the lipase from marine vibrio fischeri. *International Journal of Biology, 1*(2), 48.

Roriz, M. S., Osma, J. F., Teixeira, J. A., & Couto, S. R. (2009). Application of response surface methodological approach to optimise Reactive Black 5 decolouration by crude laccase from Trametes pubescens. *Journal of Hazardous Materials, 169*(1–3), 691–696.

Sandhya, A., Sridevi, A., Suvarnalatha, D., & Narasimha, G. (2015). Production and optimization of phytase by Aspergillus niger. *Pharmacy Letter, 7*, 148–153.

Shao, W., & Wiegel, J. (1995). Purification and characterization of two thermostable acetyl xylan esterases from Thermoanaerobacterium sp. strain JW/SL-YS485. *Applied and Environmental Microbiology, 61*(2), 729–733.

Shen, B., & Hutchinson, C. R. (1993). Tetracenomycin F1 monooxygenase: Oxidation of a naphthacenone to a naphthacenequinone in the biosynthesis of tetracenomycin C in Streptomyces glaucescens. *Biochemistry, 32*, 6656–6663.

Subhani, A., Changyong, H., Zhengmiao, Y., Min, L., & El-Ghamry, A. (2001). Impact of soil environment and agronomic practices on microbial/dehydrogenase enzyme activity in soil. A review. *Pakistan Journal of Biological Sciences, 4*(3), 333–338.

Sun, Z., Su, Q., & Rokita, S. E. (2017). The distribution and mechanism of iodotyrosine deiodinase defied expectations. *Archives of Biochemistry and Biophysics, 632*, 77–87.

Thakker, G. D., Evans, C. S., & Rao, K. K. (1992). Purification and characterization of laccase from Monocillium indicum Saxena. *Applied Microbiology and Biotechnology*, *37*(3), 321–323.

Urlacher, V. B., Lutz-Wahl, S., & Schmid, R. D. (2004). Microbial P450 enzymes in biotechnology. *Applied Microbiology and Biotechnology*, *64*(3), 317–325.

Vepsäläinen, M., Kukkonen, S., Vestberg, M., Sirviö, H., & Niemi, R. M. (2001). Application of soil enzyme activity test kit in a field experiment. *Soil Biology and Biochemistry*, *33*(12–13), 1665–1672.

Wei, M., Zhang, J. J., Liu, H., & Zhou, N. Y. (2010). para-Nitrophenol 4-monooxygenase and hydroxyquinol 1, 2-dioxygenase catalyze sequential transformation of 4-nitrocatechol in Pseudomonas sp. strain WBC-3. *Biodegradation*, *21*(6), 915–921.

Williamson, P. R. (1994). Biochemical and molecular characterization of the diphenol oxidase of Cryptococcus neoformans: Identification as a laccase. *Journal of Bacteriology*, *176*(3), 656–664.

Zeyer, J. O. S. E. F., & Kocher, H. P. (1988). Purification and characterization of a bacterial nitrophenol oxygenase which converts ortho-nitrophenol to catechol and nitrite. *Journal of Bacteriology*, *170*(4),1789–1794.

Zhang, C., & Kim, S. K. (2010). Research and application of marine microbial enzymes: Status and prospects. *Marine Drugs*, *8*(6), 1920–1934.

Zhang, J. J., Liu, H., Xiao, Y., Zhang, X.E., & Zhou, N. Y. (2009). Identification and characterization of catabolic para-nitrophenol 4-monooxygenase and para-benzoquinone reductase from Pseudomonas sp. strain WBC-3. *Journal of bacteriology*, *191*(8), 2703–2710.

Zhang, R., Xu, Y., & Xiao, R. (2015). Redesigning alcohol dehydrogenases/reductases for more efficient biosynthesis of enantiopure isomers. *Biotechnology Advances*, *33*(8), 1671–1684.

10 Application of Microbial Enzymes in Industry and Antibiotic Production

*Rishendra Kumar, Lokesh Tripathi,
and Pankaj Bhatt*

ABSTRACT

Microbial enzymes are biocatalysts that play a significant role in different industries for metabolic and biochemical reactions. Microbial enzymes are commonly used by industries for making bread, beer, cheese, dairy products, polymers, organic acids, and medicine. These enzymes have the ability to convert a substrate (compound) into different products at higher rates of reaction. The fermentation process is classified on the basis of specific parameters. Submerged fermentation (SmF) and solid-state fermentation (SSF) are two processes that involve the production of enzymes. Downstream processing methods are used for the purification and recovery of enzymes. Antibiotics are also produced by fermentation. Antibiotics encompass a broad range of antimicrobial compounds produced by living microorganisms as secondary metabolites. Most of the antibiotics are produced by *Streptomyces* species. *Streptomyces* produces bioactive compounds that are antitumor, antihypertensive, antifungal, antibacterial, and so on. Due to the constant use of antibiotics and the resistance developed by microorganisms, new research and discoveries of new antibiotics are needed to deal with the next generation of microorganisms.

Keywords: *Streptomyces*, microbial enzymes, fermentation, industry, application, antibiotics

CONTENTS

10.1 Introduction.. 208
10.2 Production of Microbial Enzymes 209
    10.2.1 Production of Enzymes.. 210
    10.2.2 Production Process of Microbial Enzymes in Industries................. 210
        10.2.2.1 Solid-State Fermentation 211
        10.2.2.2 Submerged Fermentation 211
10.3 Applications of Microbial Enzymes in Industries and Medicine 212
10.4 Antibiotic Production.. 215
    10.4.1 Antimicrobial Agents and Producing Strains 215

DOI: 10.1201/9781003202998-10

10.5 Conclusion 216
References.. 216

10.1 INTRODUCTION

The present human era considers microbial enzymes as a potential biocatalyst required by all living organisms for many breakdown reactions and syntheses. Microorganisms, like bacteria, fungi, yeast, and actinomycetes, are the preferred sources of enzymes and have the capacity to fulfill and revolutionize the current requirement of the industrial sector. Enzymes (except ribozymes) are active protein molecules that can speed up the biochemical reactions by lowering the activation energy of the reaction. Industries used microbial enzymes in brewing, baking, cheese and alcohol making, and acid production for ages (Fernandes, 2010).

In Japan, natural enzymes have been used for the fermentation of products like sake, Japanese schnapps brewed from rice, for thousand of years. Today, enzymes are widely used in various industries, such as food, tannery, beverages, textiles, feed and waste treatment, cosmetics, detergent, pulp and paper, pharmaceuticals, and confectionery (Pandey et al., 2006). Enzymes utilized today are generally isolated from animals, like pepsin, pancreatin, rennin, trypsin, and chymosin, and from plants, like papain, ficin, lipoxygenase, and bromelain, but most enzymes, such as glucoamylase, pectinase, α-amylase, and protease, are produced by microorganisms. These microbial enzymes are more active and stable, thus representing the heart of industries (Thomas et al., 2013). Sixty percent of commercial enzymes are produced by fungi, followed by 24% by bacteria, 4% by yeast, and 2% by Streptomyces. Some of the known commercial enzymes are lipases, proteases, pectin, and rennet (milk-clotting enzymes) (Demirci et al., 2014).

An enzyme has its own specificity, a specific property that helps it to recognize a particular substrate. Lipases have a significant role in the industries due to their specificity and stability under different conditions. Amylases catalyze the breakdown of starch into sugars. However, microbial enzymes used in the food and pharmaceutical industries are eco-friendly. Still, some enzymes can cause allergic reactions, and hence, protective measures have to be taken in their production and application.

Moreover, microbial enzymes are gaining interest in industries due to several driven factors, such as cost-effectiveness, nontoxicity, reduced process time, consumer demands, and higher quality, which has a positive influence on the growth rate of the global enzyme market (Miguel et al., 2013) Choi et al., 2015. The growth rate of enzymes in the global market is estimated to be about $4.2 billion, which has increased by 7% approximately over the time period till 2020, and reached $6.2 billion (2015: Industrial Enzyme Market). In India, Advanced Enzymes Technologies Ltd. is the highest manufacturer and exporter of enzymes. The demands of enzymes generally depend on their application and consumption. Enzymes include amylase, laccase, proteases, xylanase, glucoamylase, pectinase, lactoperoxidase, lipases, esterase, and phytase, which have a 30% demand in food, beverages, pharma, animal feed, dairy, textiles, polymers, and biofuel (CRISIL Research, 2013).

Despite all the successful production and advanced processes done with the help of microorganisms, infectious diseases caused by them still remain the leading problems worldwide. Bacterial infections cause about 17 million deaths annually. Bioactive compounds have a huge range of molecular targets in the control of infectious diseases. *Bacillus subtilis* and *Streptomyces griseus* can produce around 70 types of secondary metabolites, including a group of antibiotics, like b-lactams, carbapenems, alkaloids, naphthalenes, phenazines, peptides, and quinolones. Microbial fermentation is the process of producing antibiotics on microbial genera inhabiting soil that undergoes morphological differentiation (Walia et al., 2013; Garg et al., 2016).

Some antibiotics are used widely and, therefore, are being produced in large volumes, like penicillin, whereas others like vancomycin have specific applications and are produced in limited amounts. More than 10,000 different antibiotics have been isolated from fungi and gram-positive and gram-negative bacteria (Walsh, 2003.). Almost 60 to 80% of all the antibiotics used are produced by *Streptomyces*, *Bacillus*, and the filamentous fungi (Miyadoh, 1993), while 15% are obtained by groups related to actinomycetes: *Nocardia, hermoactinomyces, Micromonospora, Streptoverticillium*, and so on (Berdy, 1980).

Today, realizing the correlation between antibiotic used and resistance developed in some by microorganisms, much research devoted to new antibiotic discoveries and designs need to be done to replace and treat the next generation of resistant bacteria, such as *S. pneumoniae, K. pneumoniae, S. aureus*, and *M. tuberculosis*. Genomics has made a global approach to analyze antibiotic production (Costelloe et al., 2010). Bioinformatics and genome-sequenced pathogens help identify new antibiotics and their novel targets for fighting pathogens and their caused diseases (Bush, 2004). Molecular biology and rDNA techniques have also enabled molecular approaches to find new drugs. GlaxoSmithKline is one of such companies that have conducted studies with the antibiotic (GKS299423) acting on topoisomerase II, to prevent the resistance bacteria (Jones, 2010).

In this chapter, we find out about microbial enzymes and their production and application in industries and medicine. We also discuss antibiotic production and its future perspective.

10.2 PRODUCTION OF MICROBIAL ENZYMES

An ever-evolving branch of science and technology is enzyme technology. Enzymes are present in every living cell. Bacteria and fungi are the greatest sources of industrial enzymes. Biotechnology and bioinformatics have a great influence in improving the application of enzymes. They are easy to handle and could be grown in large tanks at a high growth rate without light. Each microbial strain can produce enzymes in large quantities and are oxidizing, metabolizing, and hydrolyzing in nature.

The global biotechnology company Novoenzymes mostly uses the fungus *Aspergillus oryzae* and the bacterium *Bacillus subtilis* for enzyme production. These are harmless to humans and have a great capacity for producing enzymes. (www.novozymes.com/en/about-us/our-business/what-are-enzymes/Pages/creat ing-the-perfect-enzyme.aspx).

The ideal microorganisms are not easy to select; some require a huge amount of nutrition, some produce undesirable by-products that may disturb the industrial process, some take a long time to grow, some are wild types and are thus not easy to domesticate, while some produce a very less amount of enzymes. The new enzymes from this huge diversity could be explored by using some tools like metagenomic screening for microbial enzymes from different habitats (e.g., volcanic vents, cow rumen, arctic tundra) (Demirci et al., 2014), genome mining in around 2,000 sequenced microbial genomes (Rodríguez et al., 2014), and exploring extremophile diversity (Liu et al., 2014).

10.2.1 Production of Enzymes

An alkaline protease called subtilisin, secreted by *Bacillus licheniformis*, was the first enzyme produced in the enzyme industry. It breaks down the proteinaceous substrate and is used as a detergent. The bacteria *Bacillus licheniformis* is also used to secrete α-amylase, which is a highly thermostable enzyme used for the breaking down of starch to oligosaccharides. Due to the very high production of enzymes from *Bacillus* strains, this genus has been considered very important for enzyme industries for decades. Glucoamylases further break down starch into glucose and are obtained from the fungal strain of *Aspergillus* (Sonenshein et al., 1993).

A *Trichoderma* strain, which is a fungus, produces an acidic cellulose complex which is used as an additive in detergents and in the treatment of textiles. The cellulases are used for bioethanol production. The enzyme glucose isomerase produced by *Streptomyces* acts as a catalyst in the conversion of glucose into fructose. The strains that have been discussed above fulfill the ideal conditions of enzymes like *Bacillus* have the tendency of surviving in the adverse conditions in a dormant state. Similarly, the fungal strains of *Aspergillus* and *Trichoderma* produce enzymes regulated with differentiation. The enzyme synthesis mechanisms include RNA synthesis, protein synthesis, and post-translational processing, which are highly conserved (Rehm and Reed, 1985).

The enzymes are different from each other in terms of their molecular structure, glycosylation degree, polypeptide chain number, and isoelectric point. Due to the differences of enzymes in their physical characteristics, their synthesis processes are done separately.

10.2.2 Production Process of Microbial Enzymes in Industries

Earlier, the commercial microbial enzymes were processed by culture methods, but now submerged culture methods are widely being used. Microorganisms used are "generally recognized as safe" strains due to their significant application in the food, bakery, and feed industries (Pandey et al., 2008; Singhania et al., 2010). The molds are cultivated on moist solid substrates (e.g., takamine uses wheat bran for producing fungal enzymes). This is referred to as the most suitable substrate as compared to other fibrous materials. The other ingredients, like acid, buffer, or nutrient salts, are added to regulate the pH and beet cosettes or soybean meal to rouse enzyme synthesis.

Boidin has invented a processed method where bacteria were cultivated in culture vessels containing liquid media with adjusted composition to production desired enzyme. Different strains of *B. subtilis* are employed on different media, depending on the desired microbial enzymes (amylase or protease). The fermentation process has been used for the production of industrial enzymes under control conditions. Maximum numbers of microbial enzymes come from a few genera, including *Aspergillus, Trichoderma, Bacillus, Streptomyces,* and *Kluyveromyces* (Sarrouh et al., 2012).

Two very important fermentation technologies are available—submerged fermentation (SmF) and solid-state fermentation (SSF)—with several benefits. Almost every industry uses the SmF process for enzyme production; however, the SSF process has gained popularity due to some specific applications in a few industries (Pandey, 2003; Singhania et al., 2009; Thomas et al., 2013b).

10.2.2.1   Solid-State Fermentation

Solid-state fermentation (SSF) has the potential for enzyme production. It involves the microorganism's cultivation on a solid substrate, like rice, wheat bran, or grains (Renge et al., 2012; Subramaniyam and Vimala, 2012). This method is alternatively used instead of submerged fermentation. It includes productivity in high volume, high product concentration and less effluents. In China, SSF has been used extensively to produce brewed food (Rodrìguez Coute and Sanromán, 2006). It is a very ancient process used to fulfill human needs.

The moisture present on the substrate plays an important role in the method. High or low water content may cause adverse effects on microbial activity. The enzymes produced in industries by the SSF method are proteases, pectinases, glucoamylases, and cellulases (Renge et al., 2012; Suganthi et al., 2011).

Bacteria, yeast, and fungi are used for producing enzymes by the SSF method. The selection of a particular strain depends on the substrate nature and the environmental conditions. The cellulases, pectinases, xylanases, and so on are the enzymes generally produced by the fungal strains *Trichoder*ma sp. and *Aspergillus* sp. (Singh and Kumar, 2019). Some potential bacterial species and filamentous fungi, such as *Bacillus licheniformis, Pseudomonas, Aspergillus* sp., and *Rhizopus* sp., respectively, produce amylolytic enzymes on a commercial basis. *A. niger* and *A. oryzae* are used to develop lactase, which is rich in infancy and is known as the brush border enzyme (Singh and Kumar, 2019; Mehaia and Cheryan, 1987). Bacterial α-amylase produced by fungal and bacterial strains is used for starch liquefaction due to its high temperature tolerance. The genetically modified strains are also used nowadays for achieving high productivity at less cost to enzyme production (Kaur and Gill, 2019).

Agricultural residues are the best substrates for SSF in the production of enzymes. Some substrates include sugar cane bagasse, grapevine trimmings, rice straw, rice husk, coconut oil cake, coconut coir pith, and starch (Pandey et al., 1999).

10.2.2.2   Submerged Fermentation

Submerged fermentation (SmF) is carried out in large vessels (1,000 m^3 volumes) in the presence of an excess of free water. It cultures microorganisms in a liquid nutrient broth. Most industries carry out enzyme production by SmF. Their bioreactors are well developed and can be controlled by several parameters, like temperature, pH, oxygen consumption, and carbon dioxide. Oxygen supply is very essential in SmF.

In SmF, four different methods of growing microorganisms are used: batch culture, fed-batch culture, perfusion-batch culture, and continuous culture. Continuous culture is the one that gives balanced growth, with less or no fluctuation in nutrients, metabolites, cell quantity, or biomass. Sterilized nutrients used in this fermentation are raw materials like maize, sugars, and soya (Renge et al., 2012).

10.3 APPLICATIONS OF MICROBIAL ENZYMES IN INDUSTRIES AND MEDICINE

Many industries, including food, feed, chemical, and pharmaceutical, depend on the enzymes for their products (Table 10.1). Most of the enzymes are biosynthesized in the industries by the microbes like bacteria, including *Pseudomonas, Clostridium,* and *Bacillus*, and fungi, including *Trichoderma, Penicillium*, and *Aspergillus*. The total market price of industrial enzymes in 2010 was $3.3 billion and was estimated to cross $4.4 billion by 2015 (BBC Research, 2011). Major enzyme-producing companies are Novozymes and Danisco in Denmark, with 45% and 17%, respectively, followed by Genencor and DSM in USA and Netherlands and BASF in Germany (Binod et al., 2008; Binod et al., 2013; BCC-Business Communications Company, Inc., 2009).

TABLE 10.1

Application of Microbial Enzymes and Its Production Method in Industries

Industry	Enzyme	Source Organism	Method of Production	Applications
Detergent	Protease, cellulase	*B. amyloliquifaciens, Trichoderma reesei, T. viride, Aspergillus* sp.	SmF	Removal of protein stains by degradation Loosening up cellulose fibers to remove dirt and color brightening
Leather	Protease, lipase	*A. oryzae, A. terreus, A. flavus, Pseudomonas* sp., *Staphylococcus* sp.	SSF, SmF	Soaking, bating, and dehairing of animal skin
Biofuel	Cellulase, xylanase	*Trichoderma reesei, T. viride, Aspergillus* sp., *Myeciliophthora thermophila, Bacillus* sp.	SmF, SSF	Hydrolyzing cellulosic biomass to generate glucose, Hydrolyzing hemicelluloses to generate pentose
Food	α- and β-amylases, glucose isomerase	*Bacillus licheniformis, Bacillus coagulans*	SSF, SmF	Production of types of syrups from starch and sucrose
Fruit juice, coffee	Cellulase, xylanase, pectinase, laccase	*Trichoderma reesei, Myeciliophthora thermophila, A. niger, Penicillium* sp.	SmF	Clarification and juice extraction, pressing, and filtration

Industry	Enzyme	Source Organism	Method of Production	Applications
Paper and pulp	Laccase, peroxidase	*A. nidulans*, *Aspergillus* sp., *Basidiomycetes*	SSF, SmF	Polymerization of materials with wood fibers
Milk	B-galactosidase	*Aspergillus oryzae*	SSF, SmF	Wheat bran and rice husk
Prebiotic	Fructosyl transferase	*Bacillus subtilis*, *Aspergillus oryzae*	SSF	Rice bran, wheat bran
Cosmetics	Endoglycosidase, lipase, laccase	*Mucor hiemalis*	SSF	Dental-related care, skin care, hair coloring
Waste management	Amyloglucosidase, cutinase, oxygenase	*A. niger*, *Fusarium solani*, *F. pisi*, *Pseudomonas* sp., *Rhodococcus* sp.	SSF, SmF	Starch hydrolysis for bioremediation Degradation of plastics and polycaprolactone Degradation of halogenated contaminants

Source: SSF, solid-state fermentation; SmF, submerged fermentation, Source (Singh and Kumar, 2019; Nizamuddin et al., 2008; Esawy et al., 2013)

By the catalytic action of enzymes like peroxidases, cellulases, oxidases, lipases, amylases, and proteases, the chemical bonds of detergents are broken down in thermophilic (more than 50°C) and alkaline (pH 7–11) conditions. The two bacterial lipases produced by *P. mendocina* and *P. alcaligenes* were introduced by Genencor International in 1995. An enzyme produced by *Bacillus* is mannanase, added to detergent, can easily remove food stains (guar gum) (Kirk et al., 2002). Lipases combined with proteases or peptidases can prepare good flavored cheese with low or no level of bitterness (Araujo et al., 2008; Tzanov et al., 2001).

The three recombinant lipases currently used in the food industry are produced by *Rhizomucor miehi*, *Thermomyces lanuginisus*, and *Fusarium oxysporum*. Nippon Paper Industries commonly use lipase enzyme produced by *Candida rugosa* (Gutiérrez et al., 2009).

The textile industry uses cellulases and laccases for denim finishing and textile effluents decolorization and bleaching (Rodriguez-Couto and Toca-Herrera, 2006). Other cellulases, like endoglucanases, break down cellulose chains randomly (Rubin, 2008); cellobiohydrolase liberates glucose from both sides of the cellulose chain (Wilson, 2009); beta-glucosidases also produce glucose from celluloses (Zhang, 2011).

Hypocrea jecorina (*Trichoderma reesei*) is mainly used by the industry as a source for cellulases and hemicellulases, which helps in the depolymerization of plant biomass to simple sugars (Zhang et al., 2006; Kumar et al., 2008; Kubicek et al., 2009).

Laccases, xylanases, and lipases are used in pulp industries for removing pitch (Gutiérrez et al., 2009). Laccases oxidize the environmental pollutants and are used for the bioremediation of herbicides and pesticides and the detoxification of liquid wastes from the textile, pulp, and petrochemical industries (Rodriguez-Couto and Toca-Herrera, 2006).

The phytases, α-galactosidases, glucanases, xylanases, and polygalacturonases are the enzymes used for poultry in the feed industry (Selle and Ravindran, 2007). Cellulases produced by *Trichoderma reesei* depolymerize the plant biomass to simple sugars (Zhang et al., 2006; Kumar et al., 2008; Kubicek et al., 2009).

Earlier calf enzymes (rennin) were used in making cheese and has been replaced by microbial enzymes (proteases) produced by *B. subtilis, Endothia parasitica*, and *Mucor miehei* (Rao et al., 1998). β-galactosidases from *Kluyveromyces lactis* hydrolyze lactose from milk. Galactosidases from *S. carlsbergensis* and xylose are used for beet sugar and fructose syrup, respectively.

In the pharmaceutical industry, enzymes have a vital role in the production of therapeutic drugs. Several microbial enzymes, such as proteolytic enzymes (for dead skin removal and burns), fibrinolytic enzymes (clot busting), dextranase (tooth decay), and rhodanase (cyanide poisoning), are used frequently in medical and involved in the synthesis of alcohol, ester, and lactones (Okafor, 2007; Cambou and Klibanov, 1984; Saxena et al., 1999).

L-dihydroxy phenylalanine (L-DOPA), a precursor for dopamine production, is a potential drug used to treat Parkinson's disease and also to control the myocardium neurogenic injury. L-DOPA is produced by tyrosinase (Ikram-ul-Haq et al., 2002; Zaidi et al., 2014). The β-lactam antibiotics like penicillins and cephalosporins are largely produced in pharmaceutical companies with the help of enzymes (Volpato, 2010).

One of the most important and effective applications of enzymes is the production of chiral medicines. The chiral alcohols, amines, epoxides, carboxylic acids, and so on are prepared with the help of lipases, esterases, proteases (Kirchner et al., 1995; Zheng and Xu, 2011; Kirk et al., 2002). Atorvastatin is an active ingredient of Lipitor, which is a cholesterol-lowering drug, produced enzymatically by ketone reductases, halohydryn dehalogenases, and glucose dehydrogenases (Ma et al., 2010).

The yeast lipases are used for the synthesis of pharmaceuticals and herbicides by catalyzing the butanolysis process, which produces enantiopure 2-chloro and 2-bromopropionic acids. Xylanase's importance has increased enormously in various applications, including clarification of juices, rumen digestion improvement, and bioconversion of lignocelluloses, agricultural residues for fuels and chemicals (Ma et al., 2010).

The thermostable plant xerophytic isoforms of laccase enzymes are used for the purpose of dyeing, pulping, textile, and bioremediation (Suenaga et al., 2007; Gomes et al., 2003). Ligninolytic enzymes are known for hydrolyzing the lignocelluloses, especially the biodegradation of lignin polymer, which is recalcitrant toward chemicals. This group of enzymes has great applications in pollution control, bioremediation, and treatment of hazardous chemicals, such as phenols, dyes, and xenobiotics (Gomes et al., 2003; Van den Burg, 2003). These enzymes are also used in the process of lignin separation and degradation from plant material in paper and pulp industries (Marrs et al., 1999).

Synthesis of 2-arylpropanoic acids (ibuprofen and naproxen) is produced by lipases from *C. antarctica* or *Pseudomonas* sp. (Gooding et al., 2010). Enzyme engineering company Codexis collaborates with Pfizer for the production of 2-methyl pentanol, an important intermediate for drug manufacturing by pharmaceutical companies (Gooding et al., 2010). β-glucosidases obtained from *Aspergillus* strains play an significant role in industrial sectors, such as food, pharmaceutical, and biofuel, due to their wide variety of glycoside substrates.

10.4 ANTIBIOTIC PRODUCTION

Chemotherapeutic agents obtained from secondary metabolism produced by microbes are the key area in medical microbiology. These chemotherapeutic agents are known as antimicrobials or antibiotics.

Alexander Fleming discovered penicillin from *Penicillium notatum* in 1929, and the efforts of Florey and Chain made in 1938 have taken as a possible step for large-scale production of antibiotics. Since from the 1940s, many newly discovered antibiotics produced by soil bacteria, such as actinomycin, neomycin, and streptomycin, and later received the Nobel Prize in Medicine for the discovery of streptomycin, the first aminoglycoside effective against tuberculosis (Schatz et al., 1944).

The early years from 1950 to 1960 were truly the golden period of antibiotic discovery and research, as one-half of the drugs used today were discovered in this period. Some discoveries were streptomycin from *S. griseus* (Waksman et al., 1946), chloramphenicol (Duggar, 1948), chlortetracycline from *S. aureofaciens* (Ehrlich et al., 1947), cephalosporin C from *C. acremonium* (Newton and Abraham, 1955), erythromycin *from S. erythraea* and vancomycin from *Amycolatopsis orientalis* (Geraci et al., 1956).

Antibiotics have two strengths of action, bacteriostatic and antibiotic. Bacteriostatic medicine does not kill the cells but only inhibits bacterial growth and gives enough time for the host to build up immunity. Bactericidal antibiotics can kill by rupturing the bacterial cell (lysis) (Waksman and Woodruff, 1940).

Antibiotics have a broad or narrow spectrum, depending on their action toward the target, including cell wall synthesis, protein synthesis, and DNA replication. Classes of antimicrobial agents are lactams, such as penicillins, cephalosporins, aminoglycosides, sulfonamides, and tetracyclines; glycopeptides, such as vancomycin; and macrolides, such as erythromycin.

Antibiotic production has increased due to the increasing number of drug-resistant pathogens. Therefore, there are new startup companies that have improved the model for antibiotic production and marketing (Reed et al., 2002). Semi-synthetic aminoglycoside antibiotics include amikacin, netilmicin, and isepamicin, which were developed to shield against resistance organisms (Miller et al., 1976; Leggett, 2015) (Van Bambeke et al., 2017). Neomycin was produced by immobilized *Streptomyces fradiae* (Srinivasulu et al., 2003), used as antibacterial agents. Most of the penicillin is converted into effective antibiotics, like ampicillin, amoxicillin, and cephalexin, through chemical synthesis. These converted antibiotics have improved the efficacy or pharmacokinetic properties.

10.4.1 ANTIMICROBIAL AGENTS AND PRODUCING STRAINS

Because of sudden breakthroughs, a large number of antibiotics have been discovered, out of which 300 antibiotics have been found for clinical use only and are currently being produced on a large scale. Improvement and production of antibiotics on a commercial scale need full knowledge of the microbes. More than 180 various types of bioactive compounds are produced by the Actinomycetes (about 75%) alone. Some of active compounds are aminoglycosides (e.g., streptomycin and kanamycin) (Nanjawade et al., 2010), ansamycins (e.g., rifampin) (Floss and Yu, 1999), anthracyclines (Kremer et al., 2001), β-lactam (e.g., cephalosporins) (Kollef, 2009), macrolides (e.g., erythromycin) (Mims et al., 2004), and tetracycline. Streptomycetes produce numerous biologically active drugs, such as chloramphenicol, amphotericin B, natamycin, nystatin, neomycin, tunicamycin, bafilomycin, ivermectin, rapamycin, tetracycline, daptomycin, and clavulanic acid (enzyme-inhibiting drug) (Yoo et al., 2015). *Streptomyces* species incude *S. coelicolor*, *S. lividans*, *S. albus*, *S. rimosus*, *S. aureofaciens*, *S. avermitilis*, and *S. venezuelae*. Some antibiotics have been extracted from *Bacillus* strains, such as moenomycins, difficidins, bacillomycins, and bacillaenes. Another genus, *Mycobacterium*, is a bacterium with very interesting antibiotic productivities. Eighty percent of *Mycobacterium* produce various bioactive compounds against microorganism (Reichenbach and Höfle, 1999).

10.5 CONCLUSION

Microorganisms provide a great range of enzymes that are used across industries, such as detergent, food, feed, leather, fine chemicals, acids, dairy, and medicine. Enzymes have special properties that help in several processes in industries and provide high specificity, improve quality, and reduce waste. Demands of microbial enzymes and antibiotics still remain high in the global market. The diversity of natural products helps in developing the future drug. However, a wide range of advances in technology, such as genomics, mutagenesis, overexpression of structural genes, proteomics, ribosome engineering, and emerging recombinant DNA techniques that facilitate the new and improved discovery of microbial enzymes and antibiotics from nature via genomes and metagenomics. Current advanced technologies, such as CRISPR/Cas, should be considered for genome editing and increasing production and can be utilized further in the field of natural microbial products, which are still a steadfast resource for novel compounds in medicine discovery.

REFERENCES

Araujo, R., Casal, M., Cavaco-Paulo, A., 2008. Application of enzymes for textiles fibers processing. *Biocatal. Biotechnol.* 26: 332–349.

BBC Research, 2011. *In Report BIO030F—Enzymes in Industrial Applications: Global Markets*. World Enzymes. Freedonia Group: Cleveland, OH, USA, 2011. www.idiverse.com/html/target_industrial_enzymes.htm (date 19.04.15).

BCC-Business Communications Company, Inc., 2009, In: *Report FOD020C-World Markets for Fermentation Ingredients*. Wellesley, MA 02481.

Berdy, J., 1980. Recent advances in and prospects for antibiotic research. *Process Biochemistry*, 15: 28–35.

Binod, P., Palkhiwala, P., Gaikaiwari, R., Nampoothiri, K.M., Duggal, A., Dey, K., et al., 2013. Industrial enzymes: Present status and future perspectives for India. *J. Sci. Ind. Res.*, 72(5): 271–286.

Binod, P., Singhania, R.R., Soccol, C.R., Pandey, A., 2008. *Industrial Enzymes. Advances in Fermentation Technology*. Asiatech Publishers, New Delhi, India.

Bush, K., 2004. Antibacterial drug discovery in the 21st century. *Clin Microbiol Infect.*, 10: 10–7.

Cambou, B., Klibanov, A.M., 1984. Preparative production of optically active esters and alcohols using esterase-catalyzed stereospecific transesterification in organic media. *J Am Chem Soc*, 106: 2687–2692.

Choi, J.M., Han, S.S., Kim, H.S., 2015. Industrial applications of enzyme biocatalysis: Current status and future aspect. *Biotechnol Adv*, 33: 1443–1454.

Costelloe, C., Metcalfe, C., Lovering, A., Mant, D., Hay, A.D., 2010. Effect of antibiotic prescribing in primary care on antimicrobial resistance in individual patients: Systematic review and meta analysis. *BMJ*. 340: c2096.

CRISIL Researchm 2013. www.crisil.com/Ratings/Brochureware/ News/CRISIL%20 Research_ipo_grading_rationale_advanced_enzymes_technologies.pdf.

Demirci, A., Izmirlioglu, G., Ercan, D., 2014. *Fermentation and Enzyme Technologies in Food Processing*. Food processing: Principles and applications. 2nd ed. New York: Wiley, pp. 107–136.

Duggar, B.M., 1948. Aureomycin; a product of the continuing search for new antibiotics. *Ann. N. Y. Acad. Sci.*, 51: 177–181.

Ehrlich, J., Bartz, Q.R., Smith, R.M., Joslyn, D.A., Burkholder, P.R., 1947. Chloromycetin, a new antibiotic from a soil actinomycete. *Science*, 106: 417.

Esawy, M.A., Abdel-Fattah, A.M., Ali, M.M., Helmy, W.A., Salama, B.M., Taie, H.A.A., Hashem, A.M., Awad, G.E.A., 2013. Levan sucrase optimization using solid state fermentation and Levan biological activities studies. *Carbohydrate Polymers*, 96: 332–341.

Fernandes, P., 2010. *Enzymes in Food Processing: A Condensed Overview on Strategies for Better Biocatalysts*. Enzyme Research. doi: 10.4061/2010/862537.

Floss, H.G., Yu, T., 1999. Lessons from the rifamycin biosynthetic gene cluster. *Curr Opin Chem Biol.*, 3(5): 592.

Garg, G., et al., 2016. Microbial pectinases: An ecofriendly tool of nature for industries. *3 Biotech*, 6(1): 47.

Geraci, J.E., Heilman, F.R., Nichols, D.R., Wellman, E.W., Ross, G.T., 1956. Some laboratory and clinical experiences with a new antibiotic, vancomycin. *Antibiot. Annu.*, 90–106.

Gomes, I., Gomes, J., Steiner, W., 2003. Highly thermostable amylase and pullulanase of the extreme thermophilic eubacterium Rhodothermus marinus: Production and partialcharacterization. *Bioresour. Technol.*, 90: 207–214.

Gooding, O., Voladri, R., Bautista, A., Hopkins, T., Huisman, G., Jenne, S., Ma, S., Mundorff, E.C., Savile, M.M., Truesdell, S.J., 2010. Development of a practical biocatalytic process for (R)-2-methylpentanol. *Org. Process Res. Dev.*, 14: 119–126.

Gutiérrez, A., del Río, J.C., Martínez, A.T., 2009. Microbial and enzymatic control of pitch in the pulp and paper industry. *Appl. Microbiol. Biotechnol.*, 82: 1005–1018.

Ikram-ul-Haq, Ali, S., Qadeer, M.A., 2002. Biosynthesis of l-DOPA by Aspergillus oryzae. *Bioresour Technol*, 85(1): 25–29.

Jones, D., 2010. The antibacterial lead discovery challenge. *Nat Rev Drug Discov.*, 9: 751–752. www.novozymes.com/en/about-us/our-business/what-are-enzymes/Pages/creating-the-perfect-enzyme.aspx (date 19.04.15).

Kaur, H., Gill, P.K., 2019. *Microbial Enzymes in Food and Beverages Processing*, in *Engineering Tools in the Beverage Industry*. Elsevier, pp. 255–282. doi: 10.1016/C2017-0-02377-7.

Kirchner, G., Scollar, M.P., Klibanov, A., 1995. Resolution of racemic mixtures via lipase catalysis in organic solvents. *J. Am. Chem. Soc.* 107: 7072–7076.

Kirk, O., Borchert, T.V., Fuglsang, C.C., 2002. Industrial enzyme applications. *Curr. Opin. Biotechnol.*, 13: 345–351.

Kollef, M.H., 2009. New antimicrobial agents for methicillin-resistant Staphylococcus aureus. *Crit Care Resusc.*, 11(4): 282–286.

Kremer, L.C., van Dalen, E., Offringa, M., et al., 2001. Anthracycline induced clinical heart failure in a cohort of 607 children: Long-term follow-up study. *J Clin Oncol.,* 19(1): 191–196.

Kubicek, C.P., Mikus, M., Schuster, A., Schmoll, M., Seiboth, B., 2009. Metabolic engineering strategies for the improvement of cellulase production by Hypocrea jecorina. *Biotechnol. Biofuels,* 2: 19–31.

Kumar, R., Singh, S., Singh, O.V., 2008. Bioconversion of lignocellulosic biomass: Biochemical and molecular perspectives. *J. Ind. Microbiol. Biotechnol.,* 35: 377–391.

Leggett, J.E., 2015. 25—Aminoglycosides. In *Mandell, Douglas, and Bennett's Principles and Practice of Infectious Diseases.* 8th ed., eds. J.E. Bennett, R. Dolin, and M.J. Blaser. Content Repository Only!, Philadelphia, PA. 310.e–321.e.

Liu, Y., et al., 2014. A novel approach for improving the yield of Bacillus subtilis transglutaminase in heterologous strains. *Journal of Industrial Microbiology & Biotechnology,* 41(8): 1227–1235.

Ma, S.K., Gruber, J., Davis, C., Newman, L., Gray, D., Wang, A., Grate, J., Huisman, G.W., Sheldon, R.A., 2010. A green-by-design biocatalytic process for atorvastatin intermediate. *Green Chem.,* 12: 81–86.

Marrs, B., Delagrave, S., Murphy, D., 1999. Novel approaches for discovering industrial enzymes. *Curr. Opin. Microbiol.,* 2: 241–245.

Mehaia, M., Cheryan, M., 1987. Production of lactic acid from sweet whey permeate concentrates. *Process Biochemistry,* 22(6): 185–188.

Miguel, A.S.M., Souza, T., Costa Figueiredo, E.V. da, Paulo Lobo, B.W., Maria, G., 2013. Enzymes in bakery: Current and future trends. In *Food Industry,* ed. I. Muzzalupo. InTech, UK, pp. 287–321.

Miller, G.H., Arcieri, G., Weinstein, M.J., Waitz, J.A., 1976. Biological activity of netilmicin, a broad-spectrum semisynthetic aminoglycoside antibiotic. *Antimicrob. Agents Chemother.,* 10: 827–836.

Mims, C., Dockrell, H.M., Goering, R.V., et al., 2004. Attacking the enemy: Antimicrobial agents and chemotherapy: Macrolides. In *Chapter Medical Microbiology.* 3rd ed. Mosby Ltd., London, p. 489.

Miyadoh, S., 1993. Research on antibiotic screening in Japan over the last decade: A producing microorganisms approach. *Actinomycetologica,* 7: 100–106.

Nanjawade, B.K., Chandrashekhara, S., Ali, M.S., Prakash, S.G., Fakirappa, V.M., 2010. Isolation and morphological characterization of antibiotic producing Actinomycestes. *Trop J Pharm Res.,* 9(3): 231–236.

Newton, G.G., Abraham, E.P., 1955. Cephalosporin C, a new antibiotic containing Sulphur and D-alpha-aminoadipic acid. *Nature,* 175: 548.

Nizamuddin, S., Sridevi, A., Narasimha, G., 2008. Production of β-galactosidase by Aspergillus oryzae in solid-state fermentation. *African Journal of Biotechnology,* 7(8): 1096–1100.

Okafor, N., 2007. *Biocatalysis: Immobilized enzymes and immobilized cells.* Modern Ind Microbiol Biotechnol, CRC Press. p 398. ISBN978-1-57808-434-0 (HC).

Pandey, A., 2003. Solid-state fermentation. *Biochem. Eng. J.,* 13(2–3): 81–84.

Pandey, A., Larroche, C., Soccol, C.R., Dussap, C.G. (Eds.), 2008. *Advances in Fermentation Technology.* Asiatech Publishers, Inc., New Delhi.

Pandey, A., Selvakumar, P., Soccol, C.R., Nigam, P., 1999. Solid-state fermentation for the production of industrial enzymes. *Curr. Sci.,* 77(1): 149–162.

Pandey, A., Webb, C., Soccol, C.R., Larroche, C. (Eds.), 2006. *Enzyme Technology.* Springer Science, USA. 740.

Rao, M.B., Tanksale, A.M., Ghatge, M.S., Deshpande, V.V., 1998. Molecular and biotechnological aspects of microbial proteases. *Microbiol. Mol. Biol. Rev.,* 62: 597–635.

Reed, S.D., Laxminarayan, R., Black, D.J., Sullivan, S.D., 2002. Economic issues and antibiotic resistance in the community. *Ann. Pharmacother.,* 36: 148–154.

Rehm, H.J., Reed, G. (Eds.), 1985. *Biotechnology, A Comprehensive Treatise in Eight Volumes*. VCH Verlagsgesellschaft, Weinheim, Germany.

Reichenbach, H., Höfle, G., 1999. Myxobacteria as producers of secondary metabolites. In *Drug Discovery from Nature*, eds. S. Grabley, R. Thiericke. Springer, Berlin, pp. 149–179.

Renge, V., S. Khedkar, Nandurkar, N.R., 2012. Enzyme synthesis by fermentation method: A review. *Sci Rev Chem Comm*, 2(4): 585e90.

Rodríguez, V., Asenjo, J.A., Andrews, B.A., 2014. Design and implementation of a high yield production system for recombinant expression of peptides. *Microbial Cell Factories*, 13(1): 65.

Rodrìguez Coute, S.R., Sanromán, M.A., 2006. Application of solid-state fermentation to food industry—A review. *Journal of Food Engineering*, 76(3): 291–302.

Rodriguez-Couto, S.R., Toca-Herrera, J.L., 2006. Industrial and biotechnological applications of laccases: A review. *Biotechnol. Adv.*, 24: 500–513.

Rubin, E.M., 2008. Genomics of cellulosic biofuels. *Nature*, 454: 841–845.

Sarrouh, B., Santos, T.M., Miyoshi, A., Dias, R., Azevedo, V., 2012. Up-to-date insight on industrial enzymes applications and global market. *J. Bioprocess Biotech.*, S4: 002.

Saxena, R.K., Ghosh, P.K., Gupta, R. et al., 1999. Microbial lipases: Potential biocatalysts for the future industry. *Curr Sci*, 77: 101–115.

Schatz, A., Bugle, E., Waksman, S.A., 1944. Streptomycin, a substance exhibiting antibiotic activity against gram-positive and gram-negative bacteria. *Proc. Soc. Exp. Biol. Med.*, 55: 66–69.

Selle, P.H., Ravindran, V., 2007. Microbial phytase in poultry nutrition. *Anim. Feed Sci. Technol.*, 135: 1–41.

Singh, P., Kumar, S., 2019. Microbial enzyme in food biotechnology. In *Enzymes in Food Biotechnology*. Elsevier, pp. 19–28.

Singhania, R.R., Patel, A.K., Pandey, A., 2010. The industrial production of enzymes. In *Industrial Biotechnology*, eds. W. Soetaert, E.J. Vandamme. Wiley-VCH Verlag, Weinheim, Germany, pp. 207–226.

Singhania, R.R., Patel, A.K., Soccol, C.R., Pandey, A., 2009. Recent advances in solid-state fermentation. *Biochem. Eng. J.*, 44(1): 13–18.

Sonenshein, A.L., Hoch, J.A., Losick, R. (Eds.), 1993. *Bacillus Subtilis and Other Gram-Positive Bacteria: Biochemistry, Physiology and Molecular Genetics*. ASM Press, Washington, DC.

Srinivasulu, B., Adinarayana, K., Ellaiah, P., 2003. Investigations on neomycin production with immobilized cells of Streptomyces marinensis NUV-5 in calcium alginate matrix. *AAPS PharmSciTech*. 4: E57.

Subramaniyam, R., Vimala, R., 2012. Solid state and submerged fermentation for the production of bioactive substances: A comparative study. *Int J Sci Nat*, 3(3): 480–486.

Suenaga, H., Ohnuki, T., Miyazaki, K., 2007. Functional screening of a metagenomic library for genes involved in microbial degradation of aromatic compounds. *Environ. Microbiol.*, 9: 2289–2297.

Suganthi, R., et al., 2011. Amylase production by Aspergillus Niger under solid state fermentation using agroindustrial wastes. *International Journal of Engineering Science and Technology*, 3(2): 1756–1763.

Thomas, L., Larroche, C., Pandey, A., 2013a. Current developments in solid-state fermentation. *Biochemical Engineering Journal*, 81: 146–161.

Thomas, L., Larroche, C., Pandey, A., 2013b. Current developments in solid-state fermentation. *Biochem. Eng. J.*, 81: 146–161.

Tzanov, T., Calafell, M., Guebitz, G.M., Cavaco-Paulo, A., 2001. Bio-preparation of cotton fabrics. *Enzyme Microb. Technol.*, 29: 357–362.

Van Bambeke, F., Mingeot-Leclercq, M.-P., Glupczynski, Y., Tulkens, P.M., 2017. 137—mechanisms of action. In *Infectious Diseases*. 4th ed., eds. J. Cohen, W.G. Powderly, S.M. Opal. Elsevier, Amsterdam, pp. 1162.e1161–1180.e1161.

Van den Burg, B., 2003. Extremophiles as a source for novel enzymes. *Curr. Opin. Microbiol.*, 6: 213–218.

Volpato, G., Rodrigues, R.C., Fernandez-Lafuente, R., 2010. Use of enzymes in the production of semi-synthetic penicillins and cephalosporins: Drawbacks and perspectives. *Curr. Med. Chem.*, 17: 3855–3873.

Waksman, S.A., Reilly, H.C., Johnstone, D.B., 1946. Isolation of streptomycin-producing strains of Streptomyces griseus. *J. Bacteriol.*, 52: 393–397.

Waksman, S.A., Woodruff, H.B., 1940. Bacteriostatic and bactericidal substances produced by a soil actinomyces. *Proc. Soc. Exp. Biol. Med.*, 45: 609–614.

Walia, A., Mehta, P., Chauhan, A., Shirkot, C.K., 2013. Antagonistic activity of plant growth promoting rhizobacteria isolated from tomato rhizosphere against soil borne fungal plant pathogens. *International Journal of Agriculture, Environment and Biotechnology*, 6(4):571–580.

Walsh, G., 2003. Pharmaceuticals, biologics and biopharmaceuticals. *Biopharmaceuticals; Biochemistry and Biotechnology*. John Wiley and Sons Limited, England, Edition 2: 1–40.

Wilson, B.D., 2009. Cellulases and biofuels. *Curr. Opin. Biotechnol.*, 20: 295–299.

Yoo, Y.J., Hwang, J.Y., Shin, H.L., Cui, H., Lee, J., Yoon, Y.J., 2015. Characterization of negative regulatory genes for the biosynthesis of rapamycin in Streptomyces rapamycinicus and its application for improved production. *J. Ind. Microbiol. Biotechnol.*, 42: 125–135.

Zaidi., K.U., Ali, A.S., Ali, S.A. et al., 2014. Microbial tyrosinases: Promising enzymes for pharmaceutical, food bioprocessing, and environmental industry. *Biochem Res Int.* doi: 10.1155/2014/854687.

Zhang, Y.H.P., 2011. What is vital (and not vital) to advance economically-competitive biofuels production. *Process Biochem.*, 46: 2091–2110.

Zhang, Y.H.P., Himmmel, M.E., Mielenz, J.R., 2006. Outlook for cellulose improvement: Screening and selection strategies. *Biotechnol. Adv.*, 24: 452–481.

Zheng, G.W., Xu, J.H., 2011. New opportunities for biocatalysis: Driving the synthesis of chiral chemicals. *Curr. Opin. Biotechnol.*, 22: 784–792.

11 ACC-Deaminase-Producing Bacteria

From Alleviating Plant Stress to their Commercial Application for Sustainable Agriculture

Anjali Pande, Vineet Singh, and Byung Wook Yun

ABSTRACT

A highly regulated and structured community of microorganisms is constantly associated with plants growing in agricultural lands and exposed to environmental conditions. Owing to the substantial role of rhizomicrobiome in enhancing plant growth and productivity, their underground colonization is of great importance to agriculture. In this regard, rhizobacteria exhibiting ACC deaminase activity are of great value in agriculture due to their role in promoting plant growth under suboptimal growth conditions. Plant growth-promoting rhizobacteria, or PGPRs, are known for their role in alleviating abiotic stress in plants and also working as biocontrol of potential phytopathogens. This chapter discusses the mechanism of action of PGPRs displaying ACC deaminase activity through the hydrolysis of 1-aminocyclopropane-1-carboxylic acid (ACC) and a summary of the steps involved in the commercialization of bio-formulations thereof.

CONTENTS

11.1 Introduction...222
11.2 Plant Microbiome 222
11.3 Plant-Growth-Promoting Rhizobacteria 223
11.4 ACC Deaminase and Its Activity During Plant Stress.......................223
11.5 Commercialization of ACC-Deaminase-Producing PGPRs for
     Sustainable Agriculture...225
11.6 Conclusion 226
References...226

DOI: 10.1201/9781003202998-11

11.1 INTRODUCTION

Environmental stress, including both biotic and abiotic stress, negatively affects plant growth and productivity, mainly due to the sessile nature of plants, which exposes them to almost all the adversities in the environment. Under field conditions, a plant grows as a complex community surrounded and partnered by microbes which together form the plant microbiome (Lundberg et al., 2012).

While microbes provide several benefits to the plant, plants also provide metabolites and reduced carbon to the microbial communities to flourish around them. One way to achieve sustainable agriculture is through the use of rhizomicrobial communities exhibiting ACC deaminase activity. Rhizobacteria exhibiting ACC deaminase activity have been extensively studied for their role in promoting plant growth (Brunetti et al., 2021; Murali et al., 2021; Naing et al., 2021) and alleviating various abiotic stress like salinity stress (Bharti and Barnawal, 2019; Gupta and Pandey, 2019; Kang et al., 2019; Anand et al., 2021; Liu et al., 2021) and drought stress (Saikia et al., 2018; Danish et al., 2020; Gowtham et al., 2020; Zhang et al., 2020). They are also useful in combating plant pathogens (Harman, 2000) and are commercially available (Velivelli et al., 2014).

Thus, PGPRs with their multifarious roles in promoting plant growth are useful to agriculture in various ways. Bioformulations containing PGPRs or their related signaling molecules offer an eco-friendly and cheaper alternative to agrochemicals. Therefore, the optimization of such formulations may enhance soil fertility and alleviate biotic (phytopathogens) and abiotic stress in plants, thus promoting plant growth and making them useful in attaining the long-term goal of sustainable agriculture (Shameer and Prasad, 2018).

11.2 PLANT MICROBIOME

A plant microbiome (also known as phytomicrobiome) typically describes those microbial communities that are directly associated with a plant and can thrive either inside or outside of the plant tissues, like roots, shoots, leaves, flowers, and seeds. Together, a plant-phytomicrobiome association is termed as holobiont (Berg et al., 2016; Theis et al., 2016; Smith et al., 2017).

The microbial community associated particularly with the roots (the rhizomicrobiome) is so far the best-studied phytomicrobiome interaction occurring in nitrogen-fixing rhizobia associated with legumes (Gray and Smith, 2005; Pande et al., 2021). The rhizomicrobiome composition is considerably controlled by the plants through the root exudates (Chaparro et al., 2012; Trabelsi and Mhamdi, 2013) and other signal compounds (Nelson and Sadowsky, 2015; Massalha et al., 2017) that recruit specific microbes over others. Another level of regulation, is the plant response to microbial quorum sensing compounds and the production of analogs for regulating the composition of rhizomicrobiome (Ortíz-Castro et al., 2009).

Consequently, plants regulate the activities of key members of the rhizomicrobiome termed as core species (Toju et al., 2018), and in turn, these core species regulate the activities within the phytomicrobiome through quorum sensing (Chauhan et al., 2015). The closer the microbial community to the root surface, the higher is the degree of influence of the plant on them (Backer et al., 2018). The zone inhabited by microorganisms inside and around the root surface is said to be the rhizosphere (Hiltner, 1904).

11.3 PLANT-GROWTH-PROMOTING RHIZOBACTERIA

Plant-growth-promoting rhizobacteria (PGPR), as the name suggests, are those bacteria that facilitate plant growth by either binding to the outer surface of the roots (the rhizosphere) or inside the roots (endophytes) or the leaves (the phyllosphere) (Davison, 1988; Kloepper et al., 1989; Firdous et al., 2019). Depending upon the plant and the associated bacterial species, a bacterial endophyte may colonize roots, stems, or entire plant tissues or may even form specific structures as nodules (symbiotic bacteria) (Oldroyd et al., 2011; Wang et al., 2012). However, irrespective of their primary location, soil bacteria that are beneficial to the plants are commonly referred to as plant-growth-promoting bacteria, or PGPB (Bashan and Holguin, 1998). The soil-borne symbiotic bacteria were among the first to be commercialized for promoting plant growth (Lucy et al., 2004).

PGPRs promote the growth of the plant directly by enhancing germination, stimulating the growth of roots or shoots and biological nitrogen fixation (*Rhizobia*), and indirectly alleviating the environmental stress in plants—for example, preventing the deleterious effects caused by a plant pathogen (Glick, 2012). The production of various antibiotics, like lipopeptides, polyketides, and antifungal metabolites, is also suggested by PGPRs in order to reduce the growth of pathogens (del Carmen Orozco-Mosqueda et al., 2018; Santoyo et al., 2019; Lau et al., 2020; Morales-Cedeno et al., 2021).

The basic mechanism of plant growth enhancement is by facilitating the plants with acquired environmental resources like fixed nitrogen, iron, and phosphate or by modulating the levels of plant hormones like auxins, cytokinins, and ethylene (Glick, 2014). Interestingly, the bacterial trait that plays a key role in promoting plant growth is the presence of the enzyme 1-aminocyclopropane-1-carboxylate (ACC) deaminase in all the PGPRs. We will discuss this enzyme in more detail in the following sections of this chapter.

11.4 ACC DEAMINASE AND ITS ACTIVITY DURING PLANT STRESS

Moist soil, rich in reduced carbon, root exudates (sugars, organic acids, and amino acids), and other plant cellular debris, provides the most suitable environment for microbial colonization. PGPRs with ACC deaminase activity regulate ethylene levels by breaking down ACC (a precursor molecule of ethylene) into α-ketobutyrate and ammonia (Singh et al., 2015), leading to suppression of ethylene concentration in stressed plants and thus preventing the ethylene-induced inhibition of plant growth or cell death (Jacobson et al., 1994; Glick et al., 1999) and several other stresses listed in Table 11.1.

Several concepts and theories have been laid by scientists regarding the mode of action of ACC deaminase in promoting plant growth. Previous studies suggested that the plant-associated bacteria synthesize and secrete the phytohormone indole-3-acetic

TABLE 11.1

Recent Examples of ACC-Deaminase-Producing Rhizobacteria and Their Role in Plant Stress Tolerance

S. No.	ACCD-Producing Rhizobacteria	Plant	Stress Tolerance	References
1.	*Pseudomonas corrugata* (DR3) and *Enterobacter soli* (DR6)	*Vitis vinifera* L.	Drought	(Liu et al., 2021)
2.	*B. amyloliquefaciens* MMR04	*Pennisetum glaucum* L.	Drought	(Murali et al., 2021)
3.	*Brevibacterium linens* (RS16)	*Oryza sativa* L.	Heat stress and combined UV-B radiation and heat stress	(Choi et al., 2021a; Choi et al., 2021b)
4.	*Microbacterium* sp. (AR-ACC2), *Methylophaga* sp. (AR-ACC3), *and* *Paenibacillus* sp. (ANR-ACC3)	*Oryza sativa* L.	Submergence	(Bal and Adhya, 2021)
5.	*Pseudomonas azotoformans*	*Solanum lycopersicum*	Salinity	(Liu et al., 2021)
6.	*Achromobacter* sp. A1	*Zea mays*	Heavy metal (cadmium)	(Sun et al., 2022)
7.	*Aneurinibacillus aneurinilyticus* (ACCo2) and *Paenibacillus* sp. (ACCo6)	*Phaseolus vulgaris*	Salinity	(Gupta and Pandey, 2019)
8.	*Achromobacter xylosoxidans* and *Enterobacter cloacae*	*Zea mays*	Drought	(Danish et al., 2020)
9.	*P. fluorescens* (P1), *P. fluorescens* (P3), *P. fluorescens* (P8), *P. fluorescens* (P14)	*Zea mays* L. var saccharata	Water deficit	(Zarei et al., 2020)
10.	*Enterobacter* sp.	*Oryza sativa*	Salinity	(Sarkar et al., 2018)

acid (IAA) in response to the root exudates containing tryptophan and other small molecules. This bacterial-IAA, along with the endogenous plant-synthesized IAA, stimulates plant cell proliferation and/or elongation or transcriptional induction of the plant enzyme ACC synthase, which then catalyzes the formation of ACC, hence emphasizing that IAA acts to stimulate the synthesis of ethylene in the plants (Glick et al., 1998).

Further supporting studies also suggested that IAA increases the extent of root exudation by loosening the plant cell walls, and the excessive ACC is exuded from the roots, shoots, or leaves (Penrose et al., 2001), which is then cleaved by ACC deaminase present in the bacteria associated with the plants (Penrose and Glick, 2003). However, on the one hand, this suggested that the bacterium acts as a sink for ACC, while on the other hand, it raises questions on the selective lowering of deleterious ethylene levels under stress without affecting the higher ethylene levels induced by stress that activates the defense response in plants. Studies suggest that the amount of ACC is low in plants at the onset of stress conditions, and so is the amount of ACC deaminase in the associated bacterium. The level of ethylene reaches its peak due to the induction of ACC-oxidase in plants, which further induces defense-related genes while also inducing bacterial ACC deaminase through the increasing levels of ACC. Since ACC oxidase has more affinity toward ACC, the ratio of ACC oxidase to ACC deaminase governs the ethylene levels in plants under such a scenario (Glick et al., 1998). Therefore, in the presence of ACC-deaminase-producing bacteria, the level of ethylene is reduced effectively before the induction of ACC oxidase levels. Overall, ACC-deaminase-producing PGPRs produce bacterial-IAA and synthesize ACC deaminase to maintain lower levels of ethylene and thus promote plant growth even under environmental stress conditions

11.5 COMMERCIALIZATION OF ACC-DEAMINASE-PRODUCING PGPRs FOR SUSTAINABLE AGRICULTURE

Bioformulations are often better and cheaper alternatives as they are eco-friendly and offer ease of handling as compared to chemical-based agricultural practices (Adoko et al., 2021). Bioformulations prepared for agricultural practices are developed based on microbial species, which may be single (monoculture) or multiple species, and some signaling molecules isolated from them are also used. Rhizobial inoculants collected from leguminous crops are the widely used agricultural inoculants and the earliest example of commercial microbial products in agriculture (Bashan and Holguin, 1998).

A few studies have observed an inconsistent effect of bacterial consortia on crop yield. For instance, a cocktail of *B. amyloliquefaciens* (a bacterium) and *Trichoderma virens* (a fungus) improves yields of corn and tomato, among other crops (Srivastava, 2021). A major challenge with such strains is managing the strains in consistent proportions. Moreover, preparing a mixture of bacterial strains is mostly advantageous because it offers interaction between the strains as compared to single strain-based formulations (Liu et al., 2018; Zarei et al., 2020).

The signaling molecules are isolated either from the PGPR (for bacteria to plant signaling) or its associated plant that triggers the PGPR genes for modulating the microbiome composition in the soil to enhance plant growth. For example, the production of lipo-chitooligosaccharides (LCOs) by rhizobia is triggered by the addition of isoflavonoids as a plant to microbe signal (Smith et al., 2015).

Commercialization of such bioformulations involves various steps, from the identification of a PGPR to the refinement of the product. This involves bacterial isolation from plant roots, stems, or leaves, their screening under controlled environment and field conditions, and on different crops, different geographic locations, and soil types. Moreover, it requires monitoring and other management practice considerations, further refinement of the final product, confirmation of no toxicological impacts, other formulation considerations for delivery, and finally, its registration for regulatory approval before its commercialization (Backer et al., 2018). Thus, keeping into consideration the aforementioned steps or modifications thereof for single species, consortia of potential bacteria, or signaling molecules associated with them may provide useful agricultural products for combating plant stress and enhancing their growth and productivity.

11.6   CONCLUSION

In order to attain the goal of agricultural sustainability, the focus should be toward the use of those biological resources which promote productivity in an eco-friendly manner. In this regard, ACC-deaminase-producing bacteria are well known for their role in promoting plant growth and productivity by lowering the detrimental levels of ethylene under environmental stress conditions. Commercialization of such bacteria or products thereof holds the potential to revolutionize agriculture and will be useful in providing food and nutritional security to future generations.

REFERENCES

Adoko, M.Y., Agbodjato, N.A., Noumavo, A.P., Amogou, O., et al. (2021). Bioformulations based on plant growth promoting rhizobacteria for sustainable agriculture: Biofertilizer or Biostimulant? *African Journal of Agricultural Research* 17(9), 1256–1260.

Anand, G., Bhattacharjee, A., Shrivas, V.L., Dubey, S., and Sharma, S. (2021). ACC deaminase positive Enterobacter-mediated mitigation of salinity stress, and plant growth promotion of Cajanus Cajan: A lab to field study. *Physiology and Molecular Biology of Plants* 27(7), 1547–1557.

Backer, R., Rokem, J.S., Ilangumaran, G., Lamont, J., Praslickova, D., Ricci, E., et al. (2018). Plant growth-promoting rhizobacteria: Context, mechanisms of action, and roadmap to commercialization of biostimulants for sustainable agriculture. *Frontiers in Plant Science* 9, 1473.

Bal, H.B., and Adhya, T.K. (2021). Alleviation of submergence stress in rice seedlings by plant growth-promoting rhizobacteria with ACC deaminase activity. *Frontiers in Sustainable Food Systems* 5, 36.

Bashan, Y., and Holguin, G. (1998). Proposal for the division of plant growth-promoting rhizobacteria into two classifications: Biocontrol-PGPB (plant growth-promoting bacteria) and PGPB. *Soil Biology & Biochemistry* 30(8–9), 1225–1228.

Berg, G., Rybakova, D., Grube, M., and Köberl, M. (2016). The plant microbiome explored: Implications for experimental botany. *Journal of Experimental Botany* 67(4), 995–1002.

Bharti, N., and Barnawal, D. (2019). Amelioration of salinity stress by PGPR: ACC deaminase and ROS scavenging enzymes activity. In *PGPR amelioration in sustainable agriculture*. Elsevier, 85–106.

Brunetti, C., Saleem, A.R., Della Rocca, G., Emiliani, G., De Carlo, A., Balestrini, R., et al. (2021). Effects of plant growth-promoting rhizobacteria strains producing ACC deaminase on photosynthesis, isoprene emission, ethylene formation and growth of Mucuna pruriens (L.) DC. in response to water deficit. *Journal of Biotechnology* 331, 53–62.

Chaparro, J.M., Sheflin, A.M., Manter, D.K., and Vivanco, J.M. (2012). Manipulating the soil microbiome to increase soil health and plant fertility. *Biology and Fertility of Soils* 48(5), 489–499.

Chauhan, H., Bagyaraj, D., Selvakumar, G., and Sundaram, S. (2015). Novel plant growth promoting rhizobacteria—prospects and potential. *Applied Soil Ecology* 95, 38–53.

Choi, J., Roy Choudhury, A., Park, S.-y., Oh, M.M., and Sa, T. (2021a). Inoculation of ACC Deaminase-Producing Brevibacterium linens RS16 Enhances Tolerance against Combined UV-B Radiation and Heat Stresses in Rice (Oryza sativa L.). *Sustainability* 13(18), 10013.

Choi, J., Roy Choudhury, A., Walitang, D.I., Lee, Y., and Sa, T. (2021b). ACC deaminase-producing Brevibacterium linens RS16 enhances heat-stress tolerance of rice (Oryza sativa L.). *Physiologia Plantarum* 174(1), e13584.

Danish, S., Zafar-ul-Hye, M., Mohsin, F., and Hussain, M. (2020). ACC-deaminase producing plant growth promoting rhizobacteria and biochar mitigate adverse effects of drought stress on maize growth. *PLoS One* 15(4), e0230615.

Davison, J. (1988). Plant beneficial bacteria. *Bio/technology* 6(3), 282–286.

del Carmen Orozco-Mosqueda, M., del Carmen Rocha-Granados, M., Glick, B.R., and Santoyo, G. (2018). Microbiome engineering to improve biocontrol and plant growth-promoting mechanisms. *Microbiological Research* 208, 25–31.

Firdous, J., Lathif, N.A., Mona, R., and Muhamad, N. (2019). Endophytic bacteria and their potential application in agriculture: A review. *Indian Journal of Agricultural Research* 53(1), 1–7.

Glick, B.R. (2012). Plant growth-promoting bacteria: Mechanisms and applications. *Scientifica*, 1–15.

Glick, B.R. (2014). Bacteria with ACC deaminase can promote plant growth and help to feed the world. *Microbiological Research* 169(1), 30–39.

Glick, B.R., Holguin, G., Patten, C., and Penrose, D.M. (1999). *Biochemical and genetic mechanisms used by plant growth promoting bacteria*. World Scientific.

Glick, B.R., Penrose, D.M., and Li, J. (1998). A model for the lowering of plant ethylene concentrations by plant growth-promoting bacteria. *Journal of Theoretical Biology* 190(1), 63–68.

Gowtham, H., Singh, B., Murali, M., Shilpa, N., Prasad, M., Aiyaz, M., et al. (2020). Induction of drought tolerance in tomato upon the application of ACC deaminase producing plant growth promoting rhizobacterium Bacillus subtilis Rhizo SF 48. *Microbiological Research* 234, 126422.

Gray, E., and Smith, D. (2005). Intracellular and extracellular PGPR: Commonalities and distinctions in the plant—bacterium signaling processes. *Soil Biology and Biochemistry* 37(3), 395–412.

Gupta, S., and Pandey, S. (2019). ACC deaminase producing bacteria with multifarious plant growth promoting traits alleviates salinity stress in French bean (Phaseolus vulgaris) plants. *Frontiers in Microbiology* 10, 1506.

Harman, G.E. (2000). Myths and dogmas of biocontrol changes in perceptions derived from research on Trichoderma harzinum T-22. *Plant Disease* 84(4), 377–393.

Hiltner, L. (1904). Über neuere erfahrungen und probleme auf dem debiete der bo denbakteriologie und unter besonderer berucksichtigung der grundund und brache. *Zbl. Bakteriol* 2, 14–25.

Jacobson, C., Pasternak, J., and Glick, B. (1994). Partial purification and characterization of ACC deaminase from the plant growth-promoting rhizobacterium Pseudomonas putida GR12-2. *Canadian Journal of Microbiology* 40(1), 19–1025.

Kang, S.-M., Shahzad, R., Bilal, S., Khan, A.L., Park, Y.-G., Lee, K.-E., et al. (2019). Indole-3-acetic-acid and ACC deaminase producing Leclercia adecarboxylata MO1 improves Solanum lycopersicum L. growth and salinity stress tolerance by endogenous secondary metabolites regulation. *BMC Microbiology* 19(1), 1–14.

Kloepper, J.W., Lifshitz, R., and Zablotowicz, R.M. (1989). Free-living bacterial inocula for enhancing crop productivity. *Trends in Biotechnology* 7(2), 39–44.

Lau, E.T., Tani, A., Khew, C.Y., Chua, Y.Q., and San Hwang, S. (2020). Plant growth-promoting bacteria as potential bio-inoculants and biocontrol agents to promote black pepper plant cultivation. *Microbiological Research* 240, 126549.

Liu, C.-H., Siew, W., Hung, Y.-T., Jiang, Y.-T., and Huang, C.-H. (2021). 1-Aminocyclopropane-1-carboxylate (ACC) Deaminase Gene in Pseudomonas azotoformans Is Associated with the Amelioration of Salinity Stress in Tomato. *Journal of Agricultural and Food Chemistry* 69(3), 913–921.

Liu, K., McInroy, J.A., Hu, C.-H., and Kloepper, J.W. (2018). Mixtures of plant-growth-promoting rhizobacteria enhance biological control of multiple plant diseases and plant-
   growth promotion in the presence of pathogens. *Plant Disease* 102(1), 67–72.
Lucy, M., Reed, E., and Glick, B.R. (2004). Applications of free living plant growth-promoting
   rhizobacteria. *Antonie van leeuwenhoek* 86(1), 1–25.
Lundberg, D.S., Lebeis, S.L., Paredes, S.H., Yourstone, S., Gehring, J., Malfatti, S., et al.
   (2012). Defining the core Arabidopsis thaliana root microbiome. *Nature* 488(7409),
   86–90.
Massalha, H., Korenblum, E., Tholl, D., and Aharoni, A. (2017). *Small molecules below-
   ground: The role of specialized metabolites in the rhizosphere.* Wiley Online Library.
Morales-Cedeno, L.R., del Carmen Orozco-Mosqueda, M., Loeza-Lara, P.D., Parra-Cota, F.I.,
   de Los Santos-Villalobos, S., and Santoyo, G. (2021). Plant growth-promoting bacterial
   endophytes as biocontrol agents of pre-and post-harvest diseases: Fundamentals, methods of application and future perspectives. *Microbiological Research* 242, 126612.
Murali, M., Gowtham, H., Singh, S.B., Shilpa, N., Aiyaz, M., Niranjana, S., et al. (2021).
   Bio-prospecting of ACC deaminase producing rhizobacteria towards sustainable agriculture: A special emphasis on abiotic stress in plants. *Applied Soil Ecology* 168, 104142.
Naing, A.H., Maung, T.T., and Kim, C.K. (2021). The ACC deaminase-producing plant growth
   promoting bacteria (PGPB): Influences of bacterial strains and ACC deaminase activities
   in plant tolerance to abiotic stress. *Physiologia Plantarum* 173(4), 1992–2012.
Nelson, M.S., and Sadowsky, M.J. (2015). Secretion systems and signal exchange between
   nitrogen-fixing rhizobia and legumes. *Frontiers in Plant Science* 6, 491.
Oldroyd, G.E., Murray, J.D., Poole, P.S., and Downie, J.A. (2011). The rules of engagement in
   the legume-rhizobial symbiosis. *Annual Review of Genetics* 45, 119–144.
Ortíz-Castro, R., Contreras-Cornejo, H.A., Macías-Rodríguez, L., and López-Bucio, J. (2009).
   The role of microbial signals in plant growth and development. *Plant Signaling &
   Behavior* 4(8), 701–712.
Pande, A., Mun, B.-G., Lee, D.-S., Khan, M., Lee, G.-M., Hussain, A., et al. (2021). NO network for plant—microbe communication underground: A review. *Frontiers in Plant Science* 12, 431.
Penrose, D.M., and Glick, B.R. (2003). Methods for isolating and characterizing ACC deaminase-containing plant growth-promoting rhizobacteria. *Physiologia Plantarum* 118(1),
   10–15.
Penrose, D.M., Moffatt, B.A., and Glick, B.R. (2001). Determination of 1-aminocycopro-
   pane-1-carboxylic acid (ACC) to assess the effects of ACC deaminase-containing bacteria on roots of canola seedlings. *Canadian Journal of Microbiology* 47(1), 77–80.
Saikia, J., Sarma, R.K., Dhandia, R., Yadav, A., Bharali, R., Gupta, V.K., et al. (2018). Alleviation of drought stress in pulse crops with ACC deaminase producing rhizobacteria
   isolated from acidic soil of Northeast India. *Scientific Reports* 8(1), 1–16.
Santoyo, G., Sánchez-Yáñez, J.M., and de los Santos-Villalobos, S. (2019). Methods for
   detecting biocontrol and plant growth-promoting traits in Rhizobacteria. In *Methods in
   rhizosphere biology research.* Springer, 133–149.
Sarkar, A., Ghosh, P.K., Pramanik, K., Mitra, S., Soren, T., Pandey, S., et al. (2018). A halotolerant Enterobacter sp. displaying ACC deaminase activity promotes rice seedling growth
   under salt stress. *Research in Microbiology* 169(1), 20–32.
Shameer, S., and Prasad, T. (2018). Plant growth promoting rhizobacteria for sustainable agricultural practices with special reference to biotic and abiotic stresses. *Plant Growth Regulation* 84(3), 603–615.
Singh, R.P., Shelke, G.M., Kumar, A., and Jha, P.N. (2015). Biochemistry and genetics of ACC
   deaminase: A weapon to "stress ethylene" produced in plants. *Frontiers in Microbiology*
   6, 937.

Smith, D.L., Gravel, V., and Yergeau, E. (2017). Signaling in the phytomicrobiome. *Frontiers in Plant Science* 8, 611.

Smith, D.L., Subramanian, S., Lamont, J.R., and Bywater-Ekegärd, M. (2015). Signaling in the phytomicrobiome: Breadth and potential. *Frontiers in Plant Science* 6, 709.

Srivastava, R.K. (2021). Biofertilizers application in agriculture: A viable option to chemical fertilizers. *Biofertilizers: Study and Impact*, 393–411.

Sun, L., Zhang, X., Ouyang, W., Yang, E., Cao, Y., and Sun, R. (2022). Lowered Cd toxicity, uptake and expression of metal transporter genes in maize plant by ACC deaminase-producing bacteria Achromobacter sp. *Journal of Hazardous Materials* 423, 127036.

Theis, K.R., Dheilly, N.M., Klassen, J.L., Brucker, R.M., Baines, J.F., Bosch, T.C., et al. (2016). Getting the hologenome concept right: An eco-evolutionary framework for hosts and their microbiomes. *Msystems* 1(2), e00028–00016.

Toju, H., Peay, K.G., Yamamichi, M., Narisawa, K., Hiruma, K., Naito, K., et al. (2018). Core microbiomes for sustainable agroecosystems. *Nature Plants* 4(5), 247–257.

Trabelsi, D., and Mhamdi, R. (2013). Microbial inoculants and their impact on soil microbial communities: A review. *BioMed Research International*, 1–11. Article ID 863240.

Velivelli, S.L., De Vos, P., Kromann, P., Declerck, S., and Prestwich, B.D. (2014). Biological control agents: From field to market, problems, and challenges. *Trends in Biotechnology* 32(10), 493–496.

Wang, D., Yang, S., Tang, F., and Zhu, H. (2012). Symbiosis specificity in the legume—rhizobial mutualism. *Cellular Microbiology* 14(3), 334–342.

Zarei, T., Moradi, A., Kazemeini, S.A., Akhgar, A., and Rahi, A.A. (2020). The role of ACC deaminase producing bacteria in improving sweet corn (Zea mays L. var saccharata) productivity under limited availability of irrigation water. *Scientific Reports* 10(1), 1–12.

Zhang, M., Yang, L., Hao, R., Bai, X., Wang, Y., and Yu, X. (2020). Drought-tolerant plant growth-promoting rhizobacteria isolated from jujube (Ziziphus jujuba) and their potential to enhance drought tolerance. *Plant and Soil* 452(1), 423–440.

12 Phytases and Their Characteristic Features and Biotechnological Applications in Animal Feed

Syed Zakir Hussain Shah, Mahroze Fatima,
Mehwish Khan, and Muhammad Bilal

ABSTRACT

In the feed industry, plant-origin ingredients are extensively used as they are cheaper and easily available. However, there are some constraints that reduce the plants' utilization in animal feeds. One of them is the anti-nutritional factors that are present in plant ingredients, such as phytate, which is a principal source of phosphorus storage in plant tissues. Phytate forms complexes with different cations, minerals, proteins, starch, and amino acids that reduce the absorption and availability of these ingredients in animals. Phytase is an enzyme that catalyzes the hydrolysis of phytate and causes the release of phosphorus and other chelated components. Phytases are classified based on their origin, optimal pH, and the site from which hydrolysis starts. Unfortunately, many ruminants and other cultured animals lack this enzyme; thus, an exogenous source of phytase is required in their feed. In poultry, aquaculture, piggery, and other cultured industries, phytase supplementation has been proven to improve the growth, feed intake, amino acid and protein utilization, and absorption of iron and minerals; increase the digestibility of other nutrients; and decrease the excretion of phosphorus.

Keywords: phytase, phytate hydrolysis, nutrients digestibility, growth performance

CONTENTS

12.1 Introduction 232
12.2 Uses of Phytase in Poultry Nutrition... 234
12.3 Nutrient Digestibility 235
12.4 Phosphorus Utilization 236
12.5 Amino Acid Availability and Dietary Energy 236

DOI: 10.1201/9781003202998-12

12.6   Broiler's Growth Performance 236
12.7   Phytase Effects on Laying Hens.. 237
       12.7.1 Egg Production.. 237
       12.7.2 Body Weight.. 237
       12.7.3 Feed Utilization... 237
       12.7.4 Egg Quality 237
       12.7.5 Blood Mineral Concentrations 237
12.8   Use of Phytase in Aquaculture Nutrition 238
12.9   Growth Performance 238
12.10 Nutrient Digestibility 238
12.11 Body Composition 239
12.12 Nutrient Retention.. 239
12.13 Use of Phytase in Pig Nutrition... 240
12.14 Mucosal Phytase... 240
12.15 Phytase from Gut Microflora 240
12.16 Intrinsic Plant Phytase... 241
12.17 Exogenous Microbial Phytase.. 241
12.18 Phytate Hydrolysis by Phytase.. 241
12.19 Digestibility and Availability of P... 241
12.20 Growth Performance of Pig.. 242
12.21 Conclusion.. 242
References... 242

12.1   INTRODUCTION

Feed formulation is the process of matching nutrient contents of locally available ingredients (raw material) to the nutrient requirements of a specific class of animals in an economical manner. In the animal feed industry, fish meal has become the most preferred component of feed because of its excellent fatty acid and amino acid profile. Fishmeal is used as an important protein source in many industries like aquaculture, poultry culture, and pig culture. Stagnant production level, high cost, and limited supply make it unsustainable for farming (Baruah *et al.*, 2004). Thus, for the production of relatively low-cost animal feeds, attention is paid toward the replacement of fishmeal with grain or plant by-products (Gatlin *et al.*, 2007). Plant-origin ingredients are cheaper and extensively available than fish meal.

In aquatic and terrestrial feeds, the major plant protein sources are corn gluten, canola meal, soybean meal, lupins, sunflower meal, and peas. Among these plant-origin protein sources, the soybean meal is most commonly used. It contains high crude protein values and is available throughout the year. For several years, nutritionists have explored the ways to employ plant-origin proteins. However, there are some constraints that reduce the plant proteins utilization in animal feed. One of them is the anti-nutritional factors that are present in plant protein sources. Phytate is an example of such anti-nutritional factors (Kumar *et al.*, 2012). The structure of phytate is given in Figure 12.1. In plant tissues, particularly seeds and bran, phytic acid, or phytate, is a basic storage form of phosphorus. It forms complexes with cations, like iron, zinc, calcium, copper, and magnesium (Figure 12.1). So it

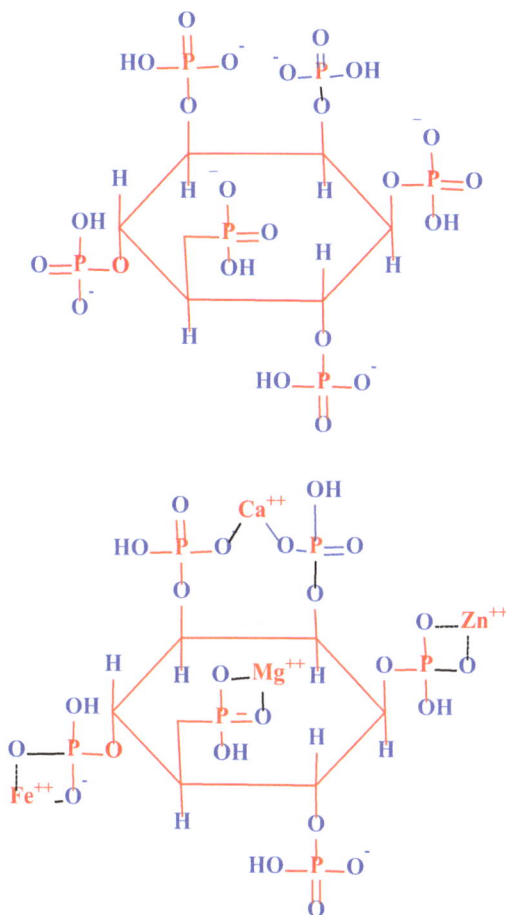

FIGURE 12.1 Chemical structure of phytate and phytate complex.

negatively affects the digestion and absorption of these minerals in the animal body (Papatryphon *et al.,* 1999). Phytate also interacts with the cations present in starch, protein, lipids, and amino acids of feedstuff, thus decreasing their nutrient digestibility in pigs, fish, and poultry.

In monogastric and agastric animals, phosphorus is not bioavailable in the form of phytate because these animals lack phytase, the intestinal digestive enzyme that is required for the hydrolysis of phytate to release phosphorus (Jackson *et al.,* 1996). Due to low phytate digestibility, the animals excrete most of phytate-P into the environment, which causes pollution in the environment, such as algal blooms in water (Baruah *et al.,* 2004). Phytase reduces phosphorus excretion by about 25–50%, depending upon the level of supplementation, species, and diet (Haefner *et al.,* 2005).

Phytase enzyme catalyzes the step-by-step breakdown of phytate into inositol and free phosphorus, making the phosphorus available for absorption in the animal

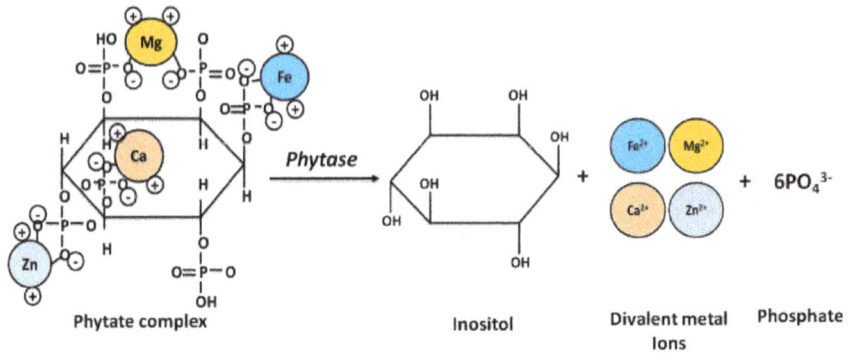

FIGURE 12.2 Release of chelated cations by phytate hydrolysis (Vashishth et al., 2017).

body (Abdulla *et al.*, 2017). It also helps to release different phytate chelated cations (Debnath *et al.*, 2005a), as shown in Figure 12.2.

Phytases are classified into two groups depending upon the site of initiation of hydrolysis, 3-phytase and 6-phytase. The 3-phytase starts hydrolysis of phytate from carbon 3- and 6-phytase from carbon 6 (Cao *et al.*, 2007). Furthermore, phytases can also be classified on the basis of optimal pH, alkaline phytase and acidic phytase. Alkaline phytase shows optimum activity at pH nearly 8 but acidic phytase at pH about 5 (Baruah *et al.*, 2007b). Additionally, phytases are also divided into plant phytases and microbial phytases on the basis of their origin (Stefan *et al.*, 2005). The phytase activity is measured in phytase units, and this is written as FTU kg^{-1} (Engelen *et al.*, 1994).

A hundred years ago, phytase was discovered in rice bran, which was excessively abundant in the environment (Suzuki *et al.*, 1907). However, nowadays, phytases are used at the industrial level in Western Europe and the United States. The applications of phytase as a feed supplement have been recently increased in Southeast Asia, India, and China (Gessler *et al.*, 2018). About 70% of swine and 90% of poultry diets are integrated with phytase to increase nutrient uptake and to control phosphorus pollution (Sharma *et al.*, 2020). In cultured animals, like pigs, fish, and poultry, phytase supplementation increases the absorption of iron and other minerals, increases body weight and feed intake, improves amino acids and proteins utilization, increases other nutrients digestibility, and decreases the phosphorus excretion in the environment. Applications of phytase in different fields are given in Figure 12.3.

12.2   USES OF PHYTASE IN POULTRY NUTRITION

Poultry farming is the form of animal husbandry, which raises domesticated birds, such as chickens, geese, turkeys, and ducks to produce meat or eggs for food. It has originated from the agricultural era. Poultry, mostly chickens, are farmed in great numbers. More than 60 billion chickens are killed for consumption around the globe annually (GASSC, 2019).

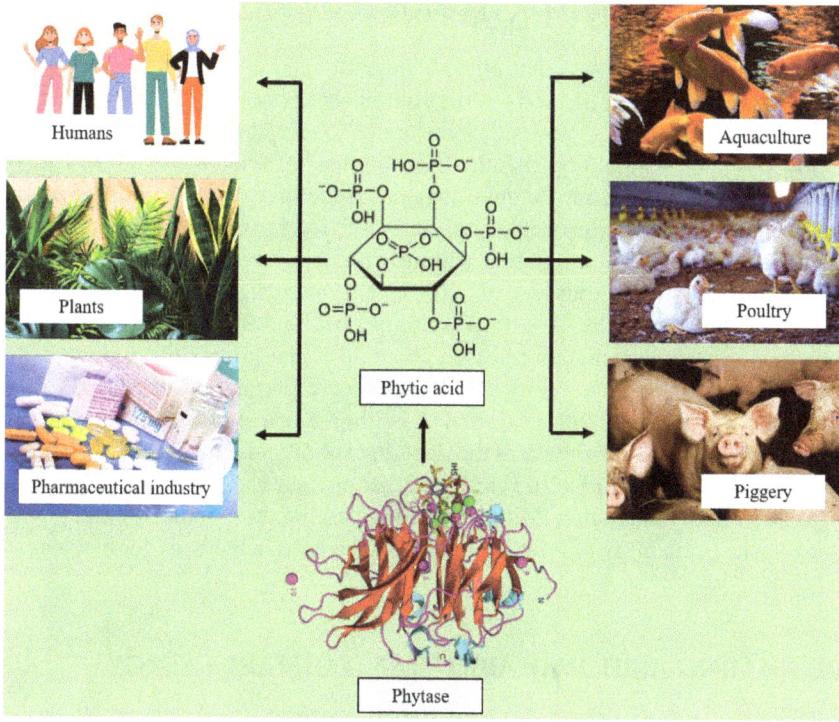

FIGURE 12.3 Role of phytase in different fields.

12.3 NUTRIENT DIGESTIBILITY

Phytase degrades the phytate present in the feed and helps to enhance the nutrient availability from plant-based feeds. Abd-Elsamee (2002) stated that usage of phytase in broiler feeds enhances the nitrogen retention and digestibility of crude protein, nitrogen-free extract, dry matter, and ether extract. In laying hens, phytase supplementation also enhanced nutrient digestibility, particularly crude protein and ether extract.

Fayza *et al.* (2003) conducted an experiment on broiler chickens and reported that the digestion of dry matter was significantly higher in the groups fed with high-P supplemented diets than the groups fed with low-P supplemented diets. In that study, phytase supplementation (600 FTU/kg) increased the digestibility of dry matter in lower-P-diet-fed hens. Similarly, Attia *et al.* (2002) revealed an increased crude fiber digestibility by phytase supplementation in the broiler diets. Moreover, Liu *et al.* (2007) reported that phytase supplementation in layer diets significantly improves amino-acid digestibility and absorption from the gut. Thus, in layers and broiler, the digestibility of various feed ingredients can be enhanced by the use of phytase.

12.4   PHOSPHORUS UTILIZATION

Various studies have proven that phytase improves the utilization of total P, phytate P, and P retention in the body. Moreover, phosphorus excretion can be reduced in the environment by phytase supplementation (Augspurger *et al.*, 2007). When the low-phytate-P-containing manure was applied to land, it contributed to the reduction of P pollution in the environment. Therefore, the combination of phytase supplementation with decreasing dietary P levels is an effective technique for enhancing phytate P utilization and also reducing the P excretion in manure (Panda *et al.*, 2005). Augspurger *et al.* (2007) reported that phytase supplementation reduced the P excretion by 42–51%. Waldroup *et al.* (2000) reported that in broiler diets, phytase supplementation released about 50% phosphorus from phytate. The calcium and phytate P availability and retention were also significantly increased by dietary phytase in hens (Zyla *et al.*, 2011). Englmaierova *et al.* (2017) reported that phytase also caused the increase in the ileal digestibility of P and Ca. Englmaierova *et al.* (2015) and Ptak *et al.* (2015) reported that phytase also plays a significant role in gut microflora modulation. Most of the recorded literature indicates that phytase improves the availability of P for the animals to be used in the body for biochemical functions.

12.5   AMINO ACID AVAILABILITY AND DIETARY ENERGY

There are quite variations in different studies related to the effect of phytase on energy utilization in poultry diets. Newkirk and Classen (2001) reported improved AME (apparent metabolizable energy) concentrations by phytase supplementation. On the contrary, Onyango *et al.* (2005) reported that AME in diets was not influenced by phytase supplementation. These variations may be due to the differences in dietary ingredients and phytase sources utilized in the studies. Rutherfurd *et al.* (2004) observed an increased availability and digestibility of amino acids and proteins by phytase supplementation in chicks. Liu and Ru (2010) reported that amino acid nutrition may be improved in hens by the addition of microbial phytase. Selle and Ravindran (2007) noted that phytase supplementation in poultry improves amino acids availability and energy utilization. Phytase improves energy utilization due to the release of free Ca ions that are necessary for α-amylase activity in starch digestion (Kies and Van Hemert, 2000).

12.6   BROILER'S GROWTH PERFORMANCE

Phytase also causes an increase in the weight gain of broiler hens. It is considered that phytase improves the performance by the hydrolysis of phytate. In this way, the anti-nutritional effects of phytate were reduced. Shirley *et al.* (2003) reported that phytase supplementation in rations contributes to the efficient utilization of feed. That, in return, causes weight gain. De Sousa *et al.* (2015) conducted research work to evaluate the phytase effects on growth performance and on blood, bone, and digestive characteristics. Results indicated that diet with phytase supplementation showed the best growth performance.

12.7 PHYTASE EFFECTS ON LAYING HENS

12.7.1 EGG PRODUCTION

Sukumar and Jalaudeen (2003) noted an increased egg production in layers by phytase supplementation in P-deficient diets. Casartelli *et al.* (2005) revealed that the egg production traits of layers were significantly affected by phytase supplementation. Ponnuvel *et al.* (2015) reported that phytase improved egg production in spite of low-energy and protein diets. But some other studies showed that egg production was not affected by phytase supplementation in feed (Musapuor *et al.*, 2005). Metwally (2005) observed that layers groups fed with phytase supplemented diets showed improved egg production (%), egg mass, and egg number compared to the groups fed with unsupplemented diets. Englmaierova *et al.* (2017) added that along with egg production, phytase also maintains the egg content and eggshell quality in older hens.

12.7.2 BODY WEIGHT

Like that of broilers, the layers' weight gain was also increased by phytase supplementation. Sukumar and Jalaudeen (2003) noted a numerically improved body weight of layers by fungal phytase supplementation in P deficient diets.

12.7.3 FEED UTILIZATION

Phytase also contributes to increased feed intake and/or feed efficiency of layers. Sukumar and Jalaudeen (2003) showed an enhanced feed efficiency by phytase supplementation. Ponnuvel *et al.* (2015) reported that without influencing feed intake, the phytase supplementation increased feed efficiency. The feed conversion and feed consumption ratio was significantly increased in layers by providing an exogenous phytase to diets (Hughes *et al.*, 2008). Thus, phytase supplementation improved feed consumption.

12.7.4 EGG QUALITY

Ponnuvel *et al.* (2015) narrated that phytase supplementation resulted in an increase in layer's egg weight. Moreover, layers groups fed with phytase supplemented diets produced eggs with greater albumen percentage and albumen weight than the control group layers were observed in a study by Metwally (2005). In contrast, Kim *et al.* (2017) did not observe any effect of phytase supplementation on the egg quality of layers.

12.7.5 BLOOD MINERAL CONCENTRATIONS

Phytate forms different complexes with other minerals like Cu, Ca, K, Zn, Fe, and Mg. Therefore, phytate is considered to be bound with important dietary minerals and make them unavailable for utilization. The phytase enzyme hydrolyzes the phytate and helps to release these minerals. In this way, these minerals become available for absorption in the body. In a study by Musapuor *et al.* (2005), the P contents of plasma in laying hens were observed to be increased, while the alkaline phosphatase activity (ALP) in plasma was significantly decreased by phytase supplementation. Kannan *et al.* (2008) noted nonsignificant differences in serum ALP, Ca, and P of various groups fed with phytase supplemented diets. A similar nonsignificant effect on serum P and Ca contents in hens was also observed by Shehab *et al.* (2012). Nevertheless, Abbasi *et al.* (2015) reported that phytase is involved in the release of iron from inositol; hence, its dietary addition helps to improve the iron reserves in hens.

12.8   USE OF PHYTASE IN AQUACULTURE NUTRITION

Aquaculture is the controlled process of cultivating aquatic organisms, especially for human consumption. It is a similar concept to agriculture, but with fish instead of plants or livestock. Aquaculture is also known as fish farming. Like other culture systems, feed cost and feed efficiency are the major factors, which control the economy of farmers (Baruah *et al.,* 2007a). The use of phytase enzyme in a cost-effective plant-protein-based diet is a relatively new trend in aquaculture compared to other culture systems.

12.9   GROWTH PERFORMANCE

It is a general consensus that the addition of phytase in the diet improves the growth performance in fish by neutralizing the negative effects of phytate present in plant-meal-based diets (Kumar *et al.*, 2012). Several studies reported significant improvements in growth performances in many fish species, including rainbow trout (Carter and Sajjadi, 2011; Dalsgaard *et al.*, 2009), catfish (Nwanna *et al.,* 2005), tilapia (Phromkunthong and Gabaudan, 2006), zebrafish (Liu *et al.*, 2013c), and common carp (Phromkunthong *et al.*, 2010) (Table 2.2). Improved growth performance is usually attributed to the phytate hydrolyzing effect of phytase resulting in enhanced availability of minerals and other nutrients to fish. However, growth performance showed no response to phytase supplementation in plant-based diets across many fish species (Fortes-Silva *et al.*, 2011; Liu *et al.*, 2012). Differences in the experimental conditions, diet composition, phytase source, and fish species are most probably the bases for any discrepancies in the results.

12.10   NUTRIENT DIGESTIBILITY

Phytase is being added in fish feed by different inclusion techniques, including pretreatment of ingredients and top spraying of pretreated diets. Top spraying retains the activity of phytase in the diet and gives a chance for phytase to perform its function inside the fish gut. The effect of phytase supplementation on mineral and other nutrient utilization has been estimated in many fish species either by pretreating the ingredients with phytase (Van Weerd *et al.*, 1999) or direct addition to the diet (Nwanna *et al.*, 2006; Vandenberg *et al.*, 2012). The effect of phytase supplementation on nutrient and mineral absorption has also been estimated. Improved protein digestibility against phytase supplementation has been reported across various fish species (Sajjadi and Carter, 2004; Zhu et al., 2014; Cheng et al., 2015; Liu et al., 2012; Vandenberg et al., 2012).

Moreover, Wang et al. (2009) reported that both the phytase treatment methods (pretreatment and top spraying) were found effective in improving mineral absorption in rainbow trout. With the pretreatment method, phytase improved the P absorption in Nile tilapia. Similar improved apparent nutrients and mineral absorption were also observed by Hussain et al. (2011) in Labeo rohita fingerlings fed on plant-based diets sprayed with graded levels of phytase. Baruah et al. (2007a) also observed improved nutrients and mineral absorption in response to phytase supplementation (through spraying) to L. rohita diets.

Recently, Liu et al. (2013a), while working on another agastric carp species, grass carp (Ctenopharyngodon idella), reported enhanced apparent digestibility coefficient (ADC%) of crude protein (CP), crude lipid, dry matter (DM), and mineral in phytase sprayed group in comparison to fish fed on a control diet. In a variety of other monogastric species, improved P availability in phytase treated diets has also been recorded (Punna and Roland, 1999; Atia et al., 2000).

12.11 BODY COMPOSITION

Reduced body lipid with phytase addition has also been reported in grass carp (Liu et al., 2014) and Nile tilapia (Cao et al., 2008). Ye et al. (2006) reported that low-P diets caused higher body lipid levels in grouper, Epinephelus coioides. However, hydrolysis of phytic acid by phytase might release sufficient P. Sufficient levels of dietary P showed a similar decrease in body lipid content in other studies (Rodehutscord, 1996; Roy and Lall, 2003). Therefore, higher P availability was the main cause of reduced lipid levels in the whole body. Phosphorus deficiency probably caused the inhibition of beta-oxidation of fatty acids (Schafer et al., 1995), resulting in higher lipid content in the control group. Similar results (increased protein) were also observed by Sardar et al. (2007) in Cyprinus carpio. Similar findings (increased ash contents) were also observed in other studies (Liu et al., 2013b; Denstadli et al., 2007; Carter and Sajjadi, 2011; Sardar et al., 2007).

Digestibility has a direct influence on the proximate composition of a fish body. Phytase supplementation has been found to improve protein contents in Cyprinus carpio (Sardar et al., 2007). Similarly, the improved body ash content in response to dietary phytase was also reported in grass carp (Liu et al., 2013b), Atlantic salmon (Denstadli et al., 2007; Carter and Sajjadi, 2011), and Cyprinus carpio (Sardar et al., 2007). Liu et al. (2014) hypothesized that lower levels of dietary P cause higher lipid deposition in grass carp. While phytase supplementation to plant-meal-based diets release enough P to lower lipid deposition in the body.

12.12 NUTRIENT RETENTION

Phytase supplementation has also been found to improve the mineral deposition in tissues and bodies of fish. Higher deposition of Ca, Na, P, Mn, Cu, K, Zn, Fe, and

Mg was observed in the bones of common carp (Nwanna *et al.*, 2007), rainbow trout (Vielma *et al.*, 1998), Nile tilapia (Liebert and Portz, 2005), and tiger puffer (Laining *et al.*, 2011) in response to phytase supplementation. Similar observations regarding the deposition of these minerals in fish bodies by feeding phytase supplemented diets has also been observed by Cheng *et al.* (2015) in yellow catfish and Baruah *et al.* (2007b) in *L. rohita*.

12.13   USE OF PHYTASE IN PIG NUTRITION

Regarding the nutrition of pigs, there are four possible bases for the activity of phytase.

1. Endogenous production by the mucosa of the small intestine
2. Large intestine, which supports the microfloral activity of enzymes
3. Intrinsic, derived from a plant-based feedstuff existing in the diet that depends on heat treatment
4. Exogenous, which contains the dietary supplementation of phytase enzyme in the feed

In the pig, phytase has been observed to degrade the phytate present in the diet (Viveros *et al.*, 2000). Eeckhout and de Paepe (1991) demonstrated that microbial phytase is more effective in pigs as compared to plant phytase. Zimmermann *et al.* (2002) confirmed that in grower pigs, the efficiency of *Aspergillus niger* is superior in the phytase from rye and wheat.

12.14   MUCOSAL PHYTASE

According to Pointillart *et al.* (1985), phosphatase and phytase capability of phytate degradation, rising from the mucosa of the small intestine, is assumed to be terminated because of its minimum and slower activity. Nevertheless, Hu *et al.* (1996) found highly noticeable mucosal phytase activity in the jejunum of pigs, as the efficiency for IP3 (myo-inositol triphosphate) was maximum and dropped when phosphorylation of molecule of phytate increased, which is linked with the decrease in solubility. The authors proposed that dephosphorylation of lower esters derived from myo-inositol phosphate might have an influence on mucosal phytase.

12.15   PHYTASE FROM GUT MICROFLORA

Animal waste contains an abundant amount of undigested phytate-P (Jendza *et al.*, 2006). Because phytate can be degraded in the hind-gut of pigs by fermentation, in the large intestine, calcium depresses phytate hydrolysis by microfloral phytase (Sandberg *et al.*, 1993). Interestingly, Leytem *et al.* (2004) observed that the pigs fed with barley-based diets have a small amount of phytate in their feces due to the hydrolysis of phytate by fermentation in the hindgut. Furthermore, Baxter *et al.* (2003) stated that 9.3% of phytate-P was hydrolyzed even in phytase-unsupplemented-diet-fed pigs by the hindgut phytase.

12.16   INTRINSIC PLANT PHYTASE

Many feed ingredients contain intrinsic plant phytase activity, including wheat (Peers, 1953). Viveros *et al.* (2000) also reported that intrinsic activity of phytate-degradation is present in wheat, barley, and some other grains containing rye and triticale, though the intrinsic activity of phytase can be reduced by steam pelleting of diets at high temperatures (Jongbloed and Kemme, 1990). On the other hand, microbial and plant phytase activity might be present in the diets containing wheat and barley pelleted at low temperatures. Zimmermann *et al.* (2003) contended that the reactions to phytases are easily preservative. In the presence of plant phytase activity, the responses to the exogenous activity of phytase would be limited (Rodehutscord *et al.,* 1996). Excitingly, responses of growth performance in weaner swine provided with phytase supplemented diets (cold-pelleted) derived from wheat were considerably more prominent as compared to parallel feeding experiment where intrinsic phytase activity was terminated by steam-pelleting (Kim *et al.*, 2005).

12.17   EXOGENOUS MICROBIAL PHYTASE

Currently, microbial phytases that are more frequently used in pig's diet are derivatives of either bacteria (*Escherichia coli*) or fungi (*A. niger, Peniophora lycii*). In an experiment, Igbasan *et al.* (2000) observed that phytases extracted from *E. coli* and *A. niger* both enhanced digestibility of Ca (18–20%) and P (33–34%) compared to weaner swine fed with unsupplemented maize-soy diets.

12.18   PHYTATE HYDROLYSIS BY PHYTASE

Schlemmer *et al.* (2001) evaluated the properties of phytate degradation of plant, mucosal, and microfloral phytases. Pigs were fed with diets containing wheat, barley, soybean meal, and rye, which were either extruded at 120°C or unprocessed where extrusion eliminated 99.5% of intrinsic phytase activity. The control and extruded diets contained 1.875 g/kg phytate-P, mainly in the form of IP6 (myo-inositol hexaphosphate, 91.0%) and IP5 (myo-inositol pentaphosphate, 6.9%). Almost complete degradation tract of phytate was observed regardless of treatments. With the regular consumption of 16.8 g of phytate, elimination was reduced to 0.4 g phytate per day, which showed that endogenous mucosal and microfloral phytase activities degraded phytate up to the level that minimized the phytate-P excretion in pigs fed with extruded diet. In extruded diet, P digestibility was 9.5% and 45.7% in the control diet, so phytase activity of intact plant supported the absorption of P.

12.19   DIGESTIBILITY AND AVAILABILITY OF P

Phytase directly affects the digestibility of P in pigs. Golovan *et al.* (2001) demonstrated that phytase supplementation increased the P digestibility up to 60–80%. Different sources of phytase affect the P availability differently. Phytases obtained from three different sources (*E. coli, A. niger,* and *P. lycii*) were compared by Augspurger *et al.* (2003). The response was measured in the form of total P released in pigs. The order of the phytate degradation rates was 40.8%, 30.6%, and 16.2% for *E. coli*, *A. niger*, and *P. lycii* phytase, respectively. Phosphorus excretion through urine represents the inefficient utilization of P in pigs. (Almeida and Stein, 2010). Golovan *et al.* (2001) observed that with phytase supplementation, the fecal phosphorus content was reduced to less than 1%, and the P digestibility was increased up to 60–80%.

12.20   GROWTH PERFORMANCE OF PIG

Ultimately, phytase enzyme supplementation to P deficient diets has the ability to improve the growth of pigs. Beers and Jongbloed (1992) reported that phytase enhances growth in pigs. They described that 1450 FTU/kg of phytase improved growth performance by 12.8%. Increases of 8.5% and 4.4% have been observed in feed intake and feed efficiency, respectively. Selle *et al.* (2003a) reported that in weaner pigs, the feed efficiency and weight gain were improved by 6.1% and 9.4%, respectively, by microbial phytase supplementation.

12.21   CONCLUSION

Phytase has efficiently proven to be an effective enzyme for hydrolysis of anti-nutrients (phytate) and improves the growth performance, nutrient digestibility, mineral absorption, nutrient retention and excretion, and enzyme activities of cultured animals, like poultry, pigs, ruminants, and fish. Phytases are acidic, basic, and neutral in nature, extracted from plants or microbes, and they are available in different forms and purity levels in the market. However, the optimum level of phytase supplementation varies with the species of cultured animal, age, sex, culture environment, species of phytase enzyme, purity of enzyme extraction, and feed ingredient used.

REFERENCES

Abbasi M, Mojtaba Z, Mehdi G, Saeed Kh. 2015. Is dietary iron requirement of broiler breeder hens at the late stage of production cycle influenced by phytase supplementation? *Journal of Applied Animal Research.* 43: 166–176.

Abd-Elsamee MO. 2002. Effect of different levels of crude protein, Sulphur amino acids, microbial phytase and their interaction on broiler chick performance. *Egyptian Poultry Science Journal.* 22: 999–1021.

Abdulla NR, Loh TC, Akit H. 2017. Effects of dietary oil sources, calcium and phosphorus levels on growth performance, carcass characteristics and bone quality of broiler chickens. *Journal of Applied Animal Research.* 45: 423–429.

Almeida FN, Stein H. 2010. Performance and phosphorus balance of pigs fed diets formulated on the basis of values for standardized total tract digestibility of phosphorus. *Journal of Animal Science.* 88:2968–2977.

Atia FA, PE Waibel, I Hermes, CW Carlson, MM Walser. 2000. Effect of dietary phosphorus, calcium, and phytase on performance of growing turkeys. *Poultry Science.* 79: 231–239.

Attia YA, Abd El-Rahman SA, Qota EMA. 2002. Effects of microbial phytase with or without cell-wall splitting enzymes on the performance of broilers fed marginal levels of dietary protein and metabolizable energy. *Egyptian Poultry Science Journal.* 21: 521–547.

Augspurger NR, Webel DM, Baker DH. 2007. An *Escherichia coli* phytase expressed in yeast effectively replaces inorganic P for finishing pigs and laying hens. *Journal of Animal Science*. 85: 1192–1198.

Augspurger NR, Webel DM, Lei XG, Baker DH. 2003. Efficacy of an *E. coli* phytase expressed in yeast for releasing phytate-bound phosphorus in young chicks and pigs. *Journal of Animal Science*. 81: 474–483.

Baruah K, Sahu NP, Pal AK, Debnath D. 2004. Dietary phytase: An ideal approach for a cost effective and low polluting aqua feed. *NAGA World Fish Centre Quarterly*. 27: 15–19.

Baruah K, Sahu NP, Pal AK, Jain KK, Debnath D, Mukherjee SC. 2007a. Dietary microbial phytase and citric acid synergistically enhances nutrient digestibility and growth performance of *Labeo rohita* (Hamilton) juveniles at sub-optimal protein level. *Aquaculture Research*. 38:109–120.

Baruah K, Sahu NP, Pal AK, Jain KK, Debnath D, Sona Y, Mukherjee SC. 2007b. Dietary microbial phytase and citric acid synergistically enhances nutrient digestibility and growth performance of *Labeo rohita* (Hamilton) juveniles at sub-optimal protein level. *Aquaculture Research*. 36: 803–812.

Baxter CA, Joern BC, Ragland D, Sands JS, Adeola O. 2003. Phytase, high-available phosphorus corn, and storage effects on phosphorus levels in pig excreta. *Journal of Environmental Quality*. 32: 1481–1489.

Beers S, Jongbloed AW. 1992. Effect of supplementary *Aspergillus niger* phytase in diets for piglets on their performance and apparent digestibility of phosphorus. *Animal Production*. 55: 425–430.

Cao L, Wang W, Yang C, Yang Y, Diana J, Yakupitiyage A, Luo Z, Li D. 2007. Application of microbial phytase in fish feed. *Enzyme and Microbial Technology*. 40: 497–507.

Cao L, Yang Y, Wang WM, Yakupitiyage A, Yuan DR, Diana JS. 2008. Effects of pre-treatment with microbial phytase on phosphorous utilization and growth performance of Nile tilapia (*Oreochromis niloticus*). *Aquaculture Nutrition*. 14: 99–109.

Carter CG, Sajjadi M. 2011. Low fishmeal diets for Atlantic salmon, *Salmo salar* L., using soy protein concentrate treated with graded levels of phytase. *Aquaculture International*. 19: 431–444.

Casartelli EM, Junqueira OM, Laurentiz AC, Lucas Junior J, Araujo LF. 2005. Effect of phytase in laying hen diet with different phosphorous sources. *British Poultry Science*. 7: 93–98.

Cheng N, Chen P, Lei W, Feng M, Wang C. 2015. The sparing effect of phytase in plant-protein-based diets with decreasing supplementation of dietary NaH_2PO_4 for juvenile yellow catfish *Pelteobagrus fulvidraco*. *Aquaculture Research*. 1–12. doi:10.1111/are.12845.

Dalsgaard J, Ekmann KS, Pedersen PB, Verlhac V. 2009. Effect of supplemented fungal phytase on performance and phosphorus availability by phosphorus-depleted juvenile rainbow trout (*Oncorhynchus mykiss*), and on the magnitude and composition of phosphorus waste output. *Aquaculture*. 286: 105–112.

Debnath D, Sahu NP, Pal AK, Jain KK, Yengkokpam S, Mukherjee SC. 2005a. Mineral status of *Pangasius pangasius* (Hamilton) fingerlings in relation to supplemental phytase: Absorption, whole body and bone mineral content. *Aquaculture Research*. 36: 326–335.

Denstadli V, Storebakken T, Svihus B, Skrede A. 2007. A comparison of online phytase pre-treatment of vegetable feed ingredients and phytase coating in diets for Atlantic salmon (*Salmo salar* L.) reared in cold water. *Aquaculture*. 269: 414–426.

De Sousa JPL, Albino LFT, Vaz RGMV, Rodrigues KF. 2015. The effect of dietary phytase on broiler performance and digestive and bone and blood biochemistry characteristics. *Revista Brasileira de Ciência Avícola (Brazilian Journal of Poultry Science)*. 17, Campinas.

Eeckhout W, de Paepe M. 1991. The quantitative effects of an industrial microbial phytase and wheat phytase on the apparent phosphorus absorbability of a mixed feed by piglets. *Mededelingen van de Faculteit Landbouwwetenschappen, Rijksuniversiteit Gent* 56: 1643–1647.

Engelen AJ, van der Heeft FC, Randsdorp PHG, Smit ELC. 1994. Simple and rapid determination of phytase activity. *Journal of AOAC International.* 77: 760–764.

Englmaierova M, Milosskrivan Skrivanova E, Cermak L. 2017. Limestone particle size and *Aspergillus niger* phytase in the diet of older hens. *Italian Journal of Animal Science.* doi: 10.10 80/1828051X.2017.1309258.

Englmaierova M, Milosskrivan Skrivanova E, Cermak L, Vlc-kova J. 2015. Effects of a low-P diet and exogenous phytase on performance, egg quality, and bacterial colonisation and digestibility of minerals in the digestive tract of laying hens. *Czech Journal of Animal Science.* 60: 542–549.

Fayza MS, El-Alaily HA, El-Medany NM, Abd El-Galil K. 2003. Improving P utilization in broiler chick diets to minimize P pollution. *Egyptian Poultry Science Journal.* 23: 201–218.

Fortes-Silva R, Sanchez-Vazquez FJ, Martinez FJ. 2011. Effects of pretreating a plant-based diet with phytase on diet selection and nutrient utilization in European sea bass. *Aquaculture.* 319: 417–422.

GASSC 2019. Global animal slaughter statistics and chart. *Faunalytics.* October 10, 2018. Retrieved November 5, 2019.

Gatlin DM, Barrows FT, Brown P, Dabrowski K, Gaylord GT, Hardy RW, Herman E, Hu G, Krogdahl A, Nelson R, Overturf K, Rust M, Sealey W, Skonberg D, Souza EJ, Stone D, Wilson R, Wurtele E. 2007: Expanding the utilization of sustainable plant products in aquafeeds: A review. *Aquaculture Research.* 38: 551–579.

Gessler NN, Serdyuk EG, Isakova EP, Deryabina YI. 2018. Phytases and the prospects for their application (Review). *Applied Biochemistry and Microbiology.* 54: 352–360.

Golovan SP, Meidinger RG, Ajakaiye A, Cottrill M, Wiederkehr MZ. 2001. Pigs expressing salivary phytase produce low-phosphorus manure. *Nature Biotechnology.* 19: 741–45.

Haefner S, Knietsch A, Scholten E, et al. 2005. Biotechnological production and applications of phytases. *Applied Microbiology and Biotechnology.* 68: 588–597.

Hu HL, Wise A, Henderson C. 1996. Hydrolysis of phytate and inositol tri-, tetra-, and penta-phosphates by the intestinal mucosa of the pig. *Nutrition Research.* 16: 781–787.

Hughes AL, Dahiya JP, Wyatt CL, Classen HL. 2008. The efficacy of QuantumTM Phytase™ in a 40 week production trial using White Leghorn laying hens fed corn-soybean meal based diets. *Poultry Science.* 87: 1156–1161.

Hussain SM, Afzal M, Rana SA, Javid A, Iqbal M. 2011. Effect of phytase supplementation on growth performance and nutrient digestibility of *Labeo rohita* fingerlings fed on corn gluten meal-based diets. *International Journal of Agriculture and Biology.* 13: 916–922.

Igbasan FA, Simon O, Milksch G, Manner K. 2000. Comparative studies of the in vitro properties of phytases from various microbial origins. *Archives of Animal Nutrition.* 53: 353–373.

Jackson L, Li MH, Robinson EH. 1996. Use of microbial phytase in channel catfish *Ictalurus punctatus* diets to improve utilization of phytate phosphorus. *Journal of the World Aquaculture Society.* 27: 309–313.

Jendza JA, Dilger RN, Sands JS, Adeola O. 2006. Efficacy and equivalency of an *Escherichia coli*-derived phytase for replacing inorganic phosphorus in the diets of broiler chicks and young pigs. *Journal of Animal Science.* 84: 3364–3374.

Jongbloed AW, Kemme PA. 1990. Effect of pelleting mixed feeds on phytase activity and the apparent absorbability of phosphorus and calcium in pigs. *Animal Feed Science and Technology.* 28: 233–242.

Kannan D, Viswanathan K, Edwin SC, Amutha R, Amutha R, Ravi R. 2008. Dietary inclusion of enzyme phytase in egg layer diet on retention of nutrients, serum biochemical characters and P excretion. *Research Journal of Agriculture and Biological Sciences.* 4: 273–277.

Kies AK, Van Hemert K. 2000. Phytase: A remarkable enzyme in: Selected topics in animal nutrition, biochemistry and physiology. In: *Reviews presented at the symposium on the occasion of the retirement of Dr. RR Marquardt.* W. Sauer and J. He. 119–136 pp.

Kim JC, Simmins PH, Mullan BP, Pluske JR. 2005. The effect of wheat phosphorus content and supplemental enzymes on digestibility and growth performance of weaner pigs. *Animal Feed Science and Technology*. 118: 139–152.

Kim JH, Pitargue FM, Jung H, Han GP, Choi HS, Kil DY. 2017. Effect of super dosing phytase on productive performance and egg quality in laying hens. *Asian-Australasian Journal of Animal Sciences*. doi: 10.5713/ajas.17.0149.

Kumar V, Sinha AK, Makkar HP, De Boeck G, Becker K. 2012. Phytate and phytase in fish nutrition. *Journal of Animal Physiology and Animal Nutrition*. 96(3): 335–364.

Laining A, Ishikawa M, Kyaw K, Gao J, Binh NT, Koshio S, Yamaguchi S, Yokoyama S, Koyama J. 2011. Dietary calcium/phosphorus ratio influences the efficacy of microbial phytase on growth, mineral digestibility and vertebral mineralization in juvenile tiger puffer, *Takifugu rubripes*. *Aquaculture Nutrition*. 17: 267–277.

Leytem AB, Turner BL, Thacker PA. 2004. Phosphorus composition of manure from swine fed low-phytate grains: Evidence for hydrolysis in the animal. *Journal of Environmental Quality*. 33: 2380–2383.

Liebert F, Portz L. 2005. Different sources of microbial phytase in plant based low phosphorus diets for Nile tilapia *Oreochromis niloticus* may provide different effects on phytate degradation. *Aquaculture*. 267: 292–299.

Liu L, Su J, Liang XF, Luo Y. 2013c. Growth performance, body lipid, brood amount, and rearing environment response to supplemental neutral phytase in zebrafish (*Danio rerio*) diet. *Zebrafish*. 10: 433–438.

Liu L, Zhou Y, Wu J, Zhang W, Abbas K, Fang LX, Luo Y. 2014. Supplemental graded levels of neutral phytase using pretreatment and spraying methods in the diet of grass carp, *Ctenopharyngodon idellus*. *Aquaculture Research*. 45: 1932–1941.

Liu LW, Luo YL, Hou HL, Pan J, Zhang W. 2013b. Partial replacement of monocalcium phosphate with neutral phytase in diets for grass carp, *Ctenopharyngodon idellus*. *Journal of Applied Ichthyology*. 29: 520–525.

Liu LW, Su JM, Luo TL. 2012. Effect of partial replacement of dietary monocalcium phosphate with neutral phytase on growth performance and phosphorus digestibility in gibel carp, *Carassius auratus gibelio* (Bloch). *Aquaculture Research*. 43: 1404–1413.

Liu LW, Su JM, Zhang T, Liang XF, Luo YL. 2013a. Apparent digestibility of nutrients in grass carp (*Ctenopharyngodon idellus*) diet supplemented with graded levels of neutral phytase using pretreatment and spraying methods. *Aquaculture Nutrition*. 19: 91–99.

Liu N, Liu GH, Li FD, Ands SJS, Zhang S, Heng ZAJ, Ru YJ. 2007. Efficacy of phytases on egg production and nutrient digestibility in layers fed reduced P diets. *Poultry Science*. 86: 2337–2342.

Liu N, Ru YJ. 2010. Effect of phytate and phytase on the ileal flows of endogenous minerals and amino acids for growing broiler chickens fed purified diets. *Animal Feed Science and Technology*. 156: 126–130.

Metwally MA. 2005. The effect of dietary P level with and without supplemental phytase or dried yeast on the performance of Dandarawi laying hens. *Egyptian Poultry Science Journal*. 26: 159–178.

Musapuor A, Pourrez J, Samie A, Moradi Shahrbabake H. 2005. The effect of phytase and different level of dietary calcium and phosphorous on phytate P utilization in laying hens. *International Journal of Poultry Science*. 8: 560–562.

Newkirk RW, Classen HL. 2001. The non-nutritional impact of phytate in canola meal fed to broiler chicks. *Animal Feed Science and Technology*. 91: 115–128.

Nwanna LC. 2007. Effect of dietary phytase on growth, enzyme activities and phosphorus load of Nile tilapia (*Oreochromis niloticus*). *Journal of Engineering and Applied Sciences*. 2: 972–976.

Nwanna LC, Ajani EK, Bamidele F. 2006. Determination of optimum dietary phytase level for the growth and mineral deposition in African catfish (*Clarias gariepinus*). *Aquatic Sciences*. 54: 75–82.

Nwanna LC, Fagbenro OA, Adeyo AO. 2005. Effects of different treatments of dietary soybean meal and phytase on the growth and mineral deposition in African catfish *Clarias gariepinus*. *Journal of Animal and Veterinary Advances*. 4: 980–987.

Onyango EM, Bedford MR, Adeola O. 2005. Efficacy of an evolved *Escherichia coli* phytase in diets of broiler chicks. *Poultry Science*. 84: 248–255.

Panda A, Rao SR, Raju M, Bhanja S. 2005. Effect of microbial phytase on production performance of White Leghorn layers fed on a diet low in non-phytate phosphorus. *British Poultry Science*. 46: 464–469.

Papatryphon E, Howell RA, Soares JH, Jr. 1999: Growth and mineral absorption by striped bass *Morone saxatilis* fed a plant feedstuff based diet supplemented with phytase. *Journal of the World Aquaculture Society*. 30: 161–173.

Peers FG. 1953. The phytase of wheat. *Biochem. J*. 53: 102–110.

Phromkunthong W, Gabaudan J. 2006. Use of microbial phytase to replace inorganic phosphorus in sex-reversed red tilapia: 1 dose response. *Songklanakarin Journal of Science and Technology*. 28: 731–743.

Phromkunthong W, Nuntapong N, Gabaudan J. 2010. Interaction of phytase RONOZYME®P (L) and citric acid on the utilization of phosphorus by common carp (*Cyprinus carpio*). *Songklanakarin Journal of Science and Technology*. 32: 547–554.

Pointillart A, Fontaine N, Thomasset M, Jay ME. 1985. Phosphorus utilization, intestinal phosphatases and hormonal control of calcium metabolism in pigs fed phytic phosphorus: Soyabean or rapeseed diets. *Nutrition Reports International*. 32: 155–167.

Ponnuvel P, Narayankutty K, Jalaludden A, Anitha P. 2015. Effect of phytase supplementation in low energy-protein diet on the production performance of layer chicken. *The Indian journal of Veterinary Science & Biotechnology*. 3: 25–27.

Ptak A, Bedford MR, Swiatkiewicz S, Zyla K, Jozefiak D. 2015. Phytase modulates ileal microbiota and enhances growth performance of the broiler chickens. *PLoS One*. 10: e0119770.

Punna S, Roland DA. 1999. Influence of supplemental microbial phytase on first cycle laying hens fed phosphorus deficient diets from day one of age. *Poultry Science*. 78: 1407–1411.

Rodehutscord M. 1996. Response of rainbow trout (*Oncorhynchus mykiss*) growing from 50 to 200 g to supplements of dibasic sodium phosphate in a purified diet. *The Journal of Nutrition*. 126: 324–331.

Rodehutscord M, Faust M, Lorenz H. 1996. Digestibility of phosphorus contained in soybean meal, barley, and different varieties of wheat, without and with supplemental phytase fed to pigs and additivity of digestibility in a wheat-soyabean-meal diet. *Journal of Animal Physiology and Animal Nutrition*. 75: 40–48.

Roy PK, Lall SP. 2003. Dietary phosphorus requirement of juvenile haddock (*Melanogrammus aeglefinus*, L.). *Aquaculture*. 221: 451–468.

Rutherfurd SM, Chung TK, Morel PCH, Moughan PJ. 2004. Effect of microbial phytase on ileal digestibility of phytate P, total P, and amino acids in a low-P diet for broilers. *Poultry Science*. 83: 61–68.

Sajjadi M, and Carter CG. 2004. Effect of phytic acid and phytase on feed intake, growth, digestibility and trypsin activity in Atlantic salmon (*Salmo salar*, L.). *Aquaculture Nutrition*. 10: 135–142.

Sandberg AS, Larsen T, Sandstrom B. 1993. High dietary calcium level decreases colonic phytate degradation in pigs fed a rapeseed diet. *The Journal of Nutrition*. 123: 559–566.

Sardar P, Randhawa HS, Abid M, Prabhakar SK. 2007. Effect of dietary microbial phytase supplementation on growth performance, nutrient utilization, body compositions and haemato-biochemical profiles of *Cyprinus carpio* (L.) fingerlings fed soyprotein-based diet. *Aquaculture Nutrition*. 13: 444–456.

Schafer A, Koppe WM, Meyer-Burgdorff KH, Gunther KD. 1995. Effects of a microbial phytase on the utilization of native phosphorus by carp in a diet based on soybean meal. *Water Science & Technology*. 31: 149–155.

Schlemmer U, Jany KD, Berk A, Schulz E, Rechkemmer G. 2001. Degradation of phytate in the gut of pigs—pathway of gastrointestinal inositol phosphate hydrolysis and enzymes involved. *Archives of Animal Nutrition.* 55: 255–280.

Selle PH, Cadogan DJ, Bryden WL. 2003a. Effects of phytase supplementation of phosphorus-adequate, lysine-deficient, wheatbased diets on growth performance of weaner pigs. *Australian Journal of Agricultural Research.* 54: 323–330.

Selle PH, Ravindran V. 2007. Microbial phytase in poultry nutrition. *Anim. Feed Sci. Technol.* 135: 1–41.

Sharma A, Ahluwalia O, Tripathi AD, Singh G, Arya SK. 2020. Phytases and their pharmaceutical applications: Mini-review. *Biocatalysis and Agricultural Biotechnology.* 23: 101439.

Shehab AE, Kamelia MZ, Khedr NE, Tahia EA, Esmaeil FA. 2012. Effect of dietary enzyme supplementation on some biochemical and hematological parameters of Japanese quails. *Journal of Animal Science Advances.* 2: 734–739.

Shirley RB, Edward HM. 2003. Graded levels of phytase past industry standards improves broilers performance. *Poultry Science.* 82: 671–680.

Stefan H, Anja K, Edzard S, Joerg B, Markus L, Oskar Z. 2005. Biotechnological production and applications of phytases. *Applied Microbiology and Biotechnology.* 68: 588–597.

Sukumar D, Jalaudeen A. 2003. Effect of supplemental phytase in diet on certain economic traits in layer chicken. *Journal of Animal Science.* 73: 1357–1359.

Suzuki U, Yoshimura K, Takaishi M. 1907. About the enzyme "phytase", which splits 'anhydro-oxy-methylene diphosphoric acid'. *Bulletin of the College of Agriculture, Tokyo Imperial University,* 7, 503–512, 1160 (in German).

Vandenberg GW, Scott SL, Noue JDL. 2012. Factors affecting nutrient digestibility in rainbow trout (*Oncorynchus mykiss*) fed a plant protein based diet supplemented with microbial phytase. *Aquaculture Nutrition.* 18: 369–379.

Van Weerd JH, Khalaf KA, Aartsen FJ, Tijssen PAT. 1999. Balance trials with African catfish *Clarias gariepinus* fed phytase-treated soybean meal-based diets. *Aquaculture Nutrition* 5: 135–142.

Vashishth A, Ram S, Beniwal V. 2017. Cereal phytases and their importance in improvement of micronutrients bioavailability. *3 Biotech.* 1; 7(1): 42.

Vielma J, Lall SP, Koskela J. 1998. Effects of dietary phytase and cholecalciferol on phosphorus bioavailability in rainbow trout (*Oncorhynchus mykiss*). *Aquaculture.* 163: 309–323.

Viveros A, Centeno C, Brenes A, Canales R, Lozano A. 2000. Phytase and acid phosphatase activities in plant feedstuffs. *Journal of Agricultural and Food Chemistry.* 48: 4009–4013.

Waldroup PW, Kersey JH, Saleh EA, Fritts A, Yan F, Stillborn HL, Crum Jr RC, Raboy V. 2000. Non phytate P requirement and P excretion of broilers chicks fed diets composed of normal or high available phosphate corn with and without microbial phytase. *Poultry Science.* 79: 1451–1459.

Wang F, Yang YH, Han ZZ, Dong HW, Yang CH, Zou ZY. 2009. Effects of phytase pretreatment of soybean meal and phytase-sprayed in diets on growth, apparent digestibility coefficient and nutrient excretion of rainbow trout (*Oncorhynchus mykiss* Walbaum). *Aquaculture International.* 17: 143–157.

Ye CX, Liu YJ, Tian LX, Mai KS, Du ZY, Yang HJ, Niu J. 2006. Effect of dietary calcium and phosphorus on growth, feed efficiency, mineral content and body composition of juvenile grouper, *Epinephelus coioides. Aquaculture.* 255: 263–271.

Zhu Y, Qiu X, Ding Q, Duan M, Wang C. 2014. Combined effects of dietary phytase and organic acid on growth and phosphorus utilization of juvenile yellow catfish *Pelteobagrus fulvidraco. Aquaculture.* 430: 1–8.

Zimmermann B, Lantzsch HJ, Mosenthin R, Biesalski HK, Drochner W. 2003. Additivity of the effect of cereal and microbial phytases on apparent phosphorus absorption in growing pigs fed diets with marginal P supply. *Animal Feed Science and Technology.* 104: 143–152.

Zimmermann B, Lantzsch HJ, Mosenthin R, Schoner FJ, Biesalski HK, Drochner W. 2002. Comparative evaluation of the efficacy of cereal and microbial phytases in growing pigs fed diets with marginal phosphorus supply. *Journal of the Science of Food and Agriculture*. 82: 1298–1304.

Zyla K, Mika M, Swiatkiewicz S, Koreleski J, Piironen J. 2011. Effects of phytase B on laying performance, eggshell quality and on P and Ca balance in laying hens fed P-deficient maize-soybean meal diets. *Czech Journal of Animal Science*. 56: 406–413.

13 Applications of Immobilized Ligninolytic Enzymes in the Degradation of Industrial Pollutants

*Muhammad Bilal, Hamza Rafiq,
Sarmad Ahmad Qamar, Asim Hussain,
Pankaj Bhatt, and Hafiz M. N. Iqbal*

ABSTRACT

In the modern world, various textile and chemical sectors demand a change from classical methodologies to novel approaches based on greener, eco-friendly, and sustainable chemicals and catalytic alternatives at both industrial and laboratory scales. In this context, herein, the catalyst based on the biological molecules shows the various features along with the high potential of environmental and biological applications. The biological catalysts are very efficient in their action as compared to the other chemical-based methods conventionally used. Moreover, there was no production of unnecessary and harmful metabolites as environmental pollutants. Modern research enabled industrial-scale production methodologies for biocatalyst development and immobilization, which is a solid base toward being tailored economically. Nonetheless, it is crucial to reprocess and recover the enzyme for feasibility at a commercial level, and this challenge can be easily accomplished using engineered catalysts by immobilization. In this chapter, we aim to concisely outline the lignin modification enzymes (LMEs) and their immobilization; manganese peroxidase (MnP), lignin peroxidase (LiP), and laccase are of key interest. Moreover, specific importance has been devoted to the recent successful experiments of different LME carriers, immobilized for the detoxification, decolorization, and degradation of the dyes/effluents of the industries.

Keywords: industrial effluents, ligninolytic enzymes, immobilization, biocatalyst, dyes degradation, environmental remediation

DOI: 10.1201/9781003202998-13

CONTENTS

13.1 Introduction..250
13.2 White Rot Fungi and Their Enzymes 251
13.3 Enzyme Immobilization for Improved Catalytic Performance....................253
      13.3.1 Immobilized LiPs..255
      13.3.2 Immobilized MnPs..255
      13.3.3 Immobilized Laccases 256
13.4 Environmental Application 257
      13.4.1 Degradation and Decolorization of Industrial Dyes 257
              13.4.1.1 LiP-Based Degradation of Dyes 258
              13.4.1.2 MnP-Based Degradation of Dyes 259
              13.4.1.3 Laccase-Based Degradation of Dyes 259
13.5 Detoxification of Waste Dyes 260
13.6 Summary and Perspectives 261
13.7 Disclosure Statement 262
References..262

13.1   INTRODUCTION

The establishment of industries is crucial for economic growth; however, it is always associated with the problem of pollution in the environment. In addition, different kinds of pollutants are released into the environment, which are highly toxic for humans and the environment. Particularly, the residual dyes from various industries and their release in the form of wastewater are a major scientific concern (Akey and Appel, 2021; Qamar et al., 2020). It has been reported that the annual production of dyes is about 7×10^5 tons. In the processing of textiles, about 15–20% of dyes are released into the environment as waste (Asgher et al., 2018; Bilal et al., 2019a, 2019c). Dyes are used in various industries in the formation of different kinds of products, such as plastics, textiles, photographs, pharmaceuticals, pulp and paper, food, tanning substances, and cosmetics, with different proportions (Issakhov et al., 2021). In addition, the removal of residual dyes in the streams significantly endangers the natural environment because of their mutagenic, carcinogenic, and genotoxic nature (Nathan et al., 2018). Therefore, to avoid ecosystem destruction, these dye-based effluents should be efficiently treated without generating crucial secondary emissions (Zhang et al., 2019). Figure 13.1 describes harmful effects of dyes/effluents on the environment, aquatic habitats, plants, and the human population.

Environmental engineers have actively been involved in designing novel technologies or improving current technologies over the last few years. Different treatment modalities are applied for effluents and dyes (Rodgers et al., 2019). Unfortunately, none of them have been industrialized due to the extensive cost, requiring high labor, harsh operating conditions, and toxic by-products (Persico and Venator, 2019). To overcome this problem, additional effort is required with high applicability of industrially adaptable (Klimek-Szczykutowicz et al., 2020). However, remediation using biobased technologies is not new. There has recently been a lot of interest in enzyme usage as green catalysts in industries for the purpose of remediation.

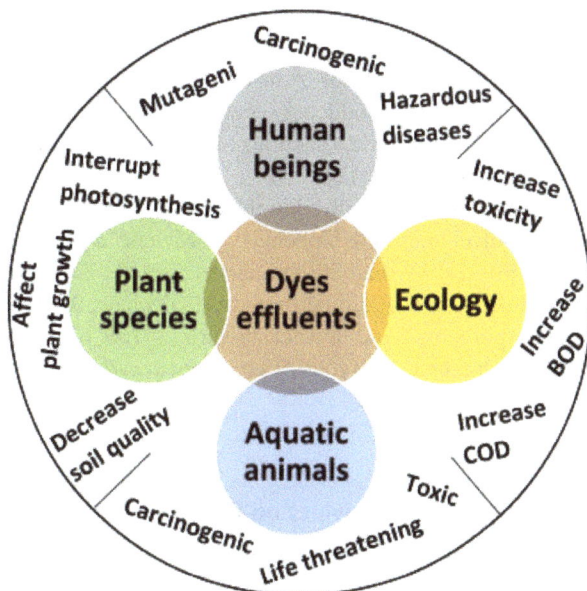

FIGURE 13.1 Harmful effects of dyes/effluents on the environment, aquatic habitats, plants, and the human population; reprinted with permission from Qamar et al. (2020).

Enzymes are natural catalysts, and they can work in various temperature and pH ranges, pollutants, and salinity concentrations (Geng et al., 2020). WRF-based LMEs are especially intriguing for environmental exploitability due to their high oxidative capacities, steric selectivity, and specificity. Enzymes are naturally active, reliable, stable, and energetic catalysts (Sánchez and Montoya, 2020).

Sometimes, these are not completely modified due to industrial exploitability. In this regard, engineering and immobilization of enzymes present remarkable and effective methods to evade problems of instability and desired biocatalyst used in the industries are obtained. Besides the issues of heat inactivation (Girelli et al., 2020), the immobilized biocatalysts show enhanced stability against salinity, organic solvents, autolysis, heavy metals, and denaturant issues (Kumari et al., 2020).

13.2 WHITE ROT FUNGI AND THEIR ENZYMES

Around 10,000 types of pigments and dyes are annually produced worldwide and are widely being used in the printing, textile, and other industries. A significant amount of the textile dyes (about 10–15%) are released in the form of wastewater. Various dyes show high stability microbial attack, temperature, and light which make them intractable compounds, which are very toxic in nature (Dao et al., 2019). Synthetically produced dyes show chemical variation and could be divided into triphenylmethane, azo, polymeric, and heterocyclic structures. Hence, the biocatalysts used in decolonization have an ability for dye degradation with carried structures (Bilal et al., 2019b). From various organisms involved in ligninolysis, white rot fungi show the highest efficiency in this work. Manganese peroxidases (EC 1.11.1.13), laccases (EC 1.10.3.2), and lignin peroxidases (EC 1.11.1.14) are enzymes with non-stereoselective and nonspecific nature, which is helpful in the degradation of lignin and possess different functions along with H_2O_2-producing secondary metabolites and oxides (Bilal et al., 2019a). Table 13.1 illustrates the functions of ligninolytic enzymes after immobilization on different support materials.

Ligninolytic enzymes are also involved in the degradation of dioxins, chlorinated phenols, pesticides, polychlorinated biphenyls, polycyclic aromatic hydrocarbons (PAHs), dyes, and explosives (Wong, 2009; Sabapathy et al., 2020). Laccases, manganese peroxidases (MnP), and lignin peroxidases (LiP) have the ability to degrade a variety of different pollutants and decolorize various dyes of different structures (Majeke et al., 2021). One or more types of enzymes are secreted by the white rot

TABLE 13.1
Immobilization of Ligninolytic Enzymes on Different Support Materials

Enzyme	Function	Immobilization Support Material	References
Manganese peroxidase	Degrade the phenolic moieties of lignin	Porous silica beads	Van et al., 2000
		Chitosan microspheres	Ran et al., 2012
Laccase	Degrades both β-1 and β-O-4 dimers via C-C cleavage, C oxidation, and alkyl-aryl cleavage	Ca-alginate beads	Bilal and Asgher, 2015
		Alginate-gelatin mix gel	Mogharabi et al., 2012
		Chitosan beads	Datta et al., 2020
		Sol-gel matrix of trimethoxysilane (TMOS) and proplytetramethoxysilane (PTMS)	Bagewadi et al., 2017
		Activated carbon fibers	Zhang et al., 2018
		Nanosized magnetic biochar	Zhang et al., 2020
		Polyacrylamide	Gokgoz and Altinok, 2012
		SiO_2 nanocarriers	Patel et al., 2014
		Nanofibrous membrane	Taheran et al., 2017
Lignin peroxidase	Cleavage of β-0-4 ether bonds and of Cα-Cβ linkages in lignins	Nano-porous gold	Qiu et al., 2009
		Fe_3O_4, SiO_2, polydopamine (PDA) nanoparticles	Guo et al., 2019
		Ca-alginate beads	Shaheen et al., 2017
		Xerogel matrix of trimethoxysilane (TMOS) and proplytetramethoxysilane (PTMS)	Asgher et al., 2012a

fungi, and the enzymes, which could be intracellular or membrane-bound, are involved in in vivo degradation of (PAHs) (Mei et al., 2020). White rot fungi are the physiological cluster of basidiomycetes that shows great ability to lignin degradation by giving an appearance like bleach in wood on which they attack (Gabhane et al., 2020). The LiP, Lac, and MnP are the LMEs, and various other enzymes are involved in the process of degradation, such as glyoxal oxidase, cytochrome P-450 monooxygenase, and oxalate decarboxylase (Schneider et al., 2020). Various strains of white rot fungi secrete different individual or multiple complexes of ligninolytic enzymes in different concentrations under the feasible environment for fermentation (He et al., 2020). Consequently, white rot fungi are classified into four classes based on the difference in composition of ligninolytic enzymes and various secretion: (i) laccase, LiP, and MnP producing: *Schyzophyllum commune, Ganoderma lucidum, Trametes versicolor, Pleurotus eryngii*, and *Pleurotus ostreatus* (Eslami et al., 2019); (ii) laccase and MnP producing: *Phlebia radiata* and *Pycnoporus cinnabarinus* (Mattila et al., 2020); (iii) MnP and LiP producing: *Phanerochaete chrysosporium* (Seo et al., 2018); and (iv) laccase and LiP producing: *Dichomitus Squalens* (Shrestha et al., 2020).

In white rot fungi, high distribution of MnP can be seen as compared to the LiP. Enzyme MnP oxidizes Mn^{2+} to Mn^{3+}, which in return produce free phenoxy radicals from the phenolic structures by oxidation (Hou et al., 2020). Therefore, LiP has greater potential than MnP and also shows significant oxidative capability in favor of in vitro phenolic substrate. Laccases are actually benzenediol oxidoreductase having the ability of lignin degradation and a wide range of other various aromatic compounds (Fang et al., 2020). They are involved in substrate oxidation by the reduction of molecular oxygen to water. VPs are also included in catalyzing the efficient oxidation of various compounds of phenol and dye molecules, which are the general peroxidases substrates (Catucci et al., 2020). This kind of enzyme oxidizes MnP and catalyzes Mn^{2+} and various other aromatic compounds, having a high redox potential like LiP. Substantial research work shows the increased interest in VPs in the previous years both as a new biocatalyst for industrial sources and models for ligninolytic enzymes.

13.3   ENZYME IMMOBILIZATION FOR IMPROVED CATALYTIC PERFORMANCE

The matrix for support (carrier) on which the catalyst may be loaded for coupling to hold the molecules in a proper direction is known as immobilization support. Due to convenient handling and by using the practical applications, it can easily be recycled, which shows the best impact on economic growth (Ariaeenejad et al., 2021). After immobilization, the enhancement in stability can be seen, and it becomes highly robust for denaturation, degradation, and aggregation. By using various techniques for enzymatic immobilization, a large number of studies have been conducted showing the greater number of applications in industrial bioprocessing (Bernal et al., 2018). Various types of support materials used for the immobilization of enzymes have been briefly explained. Nanostructured materials, which are similar in size to proteins, have gained high scientific interest among various support materials because they provide a larger surface area for enzyme attachment and allow a greater degree of freedom to the active sites of enzymes, as well as minimizing lateral contact between enzymes. The enzymatic activity could be maintained at a higher possible level. Despite their high surface immobilization areas, nanoparticles prevent the loss of enzyme particle complexes through entrainment or adsorption when used in functional microporous media, such as membranes (Hou et al., 2014). Advantageous features of immobilized enzymes using various polymeric/nanostructured support matrices have been represented in Figure 13.2.

The most important advantage of synthetic polymers as supporting materials is that the monomeric building blocks can be chosen according to enzyme specifications and the method by which the immobilizing substance is used. The composition of the polymer can be used to observe a wide range of confirmed chemical functional groups (Zdarta et al., 2018). These include carboxylic, hydroxylic, epoxy, amine, diol, and highly hydrophobic alkyl groups, as well as trialkyl amines. These groups allow effective enzyme binding and polymer surface functionality because the use of synthetic polymeric supports involves almost all forms of immobilization, the type of functional groups determines whether the enzyme is anchored into the matrix by adsorption or covalent link formation (Wouters et al., 2019). Polymers derived from nature provide excellent alternatives to synthetic polymers for enzymes immobilization purposes. Biopolymers also contain carbohydrates and proteins, such as albumin and gelatin. Gelatin, cellulose, keratins, carrageen, chitin, and alginate are examples

FIGURE 13.2 Advantageous features of immobilized enzymes using various polymeric/ nanostructured support materials; reprinted from Bilal and Iqbal (2019) with permission.

of biopolymers used for immobilization. Biopolymers have a unique set of properties ranging from biodegradable and nontoxic materials to harmless materials with excellent protein affinity, making them ideal for enzyme support (Bilal and Iqbal, 2019). Their biological origin and biocompatibility mitigate their negative effects on the structure and properties of enzymes, allowing immobilized proteins to retain strong catalytic activities. Furthermore, it allows for a direct reaction between the enzyme and the matrix and promotes surficial alteration via the availability of reactive functional units in their structures, primarily hydroxyl moieties, but also amino and carbonyl moieties (Zdarta et al., 2018).

13.3.1 IMMOBILIZED LIPS

A range of different support materials and carriers have been reported for the immobilization of LiPs obtained from different strains of white rot fungi. Surprisingly, the derivatives of immobilized enzymes exhibit higher catalytic reusability, efficiency, and thermostability as compared to free enzymes. LiPs were studied for encapsulation using carbon nanotubes (CNTs), which significantly increased the catalytic performance of enzymes (Oliveira et al., 2018). The enhancement in the catalysis efficiency may be recognized variation in the frequency of collision between enzyme and substrate in which nanoparticles are incorporated (Jiang et al., 2020). Furthermore, due to the exclusive structural features of nanoencapsulation, CNTs enhance the efficacy of enzyme, which is encapsulated, decreasing the interactions between the protein molecules (Bilal and Iqbal, 2020). In comparison to classical nanocarriers (e.g., silica, epoxy, and zirconia), CNTs show one of the important support materials for the immobilization of enzymes (Saldarriaga-Hernández et al., 2020). When enzymes are placed for immobilization with CNTs the enhanced catalytic functionality of enzymes is obtained with several folds than free enzymes (Xing et al., 2020).

The sol-gel encapsulation technology is highly efficient to use for the immobilization of LiPs. The precursor used in the sol-gel is based on the hydrolysis of alkoxide under different conditions (e.g., alkaline or acidic conditions) by following the concentrations of different hydroxylated units, which is helpful in the porous gel development (Bilal et al., 2017a). In the very first step, a metal alkoxide with low molecular weight (e.g., tetraethyl orthosilicate) can be hydrolyzed at alkaline and acidic conditions, which create (Si-OH) as a result (Kumar and Chandra, 2020). The second step the reduction reactions between the moieties of silanol, which led to the siloxane (Si-O-Si) formation, generates a matrix of sol-gel at which the enzyme is highly trapped (Pylypchuk et al., 2020). LiP based on sol-gel shows high performance and a satisfactory covalent attachment on the Sepabeads and Dilbeads NK (EC-EP3), and the other one is a siliceous foam-like matrix or efficient and stable cross-link of enzyme aggregates (CLEAs) (Zhou et al., 2020).

13.3.2 IMMOBILIZED MNPS

MnPs are often used in the bio-based decolorization, and degradation of dyes present in industrial wastewater. Consequently, various techniques for catalytic immobilization have been developed for efficient and maximum efficiency and activity (Siddeeg et al., 2020). Among these techniques, encapsulation, physical entrapment, and per-oxidase covalent linkage are most commonly used. MnP production from different microorganisms and their immobilization into sol-gel, chitosan, alginate, agar-agar, polyacrylamide, and chitosan alginate-gel or poly-alginate show the maximum activity and long storage time, maximum optical activity, thermal and pH stability, and high selectivity and sensitivity with the shorter response time and with higher reproducibility (Zhao et al., 2020). A strain of the white rot fungus, namely *T. versicolor* IBL-04, was captured in the matrix of sol-gel composed of PTMS and TMOS with a greatest 92.7% immobilization efficiency by the use of 2 mg/mL enzyme concentration (Asgher et al., 2017a). Furthermore, entrapped MNPs have been shown to preserve almost 84.5% with its original activity, after completion of the ten repetitive cycles of $MnSO_4$, compared to the enzyme that is present in soluble form, and it was also reported that it offers 65.8% high solubility and durability of MnP for 75 days at 25°C (Asgher et al., 2017b).

It has been investigated that a wide range of variables influences the encapsulation efficiency of purified MnP by *Ganoderma lucidum* IBL-05 on the beads of polyvinyl alcohol alginate (Bilal et al., 2019b). Unlike MnP in soluble form, the encapsulated PVA and their derivatives show higher catalytic performance in pH values ranging from 4 to 9 (Elkady et al., 2020). In addition to this, coupling carriers, which make the enzyme able to remain active and stable with various ranges of high temperatures, such as at 65°C, have the finest features for the support of PVA with encapsulation efficiency (Matinja et al., 2019). The immobilized enzyme shows greater stability to thermal change, a long time for storage, a wide range of pH with a high range of activation, and stability for MnP immobilization as compared to free MnP (Bilal et al., 2017). The attachment of the enzymes on the charge support material shifted the pH profile activity toward the alkaline or acidic (Yang et al., 2019). The variation in pH causes the modification of peroxidase structure and the activity of enzyme attached with the catalytic site (Sam et al., 2021). Similarly, as with other strategies for immobilization, the method used for sol-gel has contributed wonderful physicochemical features enhancement after MnP immobilization. As compared to free enzymes, allophone-carrier supported MnP shows increased thermal stability at high temperatures and a wider range of pH. It was also investigated that the three-dimensional structure of MnP is well preserved, derived from *P. chrysosporium*, and has its 23% of activity retention after immobilization on the MWCNTs in the interaction (Bosco et al., 2017).

13.3.3   IMMOBILIZED LACCASES

For better functionality and catalytic performance, enzymes usually have been improved earlier than their application in industries where recyclability and improved yield are required (Behbahani et al., 2020). Due to multiple applications of laccases from fungal sources, many of the researchers have attempted to develop a number of different carriers helpful in the development of nano-biocatalyst with high efficiency (Parra-Arroyo et al., 2020). Attachment of multiple enzymes results in the increased stability and binding capacity in which the use of hetero-functional support takes place (Yang et al., 2020). Laccase immobilization on the surface of microcapsules and gel not only modifies the immobilization degree but also enhances recoverability with catalytic features and functional stability (Sarkar et al., 2020). Another laccase-producing fungus, *T. versicolor*, has been immobilized on carbon nanomaterials, such as graphene oxides (GOs), carboxylated multi-walled carbon nanotubes (MWNT-COOHs), and multi-walled carbon nanotubes (MWCNTs), by using different techniques (Bilal et al., 2020).

In the application in the environment, laccase immobilized on the CNMs act as usable catalyst depending on pH and offer high stability against varied temperature (Bilal et al., 2020). CNTs are placed on the membrane, which enhances the desired features of immobilized laccase, such as the maintenance activity loading of enzymes, thermostability, shelf-life extension, and operative stabilities under the different conditions of various chemical agents (Lara-Espinoza et al., 2021). The carbon nanotubes provide the microenvironment with biocompatibility, and increasing the electrical conductance, the improvement in features can also take place (Nyika, 2021).

The immobilized laccase and its derivatives show fivefold enhanced efficiency and performance as compared to the laccase present in the free form. Moreover, it offers the ability to withstand a wide range of temperatures, from 50 to 80°C, when compared to free enzymes, which makes it the finest candidate for various applications in industrial sectors. Another technique for the entrapment of laccase enzymes on the sol-gel matrix from strain *P. ostreatus* with purified form is successfully immobilized (Asgher et al., 2012b). The immobilized enzyme shows optimum thermostability and activity at 60°C and pH 5 with a 230-minute time of half-life and shows the higher V_{max} and K_m as compared to the free enzyme. Secondary interaction and method of immobilization may be the exposition for pH optimal displacement from the lower acidic to higher (Aricov et al., 2020). Likewise, solid-phase catalyst shows greater stability against various denaturants as compared to free enzyme. A method of physical absorption in which immobilization of laccase on different sizes of nano-porous gold was performed (Salaj-Kosla et al., 2013). The similar effect can be seen for polyacrylonitrile (PAN)-based support, which possess excellent solvent and mechanical properties, resistance to abrasion with high tensile strength. The encapsulation of laccase with support material of Ca-alginate beads shows activity of about 61%, and the recovery with Ca-alginates beads is about 42.5%. Similarly, the immobilization of laccases on various kinds of support materials shows resistance for conformational change, hence highly increasing the life span and steadiness (Deepa et al., 2020).

13.4 ENVIRONMENTAL APPLICATION

13.4.1 DEGRADATION AND DECOLORIZATION OF INDUSTRIAL DYES

Among other microbes, white rot fungi are the best to produce enzymes involved in the process of decolorization or degradation of dyes, which are released into the environment as an effluent. These organisms show greater potential in the transfer of a wide range of organic compounds (Varjani et al., 2020). The present scientific literature related to dye-based pollutants aiming to decolorize/degrade summarizes the possible use of the enzymes from white rot fungi in the lysis of lignocellulose, which is really helpful in the metabolism of dyes (Dileepkumar et al., 2020). Lignin and dyes have similarities in structure; therefore, dyes are easily degraded by these enzymes. The potential of environmental and biological significance has controlled on substantially high demand for enzyme use in ligninolysis (Routoula and Patwardhan, 2020). Nevertheless, still, there are many difficulties in industrial applications of ligninolytic enzymes, which are involved in the degradation of dyes and enzyme non-reusability. Therefore, ligninolytic enzyme immobilization has shown high practicality as compared with older chemical-based industrial processes (Ojha and Thareja, 2020).

13.4.1.1   LiP-Based Degradation of Dyes

To explain the decolorization of the dyes or effluents of industries based on the assisted LiP, some degradation of the organic molecules with aromatic compounds degraded by the LiP has been discussed in this portion. Various different aromatic compounds are mineralizing with high resistance and can be catalyzed using the immobilized LiP (Singh and Dwivedi, 2020). The catalytic cycle of LiP-assisted degradation is shown in Figure 13.3 (Singh et al., 2021). Furthermore, it is also involved in the oxidation of many phenolic and polycyclic compounds (Bento et al., 2020). The LiP immobilization on the absorbent ceramic support completely increases the stability of enzymes and offers a greater potential for degradation of environmental persistent aromatic pollutants (Gianolini et al., 2020). Likewise, it was investigated that LiP immobilization on activated CNBr Sepharose 4B is used in the removal of color from the kraft effluents. The treatment of about 3 hours with LiP immobilized on the Amberlite (IRA-400) showed greater results in decolorization (70%), phenol (55%), and total organic carbon removal from the effluents of pulp mill (Husain and

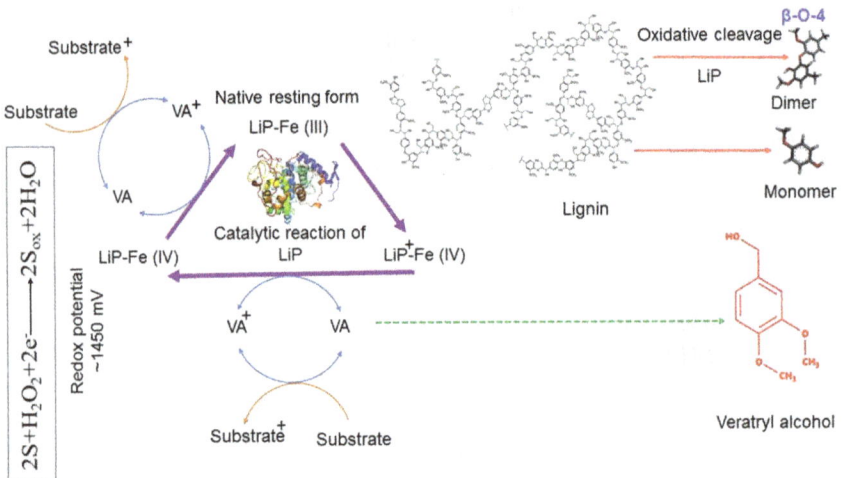

FIGURE 13.3 Extended schematic illustration of LiP catalytic cycle, comprising the oxidative catalysis of VA and oxidative of lignin monomer and dimer compounds (β-O-4 linkage-based dimer compound) (Singh et al., 2021).

Ulber, 2011). There is another system of removal of dyes in which LiP is immobilized on the nano-porous gold was used for the decolorization of pyrogallol red, fuchsine, and rhodamine B (Mawad et al., 2020).

13.4.1.2 MnP-Based Degradation of Dyes

The magnetic nanoparticles extracted by *G. lucidum* IBL-05 and the nanoparticles entrapped by polyvinyl alcohol alginate (PVA) beads have the highest efficiency for decolorization against a new group of dyes that react with Sandal and a wide range of industrial wastewater, which have a range of 77.96% to 91.78% and 60 to 79%, respectively (Bilal et al., 2019a). Amazingly, after the completion of the six batches of decolorization, the PAV-alginate magnetic nanoparticles preserve 63.9% of their original activity for the Foron Blue (E$_2$BLN) dye (Rather et al., 2019). But the reduction activity loss was irrelevant for the start of three cycles and gradually moved toward decomposition (Bhat et al., 2020). Free radical accumulation is very active and causes the reduction in practical activity such as the lipid, peroxyl, Mn^{3+}, and hydroxyl radicals, and regulator in microsphere environment that knotted the enzyme active site, resulting in the activation of enzymes (Bilal and Iqbal, 2019). It was reported that the strain *G. lucidum* IBL-05, in which magnetic nanoparticles were entrapped by the sol-gel method, showed a decolorization efficiency of about 90% after 4 hours (Bilal, Asgher and Ramzan, 2015; Khalid et al., 2020). The parameter for the toxicity, such as the nitroimines and formaldehyde, is helpful in the effluent degradation within the accepted range (Shaheen et al., 2017). The decolorization or degradation of three structurally dissimilar dyes, such as the Reactive Blue 21 (RB-21), Reactive Yellow (RY-145A), and Reactive Red 195A (RR- 195A), have also been described by agar-agar encapsulation of manganese peroxidase of *G. lucidum* IBL-05. The decolorization of various dyes with maximum efficiency was within the range of 78.6–84.7% (Bilal et al., 2016).

13.4.1.3 Laccase-Based Degradation of Dyes

Decolorization or degradation of dyes based on laccases has widely been reported in the literature using various techniques (Figure 13.4) (Pulicharla et al., 2018; Britos et al., 2018). Immobilization of laccases of *Trametes modesta* on the oxide pellets of aluminum is involved in the decolorization and degradation of different dyes (Morsi et al., 2020). The alumina is activated and functionalized using a glutaric dialdehyde and 3-aminopropyltriethoxysilane (APTES), and a sensor of spectroscopic was used for the monitoring of decolorization. It is also investigated that APTES is used for salinized pellets of alumina, which is followed by activation by using glutaraldehyde cross-linked enzyme derived from *T. hirsute* and used in degradation and decolorization of methyl green and Remazol Brilliant Blue R (RBBR) (Mate and Mishra, 2020). The actual degradation/decolorization of the RBBR dye was recognized due to the high activity of enzymes rather than carrier support for the adsorption of the dyes (Begum et al., 2020). A greater rate of decolorization with 55% and 80% were achieved for RB-4 and RY-2 by the use of MCMs immobilized laccase in the batch reactors. The removal of two different-colored dyes, Acid Orange 7 and Acid Blue 25, with the result of 64% and 76% after the contact time of 65 minutes. It was investigated that in the 1-hour contact time, 96% of azo phenolic dye decolorization was

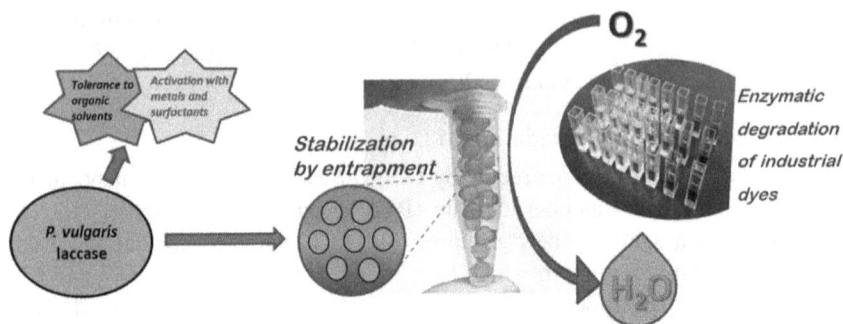

FIGURE 13.4 Degradation of industrial dyes by entrapment stabilized laccases produced by *P. vulgaris*; reprinted from Britos et al. (2018) with permission.

executed, but there is the highest operational stability, and good storage of laccase can be seen (Yamak et al., 2009; Ramírez-Montoya et al., 2015). The removal of dyes from the aqueous solution using the immobilized laccase from *T. versicolor* on the chitosan/microspores shows the efficiency of dye removal (Asgher et al., 2018).

13.5   DETOXIFICATION OF WASTE DYES

The industry effluents containing hazardous dyes and their various intermediates are directly or indirectly harmful to the natural environment (Amin et al., 2020). Hence, the status of treated or untreated effluents and dyes must be understood, in addition to efficient tool use with biological treatment methodologies. These effluents and dyes are pollutants and have different toxic intermediates, which greatly threaten humans and aquatic organisms (Maqbool et al., 2020). It was investigated that the high release of the industrial effluents and dyes cause irregular pigmentation on the surface of the water, affect fauna and flora directly or indirectly, and cause great concern from scientists about the safety of the environment and quality of water for aquatic life and for other purposes (Sharif et al., 2020). A large number of azo dyes show indirect or direct effects, which cause serious allergies, tumor formation, or cancers (Figure 13.5) (Bayramoglu et al., 2021). Numerous biological systems play an important role in the decolorization, mineralization, and degradation of dyes from the textile industries and making them less effective and toxic (Haque et al., 2021). In previous years, various scientific reports showed the LME carrier support of white rot fungi not only degrade or decolorize but also play an important role in the detox-ification of the highly polluting compounds. Therefore, immobilization has proven highly responsive environmentally friendly technology rather than chemical-based methodologies (Chankhanittha et al., 2021).

For the treatment of azo dyes, immobilization of enzymes from different sources has been studied and shows the difference in the detoxifications raised by the toxicity and sensitivity assays, and using the beads of the CPC-silica for the laccase immobi-lization of *T. versicolor* is highly involved in the decolorization of sulfamethoxazole (SMZ) and sulfathiazole (STZ), confirmed by using micro-toxicity assay (Al-Tohamy

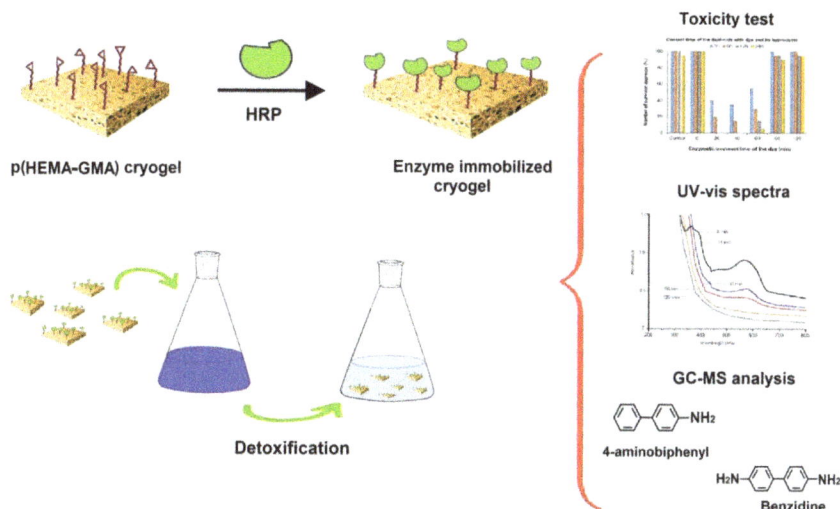

FIGURE 13.5 Immobilized horseradish peroxidase-based detoxification of a benzidine-based azo dye (Bayramoglu et al., 2021).

et al., 2020). The presence of some untreated azo dyes, which are involved in the D-magna complete mortality, on the other hand, has a higher survival rate of crustaceans, which can be seen while treating with the treated sample of a microbial consortium with immobilized polyurethane foam (Ameenudeen et al., 2021). The study of phytotoxicity revealed that when the seeds of *Triticum aestivum* were germinated and exposed to a solution of untreated dyes, they showed growth inhibition (Lian et al., 2020). On the other hand, encapsulated agar-agar MnP enhanced the rate of germination and improved the radical growth and plumule in which the growth of plants takes place in the solution of dyes. Likewise, before applying the treatment, the mortality rate of *Artemia salina* was 32.5–38.2%, which was decreased to 2.03%–6.7% after the immobilized MnP treatment (Ahammed et al., 2020). Phytotoxicity assay confirmed the dyes transformation into the simpler non-toxic compounds. Therefore, the immobilized enzyme agar-MnP showed improved degradation and detoxification of industrials dyes (Vieira et al., 2021).

13.6 SUMMARY AND PERSPECTIVES

In conclusion, the reviewed data above show significant evidence related to the usage of white rot fungi and their exclusive LMEs to challenge a bigger class of dye pollutants released from the industries. However, biocatalytic engineering of enzymes is not new, but a broad range of enzymes related to industries, including LMEs, shows a great working ability for bioremediation and was not used more in the past years. Nevertheless, the working strategies, immobilization, and engineering has been applied to the LMEs, which proves itself the best technique for enzyme stability, and this kind of enzyme shows many applications in the different types of sectors in the modern world. In the past few years, numerous noteworthy efforts and steps have been taken to make it possible to apply the different techniques. In addition, this system involves the immobilization of enzymes that are naturally ever-green, indicating the high potential tool for various biotechnological and industrial applications. Although aside from the ligninolytic enzyme engineering and immobilization, these enzymes also have established exclusive properties in degradation, decolorization, and detoxification of different types of pollutants at laboratories levels and plot and semi-plot scales. All at once, it is necessary to develop the available training and additional scientific studies for a good understanding of the mechanism of degradation for various nonphenolic substrates. Moreover, the bioremediation using immobilized enzymes provide biobased and eco-friendly technology. The described processes are eco-friendly and nontoxic, which use enzyme-based green catalyst and their immobilization for different environmental and biotechnological applications.

13.7 DISCLOSURE STATEMENT

The authors declare no competing financial interests.

REFERENCES

Ahammed, G. J., Wang, Y., Mao, Q., Wu, M., Yan, Y., Ren, J., . . . Chen, S. (2020). Dopamine alleviates bisphenol A-induced phytotoxicity by enhancing antioxidant and detoxification potential in cucumber. *Environmental Pollution*, 259, 113957.

Akey, P., & Appel, I. (2021). The limits of limited liability: Evidence from industrial pollution. *The Journal of Finance*, 76(1), 5–55.

Al-Tohamy, R., Sun, J., Fareed, M. F., Kenawy, E. R., & Ali, S. S. (2020). Ecofriendly biodegradation of Reactive Black 5 by newly isolated Sterigmatomyces halophilus SSA1575, valued for textile azo dye wastewater processing and detoxification. *Scientific Reports*, 10(1), 1–16.

Ameenudeen, S., Unnikrishnan, S., & Ramalingam, K. (2021). Statistical optimization for the efficacious degradation of reactive azo dyes using Acinetobacter baumannii JC359. *Journal of Environmental Management*, 279, 111512.

Amin, S., Rastogi, R. P., Chaubey, M. G., Jain, K., Divecha, J., Desai, C., & Madamwar, D. (2020). Degradation and toxicity analysis of a reactive textile diazo dye-Direct Red 81 by newly isolated Bacillus sp. DMS2. *Frontiers in Microbiology*, 11, 2280.

Ariaeenejad, S., Motamedi, E., & Salekdeh, G. H. (2021). Immobilization of enzyme cocktails on dopamine functionalized magnetic cellulose nanocrystals to enhance sugar bioconversion: A biomass reusing loop. *Carbohydrate Polymers*, 256, 117511.

Aricov, L., Leonties, A. R., Gîfu, I. C., Preda, D., Raducan, A., & Anghel, D. F. (2020). Enhancement of laccase immobilization onto wet chitosan microspheres using an iterative protocol and its potential to remove micropollutants. *Journal of Environmental Management*, 276, 111326.

Asgher, M., Iqbal, H. M. N., & Irshad, M. (2012a). Characterization of purified and xerogel immobilized novel lignin peroxidase produced from Trametes versicolor IBL-04 using solid state medium of corncobs. *BMC Biotechnology*, 12(1), 1–8.

Asgher, M., Kamal, S., & Iqbal, H. M. N. (2012b). Improvement of catalytic efficiency, thermostability and dye decolorization capability of Pleurotus ostreatus IBL-02 laccase by hydrophobic sol gel entrapment. *Chemistry Central Journal*, 6(1), 1–10.

Asgher, M., Noreen, S., & Bilal, M. (2017a). Enhancement of catalytic, reusability, and long-term stability features of Trametes versicolor IBL-04 laccase immobilized on different polymers. *International Journal of Biological Macromolecules*, 95, 54–62.

Asgher, M., Noreen, S., & Bilal, M. (2017b). Enhancing catalytic functionality of Trametes versicolor IBL-04 laccase by immobilization on chitosan microspheres. *Chemical Engineering Research and Design*, 119, 1–11.

Asgher, M., Wahab, A., Bilal, M., & Iqbal, H. M. (2018). Delignification of lignocellulose biomasses by alginate—chitosan immobilized laccase produced from Trametes versicolor IBL-04. Waste and Biomass *Valorization*, 9(11), 2071–2079.

Bagewadi, Z. K., Mulla, S. I., & Ninnekar, H. Z. (2017). Purification and immobilization of laccase from Trichoderma harzianum strain HZN10 and its application in dye decolorization. *Journal of Genetic Engineering and Biotechnology*, 15(1), 139–150.

Bayramoglu, G., Akbulut, A., & Arica, M. Y. (2021). Utilization of immobilized horseradish peroxidase for facilitated detoxification of a benzidine based azo dye. *Chemical Engineering Research and Design*, 165, 435–444.

Begum, S., Narwade, V. N., Halge, D. I., Jejurikar, S. M., Dadge, J. W., Muduli, S., . . . Bogle, K. A. (2020). Remarkable photocatalytic degradation of Remazol Brilliant Blue R dye using bio-photocatalyst 'nano-hydroxyapatite'. *Materials Research Express*, 7(2), 025013.

Behbahani, M., Nosrati, M., Moradi, M., & Mohabatkar, H. (2020). Using Chou's general pseudo amino acid composition to classify laccases from bacterial and fungal sources via Chou's five-step rule. *Applied Biochemistry and Biotechnology*, 190(3), 1035–1048.

Bento, R. M., Almeida, M. R., Bharmoria, P., Freire, M. G., & Tavares, A. P. (2020). Improvements in the enzymatic degradation of textile dyes using ionic-liquid-based surfactants. *Separation and Purification Technology*, 235, 116191.

Bernal, C., Rodriguez, K., & Martinez, R. (2018). Integrating enzyme immobilization and protein engineering: An alternative path for the development of novel and improved industrial biocatalysts. *Biotechnology Advances*, 36(5), 1470–1480.

Bhat, S. A., Rashid, N., Rather, M. A., Bhat, S. A., Ingole, P. P., & Bhat, M. A. (2020). Highly efficient catalytic reductive degradation of Rhodamine-B over Palladium-reduced graphene oxide nanocomposite. *Chemical Physics Letters*, 754, 137724.

Bilal, M., Adeel, M., Rasheed, T., Zhao, Y., & Iqbal, H. M. (2019c). Emerging contaminants of high concern and their enzyme-assisted biodegradation—a review. *Environment International*, 124, 336–353.

Bilal, M., & Asgher, M. (2015). Dye decolorization and detoxification potential of Ca-alginate beads immobilized manganese peroxidase. *BMC Biotechnology*, 15(1), 1–14.

Bilal, M., Asgher, M., Iqbal, H. M., Hu, H., Wang, W., & Zhang, X. (2017a). Bio-catalytic performance and dye-based industrial pollutants degradation potential of agarose-immobilized MnP using a Packed Bed Reactor System. *International Journal of Biological Macromolecules*, 102, 582–590.

Bilal, M., Asgher, M., Iqbal, H. M., Hu, H., & Zhang, X. (2017b). Biotransformation of lignocellulosic materials into value-added products—A review. *International Journal of Biological Macromolecules*, 98, 447–458.

Bilal, M., Asgher, M., & Ramzan, M. (2015). Purification and biochemical characterization of extracellular manganese peroxidase from Ganoderma lucidum IBL-05 and its application. *Scientific research and Essays*, 10(14), 456–464.

Bilal, M., Asgher, M., Shahid, M., & Bhatti, H. N. (2016). Characteristic features and dye degrading capability of agar agar gel immobilized manganese peroxidase. *International Journal of Biological Macromolecules*, 86, 728–740.

Bilal, M., Ashraf, S. S., Cui, J., Lou, W. Y., Franco, M., Mulla, S. I., & Iqbal, H. M. (2020). Harnessing the biocatalytic attributes and applied perspectives of nanoengineered laccases—A review. *International Journal of Biological Macromolecules*, 166, 352–373.

Bilal, M., Ashraf, S. S., & Iqbal, H. M. (2020). Laccase-mediated bioremediation of dye-based hazardous pollutants. In *Methods for Bioremediation of Water and Wastewater Pollution* (pp. 137–160). Springer, Cham.

Bilal, M., Barceló, D., & Iqbal, H. M. (2020). Persistence, ecological risks, and oxidoreductases-assisted biocatalytic removal of triclosan from the aquatic environment. *Science of The Total Environment*, 139194.

Bilal, M., & Iqbal, H. M. (2019). Naturally-derived biopolymers: Potential platforms for enzyme immobilization. *International Journal of Biological Macromolecules*, 130, 462–482.

Bilal, M., & Iqbal, H. M. (2020). Ligninolytic enzymes mediated ligninolysis: An untapped biocatalytic potential to deconstruct lignocellulosic molecules in a sustainable manner. *Catalysis Letters*, 150(2), 524–543.

Bilal, M., Jing, Z., Zhao, Y., & Iqbal, H. M. (2019a). Immobilization of fungal laccase on glutaraldehyde cross-linked chitosan beads and its bio-catalytic potential to degrade bisphenol A. *Biocatalysis and Agricultural Biotechnology*, 19, 101174.

Bilal, M., Rasheed, T., Nabeel, F., Iqbal, H. M., & Zhao, Y. (2019b). Hazardous contaminants in the environment and their laccase-assisted degradation—a review. *Journal of Environmental Management*, 234, 253–264.

Bosco, F., Mollea, C., & Ruggeri, B. (2017). Decolorization of Congo Red by Phanerochaete chrysosporium: The role of biosorption and biodegradation. *Environmental Technology*, 38(20), 2581–2588.

Britos, C. N., Gianolini, J. E., Portillo, H., & Trelles, J. A. (2018). Biodegradation of industrial dyes by a solvent, metal and surfactant-stable extracellular bacterial laccase. *Biocatalysis and Agricultural Biotechnology*, 14, 221–227.

Catucci, G., Valetti, F., Sadeghi, S. J., & Gilardi, G. (2020). Biochemical features of dye-decolorizing peroxidases: Current impact on lignin degradation. *Biotechnology and Applied Biochemistry*, 67(5), 751–759.

Chankhanittha, T., Somaudon, V., Watcharakitti, J., & Nanan, S. (2021). Solar light-driven photocatalyst based on bismuth molybdate (Bi 4 MoO 9) for detoxification of anionic azo dyes in wastewater. *Journal of Materials Science: Materials in Electronics*, 1–15.

Dao, A. T., Vonck, J., Janssens, T. K., Dang, H. T., Brouwer, A., & de Boer, T. E. (2019). Screening white-rot fungi for bioremediation potential of 2, 3, 7, 8-tetrachlorodibenzo-p-dioxin. *Industrial Crops and Products*, 128, 153–161.

Datta, S., Veena, R., Samuel, M. S., & Selvarajan, E. (2020). Immobilization of laccases and applications for the detection and remediation of pollutants: A review. *Environmental Chemistry Letters*, 1–18.

Deepa, T., Gangwane, A. K., Sayyed, R. Z., & Jadhav, H. P. (2020). Optimization and scale-up of laccase production by Bacillus sp. BAB-4151 isolated from the waste of the soap industry. *Environmental Sustainability*, 3(4), 471–479.

Dileepkumar, V. G., Surya, P. S., Pratapkumar, C., Viswanatha, R., Ravikumar, C. R., Kumar, M. A., . . . Santosh, M. S. (2020). NaFeS2 as a new photocatalytic material for the degradation of industrial dyes. *Journal of Environmental Chemical Engineering*, 8(4), 104005.

Elkady, M., Salama, E., Amer, W. A., Ebeid, E. Z. M., Ayad, M. M., & Shokry, H. (2020). Novel eco-friendly electrospun nanomagnetic zinc oxide hybridized PVA/alginate/chitosan nanofibers for enhanced phenol decontamination. *Environmental Science and Pollution Research*, 27(34), 43077–43092.

Eslami, H., Shariatifar, A., Rafiee, E., Shiranian, M., Salehi, F., Hosseini, S. S., . . . Ebrahimi, A. A. (2019). Decolorization and biodegradation of reactive Red 198 Azo dye by a new Enterococcus faecalis—Klebsiella variicola bacterial consortium isolated from textile wastewater sludge. *World Journal of Microbiology and Biotechnology*, 35(3), 38.

Fang, W., Zhang, X., Zhang, P., Morera, X. C., van Lier, J. B., & Spanjers, H. (2020). Evaluation of white rot fungi pretreatment of mushroom residues for volatile fatty acid production by anaerobic fermentation: Feedstock applicability and fungal function. *Bioresource Technology*, 297, 122447.

Gabhane, J., Kumar, S., & Sarma, A. K. (2020). Effect of glycerol thermal and hydrothermal pretreatments on lignin degradation and enzymatic hydrolysis in paddy straw. *Renewable Energy*, 154, 1304–1313.

Gao, Y., Wang, G., Gu, H., Zhang, J., Li, W., & Fu, Y. (2021). Cooperatively controlling the enzyme mimicking Pt nanomaterials with nucleotides and solvents. *Colloids and Surfaces A: Physicochemical and Engineering Aspects*, 613, 126070.

Geng, Z., Zhang, Y., Li, C., Han, Y., Cui, Y., & Yu, B. (2020). Energy optimization and prediction modeling of petrochemical industries: An improved convolutional neural network based on cross-feature. *Energy*, 194, 116851.

Gianolini, J. E., Britos, C. N., Mulreedy, C. B., & Trelles, J. A. (2020). Hyperstabilization of a thermophile bacterial laccase and its application for industrial dyes degradation. *3 Biotech*, 10, 1–7.

Girelli, A. M., Astolfi, M. L., & Scuto, F. R. (2020). Agro-industrial wastes as potential carriers for enzyme immobilization: A review. *Chemosphere*, 244, 125368.

Gokgoz, M., & Altinok, H. (2012). Immobilization of laccase on polyacrylamide and polyacrylamide-kappa-carragennan-based semi-interpenetrating polymer networks. *Artificial Cells Blood Substitutes And Biotechnology*, 40(5), 326–330.

Guo, J., Liu, X., Zhang, X., Wu, J., Chai, C., Ma, D., . . . Ge, W. (2019). Immobilized lignin peroxidase on Fe3O4@ SiO2@ polydopamine nanoparticles for degradation of organic pollutants. *International Journal of Biological Macromolecules*, 138, 433–440.

Haque, M. M., Haque, M. A., Mosharaf, M. K., & Marcus, P. K. (2021). Decolorization, degradation and detoxification of carcinogenic sulfonated azo dye methyl orange by newly developed biofilm consortia. *Saudi Journal of Biological Sciences*, 28(1), 793–804.

He, J., Huang, C., Lai, C., Huang, C., Li, M., Pu, Y., . . . Yong, Q. (2020). The effect of lignin degradation products on the generation of pseudo-lignin during dilute acid pretreatment. *Industrial Crops and Products*, 146, 112205.

Hou, J., Dong, G., Ye, Y., & Chen, V. (2014). Laccase immobilization on titania nanoparticles and titania-functionalized membranes. *Journal of Membrane Science*, 452, 229–240.

Hou, L., Ji, D., Dong, W., Yuan, L., Zhang, F., Li, Y., & Zang, L. (2020). The synergistic action of electro-Fenton and white-rot fungi in the degradation of lignin. *Frontiers in Bioengineering and Biotechnology*, 8, 99.

Husain, Q., & Ulber, R. (2011). Immobilized peroxidase as a valuable tool in the remediation of aromatic pollutants and xenobiotic compounds: A review. *Critical Reviews in Environmental Science and Technology*, 41(8), 770–804.

Issakhov, A., Alimbek, A., & Zhandaulet, Y. (2021). The assessment of water pollution by chemical reaction products from the activities of industrial facilities: Numerical study. *Journal of Cleaner Production*, 282, 125239.

Jiang, Z., Zhu, W., Xu, G., Xu, X., Wang, M., Chen, H., . . . Lin, M. (2020). Ni-nanoparticle-bound boron nitride nanosheets prepared by a radiation-induced reduction-exfoliation method and their catalytic performance. *Journal of Materials Chemistry A*, 8(18), 9109–9120.

Khalid, N., Asgher, M., & Qamar, S. A. (2020). Evolving trend of Boletus versicolor IBL-04 by chemical mutagenesis to overproduce laccase: Process optimization, 3-step purification, and characterization. *Industrial Crops and Products*, 155, 112771.

Klimek-Szczykutowicz, M., Szopa, A., & Ekiert, H. (2020). Citrus limon (Lemon) phenomenon—a review of the chemistry, pharmacological properties, applications in the modern pharmaceutical, food, and cosmetics industries, and biotechnological studies. *Plants*, 9(1), 119.

Kumar, A., & Chandra, R. (2020). Ligninolytic enzymes and its mechanisms for degradation of lignocellulosic waste in environment. *Heliyon*, 6(2), e03170.

Kumari, E., Görlich, S., Poulsen, N., & Kröger, N. (2020). Genetically programmed regioselective immobilization of enzymes in biosilica microparticles. *Advanced Functional Materials*, 30(25), 2000442.

Lara-Espinoza, C., Sanchez-Villegas, J. A., Lopez-Franco, Y., Carvajal-Millan, E., Troncoso-Rojas, R., Carvallo-Ruiz, T., & Rascon-Chu, A. (2021). Composition, physicochemical features, and covalent gelling properties of ferulated pectin extracted from three sugar beet (Beta vulgaris L.) cultivars grown under desertic conditions. *Agronomy*, 11(1), 40.

Lian, J., Zhao, L., Wu, J., Xiong, H., Bao, Y., Zeb, A., . . . Liu, W. (2020). Foliar spray of TiO_2 nanoparticles prevails over root application in reducing Cd accumulation and mitigating Cd-induced phytotoxicity in maize (Zea mays L.). *Chemosphere*, 239, 124794.

Majeke, B. M., Collard, F. X., Tyhoda, L., & Görgens, J. F. (2021). The synergistic application of quinone reductase and lignin peroxidase for the deconstruction of industrial (technical) lignins and analysis of the degraded lignin products. *Bioresource Technology*, 319, 124152.

Maqbool, Z., Nadeem, H., Mahmood, F., Siddique, M. H., Shahzad, T., Azeem, F., . . . Hussain, S. (2020). Environmental effects and microbial detoxification of textile dyes. In *Methods for Bioremediation of Water and Wastewater Pollution* (pp. 289–326). Springer, Cham.

Mate, C. J., & Mishra, S. (2020). Synthesis of borax cross-linked Jhingan gum hydrogel for remediation of Remazol Brilliant Blue R (RBBR) dye from water: Adsorption isotherm, kinetic, thermodynamic and biodegradation studies. *International Journal of Biological Macromolecules,* 151, 677–690.

Matinja, A. I., Zain, N. A. M., Suhaimi, M. S., & Alhassan, A. J. (2019). Optimization of biodiesel production from palm oil mill effluent using lipase immobilized in PVA-alginate-sulfate beads. *Renewable Energy,* 135, 1178–1185.

Mattila, H. K., Mäkinen, M., & Lundell, T. (2020). Hypoxia is regulating enzymatic wood decomposition and intracellular carbohydrate metabolism in filamentous white rot fungus. *Biotechnology for Biofuels*, 13(1), 1–17.

Mawad, A. M., Abd Hesham, E. L., Yousef, N. M., Shoreit, A. A., Gathergood, N., & Gupta, V. K. (2020). Role of bacterial-fungal consortium for enhancement in the degradation of industrial dyes. *Current Genomics*, 21(4), 283–294.

Mei, J., Shen, X., Gang, L., Xu, H., Wu, F., & Sheng, L. (2020). A novel lignin degradation bacteria-Bacillus amyloliquefaciens SL-7 used to degrade straw lignin efficiently. *Bioresource Technology*, 310, 123445.

Mogharabi, M., Nassiri-Koopaei, N., Bozorgi-Koushalshahi, M., Nafissi-Varcheh, N., Bagherzadeh, G., & Faramarzi, M. A. (2012). Immobilization of laccase in alginate-gelatin mixed gel and decolorization of synthetic dyes. *Bioinorganic Chemistry and Applications*, 2012, 823830.

Morsi, R., Bilal, M., Iqbal, H. M., & Ashraf, S. S. (2020). Laccases and peroxidases: The smart, greener and futuristic biocatalytic tools to mitigate recalcitrant emerging pollutants. *Science of The Total Environment*, 714, 136572.

Nathan, V. K., Kanthimathinathan, S. R., Rani, M. E., Rathinasamy, G., & Kannan, N. D. (2018). Biobleaching of waste paper using lignolytic enzyme from Fusarium equiseti VKF2: A mangrove isolate. *Cellulose*, 25(7), 4179–4192.

Nyika, J. M. (2021). The use of microorganism-derived enzymes for bioremediation of soil pollutants. In *Recent Advancements in Bioremediation of Metal Contaminants* (pp. 54–71). IGI Global. Technical University of Kenya, Kenya.

Ojha, A., & Thareja, P. (2020). Graphene-based nanostructures for enhanced photocatalytic degradation of industrial dyes. *Emergent Materials*, 1–12.

Oliveira, S. F., da Luz, J. M. R., Kasuya, M. C. M., Ladeira, L. O., & Junior, A. C. (2018). Enzymatic extract containing lignin peroxidase immobilized on carbon nanotubes: Potential biocatalyst in dye decolourization. *Saudi Journal of Biological Sciences*, 25(4), 651–659.

Parra-Arroyo, L., Parra-Saldivar, R., Ramirez-Mendoza, R. A., Keshavarz, T., & Iqbal, H. M. (2020). Laccase-assisted cues: State-of-the-art analytical modalities for detection, quantification, and redefining "removal" of environmentally related contaminants of high concern. In *Laccases in Bioremediation and Waste Valorisation* (pp. 173–190). Springer, Cham.

Patel, S. K., Kalia, V. C., Choi, J. H., Haw, J. R., Kim, I. W., & Lee, J. K. (2014). Immobilization of laccase on $ SiO_2 $ nanocarriers improves its stability and reusability. *Journal of Microbiology and Biotechnology*, 24(5), 639–647.

Persico, C. L., & Venator, J. (2019). The effects of local industrial pollution on students and schools. *Journal of Human Resources,* 0518–9511R2.

Pulicharla, R., Das, R. K., Brar, S. K., Drogui, P., & Surampalli, R. Y. (2018). Degradation kinetics of chlortetracycline in wastewater using ultrasonication assisted laccase. *Chemical Engineering Journal*, 347, 828–835.

Pylypchuk, I. V., Daniel, G., Kessler, V. G., & Seisenbaeva, G. A. (2020). Removal of diclofenac, paracetamol, and carbamazepine from model aqueous solutions by magnetic Sol—Gel encapsulated horseradish peroxidase and lignin peroxidase composites. *Nanomaterials*, 10(2), 282.

Qamar, S. A., Ashiq, M., Jahangeer, M., Riasat, A., & Bilal, M. (2020). Chitosan-based hybrid materials as adsorbents for textile dyes—A review. *Case Studies in Chemical and Environmental Engineering*, 2, 100021.

Qiu, H., Li, Y., Ji, G., Zhou, G., Huang, X., Qu, Y., & Gao, P. (2009). Immobilization of lignin peroxidase on nanoporous gold: Enzymatic properties and in situ release of H2O2 by co-immobilized glucose oxidase. *Bioresource Technology*, 100(17), 3837–3842.

Ramírez-Montoya, L. A., Hernández-Montoya, V., Montes-Morán, M. A., & Cervantes, F. J. (2015). Correlation between mesopore volume of carbon supports and the immobilization of laccase from Trametes versicolor for the decolorization of Acid Orange 7. *Journal of Environmental Management*, 162, 206–214.

Ran, Y. H., Che, Z. F., & Chen, W. Q. (2012). Co-immobilized lignin peroxidase and manganese peroxidase from coriolus versicolor capable of decolorizing molasses waste water. In *Applied Mechanics and Materials* (Vol. 138, pp. 1067–1071). Switzerland: Trans Tech Publications Ltd.

Rather, M. A., Bhat, S. A., Pandit, S. A., Bhat, F. A., Rather, G. M., & Bhat, M. A. (2019). As catalytic as silver nanoparticles anchored to reduced graphene oxide: Fascinating activity of imidazolium based surface active ionic liquid for chemical degradation of rhodamine B. *Catalysis Letters*, 149(8), 2195–2203.

Rodgers, K., McLellan, I., Peshkur, T., Williams, R., Tonner, R., Hursthouse, A. S., . . . Henriquez, F. L. (2019). Can the legacy of industrial pollution influence antimicrobial resistance in estuarine sediments? *Environmental Chemistry Letters,* 17(2), 595–607.

Routoula, E., & Patwardhan, S. V. (2020). Degradation of anthraquinone dyes from effluents: A review focusing on enzymatic dye degradation with industrial potential. *Environmental Science & Technology*, 54(2), 647–664.

Sabapathy, P. C., Devaraj, S., Meixner, K., Anburajan, P., Kathirvel, P., Ravikumar, Y., . . . Qi, X. (2020). Recent developments in Polyhydroxyalkanoates (PHAs) production—A review. *Bioresource Technology,* 306, 123132.

Salaj-Kosla, U., Pöller, S., Schuhmann, W., Shleev, S., & Magner, E. (2013). Direct electron transfer of Trametes hirsuta laccase adsorbed at unmodified nanoporous gold electrodes. *Bioelectrochemistry*, 91, 15–20.

Saldarriaga-Hernández, S., Velasco-Ayala, C., Flores, P. L. I., de Jesús Rostro-Alanis, M., Parra-Saldivar, R., Iqbal, H. M., & Carrillo-Nieves, D. (2020). Biotransformation of lignocellulosic biomass into industrially relevant products with the aid of fungi-derived lignocellulolytic enzymes. *International Journal of Biological Macromolecules,* 161, 1099–1116.

Sam, S. P., Adnan, R., & Ng, S. L. (2021). Statistical optimization of immobilization of activated sludge in PVA/alginate cryogel beads using response surface methodology for p-nitrophenol biodegradation. *Journal of Water Process Engineering*, 39, 101725.

Sánchez, Ó. J., & Montoya, S. (2020). Assessment of polysaccharide and biomass production from three white-rot fungi by solid-state fermentation using wood and agro-industrial residues: A kinetic approach. *Forests*, 11(10), 1055.

Sarkar, S., Banerjee, A., Chakraborty, N., Soren, K., Chakraborty, P., & Bandopadhyay, R. (2020). Structural-functional analyses of textile dye degrading azoreductase, laccase and peroxidase: A comparative in silico study. *Electronic Journal of Biotechnology*, 43, 48–54.

Singh, A. K., Bilal, M., Iqbal, H. M., & Raj, A. (2021). Lignin peroxidase in focus for catalytic elimination of contaminants—A critical review on recent progress and perspectives. *International Journal of Biological Macromolecules*, 177, 58–82.

Schneider, W. D. H., Fontana, R. C., Baudel, H. M., de Siqueira, F. G., Rencoret, J., Gutiérrez, A., . . . Camassola, M. (2020). Lignin degradation and detoxification of eucalyptus wastes by on-site manufacturing fungal enzymes to enhance second-generation ethanol yield. *Applied Energy*, 262, 114493.

Seo, H., Kim, K. J., & Kim, Y. H. (2018). In silico-designed lignin peroxidase from Phanerochaete chrysosporium shows enhanced acid stability for depolymerization of lignin. *Biotechnology for Biofuels*, 11(1), 1–13.

Shaheen, R., Asgher, M., Hussain, F., & Bhatti, H. N. (2017). Immobilized lignin peroxidase from Ganoderma lucidum IBL-05 with improved dye decolorization and cytotoxicity reduction properties. *International Journal of Biological Macromolecules*, 103, 57–64.

Sharif, A., Nasreen, Z., Bashir, R., & Kalsoom, S. (2020). 8. Microbial degradation of textile industry effluents: A review. *Pure and Applied Biology (PAB)*, 9(4), 2361–2382.

Shrestha, S., Kognou, A. L. M., Zhang, J., & Qin, W. (2020). Different facets of lignocellulosic biomass including pectin and its perspectives. *Waste and Biomass Valorization*, 1–19.

Siddeeg, S. M., Tahoon, M. A., Mnif, W., & Ben Rebah, F. (2020). Iron oxide/chitosan magnetic nanocomposite immobilized manganese peroxidase for decolorization of textile wastewater. *Processes*, 8(1), 5.

Singh, G., & Dwivedi, S. K. (2020). Decolorization and degradation of Direct Blue-1 (Azo dye) by newly isolated fungus Aspergillus terreus GS28, from sludge of carpet industry. *Environmental Technology & Innovation*, 18, 100751.

Taheran, M., Naghdi, M., Brar, S. K., Knystautas, E. J., Verma, M., & Surampalli, R. Y. (2017). Covalent immobilization of laccase onto nanofibrous membrane for degradation of pharmaceutical residues in water. *ACS Sustainable Chemistry & Engineering*, 5(11), 10430–10438.

Tian, F., Wang, Y., Guo, G., Ding, K., Yang, F., Wang, H., . . . Liu, C. (2021). Enhanced azo dye biodegradation at high salinity by a halophilic bacterial consortium. *Bioresource Technology*, 124749.

Van Aken, B., Ledent, P., Naveau, H., & Agathos, S. N. (2000). Co-immobilization of manganese peroxidase from Phlebia radiata and glucose oxidase from Aspergillus niger on porous silica beads. *Biotechnology Letters*, 22(8), 641–646.

Varjani, S., Rakholiya, P., Ng, H. Y., You, S., & Teixeira, J. A. (2020). Microbial degradation of dyes: An overview. *Bioresource Technology*, 123728.

Vieira, G. A. L., Cabral, L., Otero, I. V. R., Ferro, M., de Faria, A. U., de Oliveira, V. M., . . . Sette, L. D. (2021). Marine associated microbial consortium applied to RBBR textile dye detoxification and decolorization: Combined approach and metatranscriptomic analysis. *Chemosphere*, 267, 129190.

Wong, D. W. (2009). Structure and action mechanism of ligninolytic enzymes. *Applied Biochemistry and Biotechnology*, 157(2), 174–209.

Wouters, B., Pirok, B. W., Soulis, D., Perticarini, R. C. G., Fokker, S., van den Hurk, R. S., . . . Schoenmakers, P. J. (2019). On-line microfluidic immobilized-enzyme reactors: A new tool for characterizing synthetic polymers. *Analytica chimica acta*, 1053, 62–69.

Xing, B. S., Han, Y., Cao, S., Wen, J., Zhang, K., Yuan, H., & Wang, X. C. (2020). Cosubstrate strategy for enhancing lignocellulose degradation during rumen fermentation in vitro: Characteristics and microorganism composition. *Chemosphere*, 250, 126104.

Yamak, O., Kalkan, N. A., Aksoy, S., Altinok, H., & Hasirci, N. (2009). Semi-interpenetrating polymer networks (semi-IPNs) for entrapment of laccase and their use in Acid Orange 52 decolorization. *Process Biochemistry*, 44(4), 440–445.

Yang, Q., Yan, Y., Yang, X., Liao, G., Wang, D., & Xia, H. (2019). Enzyme immobilization in cage-like 3D-network PVA-H and GO modified PVA-H (GO@ PVA-H) with stable conformation and high activity. *Chemical Engineering Journal*, 372, 946–955.

Yang, X., Wu, Y., Zhang, Y., Yang, E., Qu, Y., Xu, H., . . . Yan, J. (2020). A thermo-active laccase isoenzyme from Trametes trogii and its potential for dye decolorization at high temperature. *Frontiers in Microbiology*, 11, 241.

Zdarta, J., Meyer, A. S., Jesionowski, T., & Pinelo, M. (2018). A general overview of support materials for enzyme immobilization: Characteristics, properties, practical utility. *Catalysts*, 8(2), 92.

Zhang, C., Gong, L., Mao, Q., Han, P., Lu, X., & Qu, J. (2018). Laccase immobilization and surface modification of activated carbon fibers by bio-inspired poly-dopamine. *RSC Advances*, 8(26), 14414–14421.

Zhang, J., Wu, Q., & Zhou, Z. (2019). A two-stage DEA model for resource allocation in industrial pollution treatment and its application in China. *Journal of Cleaner Production*, 228, 29–39.

Zhang, Y., Piao, M., He, L., Yao, L., Piao, T., Liu, Z., & Piao, Y. (2020). Immobilization of laccase on magnetically separable biochar for highly efficient removal of bisphenol A in water. *RSC Advances*, 10(8), 4795–4804.

Zhao, J., Li, J., Liu, H., Zhang, X., Zheng, K., Yu, H., . . . Huo, J. (2020). Cesium immobilization in perovskite-type Ba1-x (La, Cs) xZrO3 ceramics by sol-gel method. *Ceramics International,* 46(7), 9968–9971.

Zhou, W., Zhang, W., & Cai, Y. (2020). Laccase immobilization for water purification: A comprehensive review. *Chemical Engineering Journal*, 126272.

14 Role of Streptokinase as a Thrombolytic Agent for Medical Applications

Hamza Rafeeq, Muhammad Anjum Zia,
Asim Hussain, Ayesha Safdar, Muhammad Bilal
and Hafiz M. N. Iqbal

ABSTRACT

In the human body, blood circulation is affected by the formation of blood clots caused by serious heart-related diseases and heart-related problems, such as acute myocardial infarction (AMI). There are many thrombolytic agents, such as streptokinase, tPa, and urokinase. Streptokinase is the most useful agent, even in underdeveloped countries, because of its low cost and easy availability. This agent is naturally synthesized by using different strains of streptococci of groups C, A, and G, and its main function is to convert plasminogen into the plasmin by activating it and to break the fibrin, which is the main cause of blood clots deposited in the veins. Streptokinase therapy is most useful in heart attacks. It has been internationally utilized in therapy, mostly in myocardial infarction, because it shows strong activity to dissolve fibrin. In the circulatory system, the inactive precursor of plasmin, also known as plasminogen, activates the plasmin. Plasminogen is activated by streptokinase by making a complex that in turn breaks the bond in plasminogen molecule giving rise to plasmin/inactive plasminogen. This is a good method for the treatment of occult pleural effusion and is a good alternative for surgical femoropopliteal procedures.

Keywords: streptokinase, plasminogen, blood clot, myocardial infarction, fibrin

CONTENTS

14.1   Introduction 272
14.2   Thrombus 272
14.3   Plasminogen and Plasmin 274
14.4   Activators of Plasminogen 275
14.5   Streptokinase 276
14.6   Natural Source of Streptokinase... 276
14.7   Beta-Hemolytic Streptococci 276
14.8   Streptokinase Mechanism and Structure.. 277
14.9   Discovery of Streptokinase 278

DOI: 10.1201/9781003202998-14

14.10 Commonly Used Fibrinolytic Agents... 280
14.11 Structure and Properties of Streptokinase 280
14.12 Production, Purification, and Characterization of Streptokinase 283
14.13 Applications of Streptokinase 284
14.14 Conclusion.. 285
14.15 Acknowledgments 286
14.16 Conflict of Interests 286
References.. 286

14.1 INTRODUCTION

Different heart diseases, like atherosclerosis, deep-vein thrombus, coronary heart disease, and thrombosis have alarmed the world. These diseases are responsible for many deaths. According to an estimate, about 12 million people die every year because of heart diseases, and about 80 million are suffering from them (Yusuf et al., 2021). After comprehending the severe risk factors related to this health issue, researchers have started to pay great attention to overcome the problem (Capak et al., 2007).

In developing and underdeveloped countries, death is the most prominently caused by myocardial infarction (MI). According to the 2013 report of WHO, about 17.3 million casualties from all over the world were due to cardiovascular diseases and cardiac complications. Furthermore, it is found that the death rate is more in developing and low-income countries (Rizzoli et al., 2013).

There are several drugs that have the capability to dissolve the clots. Hence, these drugs are frequently used to cure heart diseases, whereas streptokinase is the most prominent of all other drugs. Streptokinase is the extracellular protein that is used to break fibrin (Raee et al., 2017). Enzymes from microbial sources used to cure heart diseases are involved urokinase (uPA), streptokinase (SK), and tissue plasminogen activators (tPA). Streptokinase has great importance in curing heart diseases among all other enzymes due to its cheap cost and satisfactory outcomes (Hasanpour et al., 2021). It has a bacterial origin, produced through fermentation (Mohanty and Khasa, 2021). Microorganisms produce streptokinase. Their applications, sources, and fermentation media optimize the production of streptokinase and determine streptokinase activity, and various techniques are reviewed in the literature.

14.2 THROMBUS

Thrombosis is blood clot formation in the blood vessels, and this is actually the result of an imbalance of hemostasis and its inhibitors (Rizzoli et al., 2013). Changes in the blood vessels result in thrombosis, especially if the blood flow is slow (Kikkert et al., 2014). Atherosclerosis is due to changes in blood vessels. The plaques are protrusions in blood vessels, and these plaques are made up of lipids and collagens made up of smooth muscles (Tadayon et al., 2015). Different cells, such as free fatty acids, cholesterol, and scar tissues present in the arterial walls, make the arteries constrict, making it difficult for blood to flow to the heart. Chest pain is caused by slowed blood flow, and heart attack results from complete blockage of blood. And the most prominent cause of death in the developed countries is thrombosis. Forty-five percent of total deaths are due to vascular diseases in the United States (Ali et al., 2016). Lifestyle and diet changes prevent thrombosis. High plasma lipid levels are due to an imbalanced diet, lack of physical activity, high blood pressure, and cigarette smoking (Ali et al., 2016). Figure 14.1 represents the mechanism of thrombus formation in a circulatory system.

Thrombolytic drugs enhance fibrin clot dissolution. Anticoagulants have the ability to avert thrombosis and also pre-existing thrombosis enlargement (Litvinov et al., 2019). The thrombolytic agents promote the degradation of thrombi. The hemostatic system normally prevents blood clot formation in vessels (Cone et al., 2020). Hemostasis leads to myocardial infarction and pulmonary embolism besides stroke (Virani et al., 2020). Clinical involvement is required in cases of the failure of hemostasis (Yepes et al., 2009). Intrinsic pathway factors of the fibrinolytic systems can be activated by fibrinolytic agents exclusively. These are also responsible for the lysis of blood clots, which results in the restoration of blood flow in the occluded vessels (Yamashita et al., 2009).

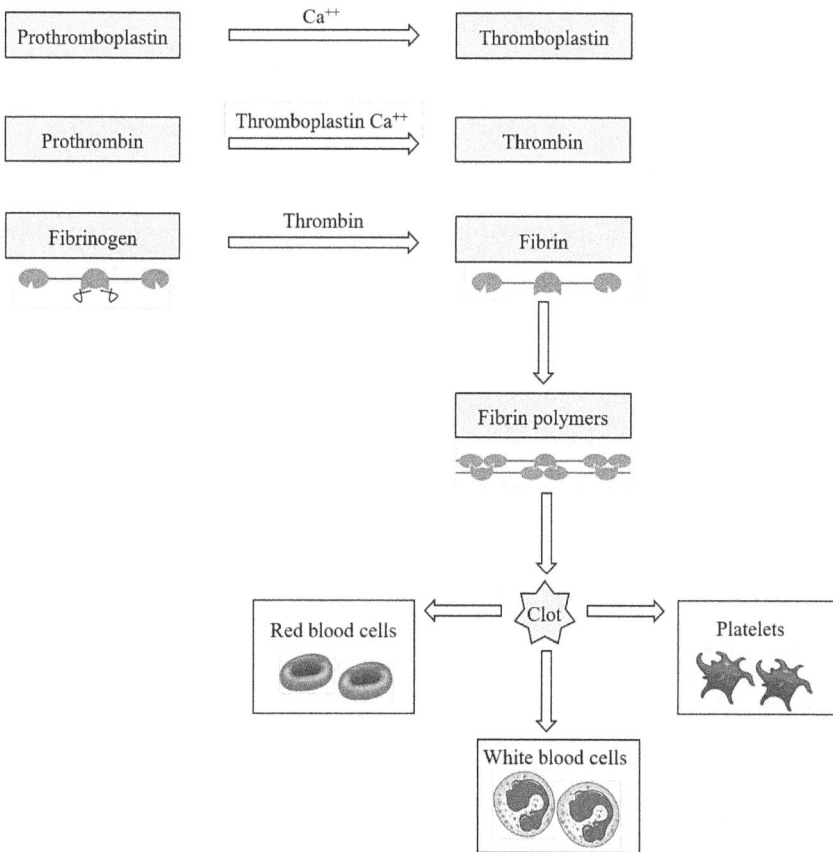

FIGURE 14.1 Mechanism of thrombus formation in a circulatory system.

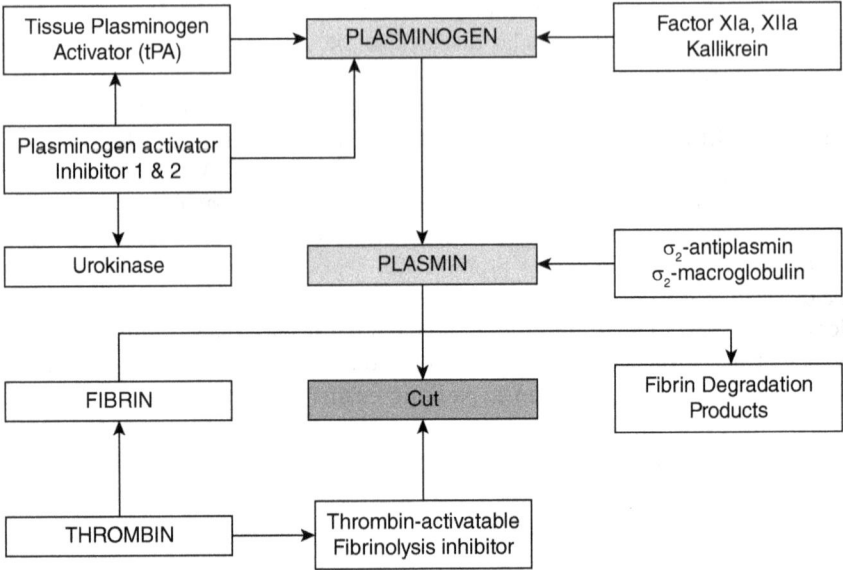

FIGURE 14.2 Mechanism of fibrin degradation in the circulatory system.

Commonly used thrombolytic agents in clinical practice are urokinase (uPA), strep-tokinase (SK), and tissue plasminogen activators (tPA). These thrombolytic agents convert the inert plasminogen to its active protease form that dissolves thrombi, and the degraded products are removed by phagocytosis (Kunamneni et al., 2007). The most widely used agent of this type is streptokinase isolated from hemolytic strep-tococci. However, urokinase is isolated from the male urine, and it is an additional example of a thrombolytic agent. Streptokinase is a protein that makes a complex of 1:1 molar ratio with plasminogen to make its enzyme activity effective (which is shown in conformational changes as a result of the interaction). The rapid generation of plasmin is caused by this interaction. Plasmin binds to fibrin and leads to the dis-solution of the thrombus, as shown in Figure 14.2. Streptokinase is largely accepted because it is cheap in developing countries (Cortes-Canteli et al., 2010). Meanwhile, this thrombolytic agent also causes allergic reactions (Karnabatidis et al., 2011). The mechanism of the fibrinolytic system is presented in Figure 14.2.

14.3 PLASMINOGEN AND PLASMIN

Plasma proenzyme is a plasminogen with a molecular weight 90 kDa. By proteo-lytic activation, it is transformed in its active serine protease plasmin (Mehra et al., 2020). In the plasminogen molecule, cleavage of peptide bond takes place between Val.561 and the Arg.560 and by the mammalian plasminogen activators, leaving behind the plasmin molecule containing the light and heavy chains by two disulfide bonds. The molecular weight of the light chain is 25 kDa, and ports are active sites with Asp.645, His602, and Ser.740 and make catalytic characteristic trails of a serine protease (Bansal, 2020). However, the heavy chain is 65 kDa and contains five disulfide double bonds, triple loop kringle structure. Plasmin cleaves the peptide bonds next to lysine or arginine residues as it is a broad-spectrum protease (Ruf et al., 2011). The native plasminogen consists of 791 amino acids, and the amino acid terminal is a glutamic acid residue. So it is known as glutamic acid plasminogen. By treating it with plasmin between the residue Arg67-Met.68-Lys.76-Lys.77-Val.78, plasminogen can be modified and leave behind an activated peptide having the molecular weight of 8 kDa (De Leonardis et al., 2015).

The truncated type of zymogens is denoted by lys-plasminogen. There is stronger interaction between receptors and target molecules of plasminogen due to the conformation of lys-plasminogen, and it also helps in the conversion of plasminogen to plasmin (McMahon and Kwaan, 2015). The three-dimensional structure of plasminogen has not been reported in the crystal structure of micro plasmin, micro plasminogen, and truncated form that contains 20 amino acids of long fragments of heavy chains associated with the light chain by two disulfide linkage have been determined. The plasmin catalytic domain is a solid component, which can recruit factors, adaptor molecules, and bacterial and plasminogen activators, such as staphylokinase and streptokinase, which alter the substrate demonstration in the catalytic domain (Baharifar and Amani, 2016).

14.4 ACTIVATORS OF PLASMINOGEN

Bacterial plasminogen activators have been evolved from several invasive human pathogens. These plasminogen (plg) activators are either surface bound or secreted proteins. Plasminogen activators are particularly divided into two groups: streptokinase (SK) and staphylococcus (SAK) (Elmi et al., 2019). Both of these are not enzymes but make complexes with plasminogen and plasmin in a 1:1 ratio and are responsible for causing the alteration in plasminogen conformation and its specificity. The SK-plasmin and SAK-plasmin complex obtain a significant efficacy to trigger plasminogen according to contraindications of plasmin alone (Zuo et al., 2021). It has been observed that both staphylococcus and streptococcus are minute sequence homology.

The classification of thrombolytic agents is given in Table 14.1. However, in their crystal configurations, they have implemented a similar fold, called beta grasp, in which staphylokinase and three streptokinase domains each have five- or four-stranded beta sheets and a central alpha helix or twisted coil (Zhou et al., 2016) Both streptokinase and staphylokinase show similar mechanisms of activation, but in certain crucial features, like fibrin dependency of initiation in human plasma, there is a difference between their mechanism of plasminogen activation (Angelucci et al., 2019). In a research, it was found that when they treated with the tPA, Uk, and SK, it was observed that there was no difference in mortality rate. The enzyme used for the breakdown of fibrin is streptokinase having proteases characteristics in nature. Anticoagulant streptokinase, termed fibrinolysin, is a well-known bacterial protein that reacts with the plasminogen (Mahmoud et al., 2020).

TABLE 14.1

Classification of Thrombolytic Agents Based on Fibrin Specificity

Generation of Thrombolytic Agents	Fibrin-Specific	Non-Fibrin-Specific	
First generation	–	Streptokinase Urokinase	(Shibata et al., 2019)
Second generation	Altepase Tissue plasminogen activator (tPA)	Streptokinase plasminogen activating complex (APSAC) Prourokinase	(Vishnu and Srivastava, 2019)
Third generation	Tenecteplase Lanoteplase Reteplase Pamiteplase Monteplse	– – – – –	(Nedaeinia et al., 2020)

14.5   STREPTOKINASE

It is an extracellular enzyme produced by different strains of beta-hemolytic streptococci. Streptokinase is the protease, which consists of single-chain polypeptide and performs the function of activating plasminogen in blood circulation. Its molecular weight
is 45–50 kDa (Tadayon et al., 2015), and it contains 414 amino acids, having isoelectric
pH 7.3–7.6. It shows maximum activity at pH 7.5 (Chapurina et al., 2016). Streptokinase
does not contain compounds such as cysteine, phosphorus, lipids, and carbohydrates.
There is a great difference between the structure of streptokinase produced by different
strains of cocci (Basak et al., 2016). In 1982, two scientists named Jackson and Tang
reported the whole sequence of amino acids of streptokinase that also shows the difference in structure (Huish et al., 2017). Contrasting other fibrinolytic agents, streptokinase
does not show its own proteolytic activity. It activates the plasminogen as it forms the
complex with plasminogen. Streptokinase plasminogen complex activates the other free
proenzymes molecule plasminogen into proteolytic plasmin (pl) (Wang et al., 2020).
The mechanism action of streptokinase is presented in Figure 14.3.

14.6   NATURAL SOURCE OF STREPTOKINASE

Microorganisms have the ability to grow on a variety of substrates leading to produce
significantly diverse products. New methods have been developed for increasing the
production of existing products to improve their quality (Keramati et al., 2020).
Streptokinase production from streptococci was first achieved by Billroth in 1874 in
secretions from infected tissues. Later, streptococcus was also found in the blood of
patients suffering from scarlet fever (Mahara et al., 2016).

14.7   BETA-HEMOLYTIC STREPTOCOCCI

Streptokinase is naturally produced by different species of beta-hemolytic streptococci. In 1933, Lancefield differentiated beta-hemolytic streptococcal strain from

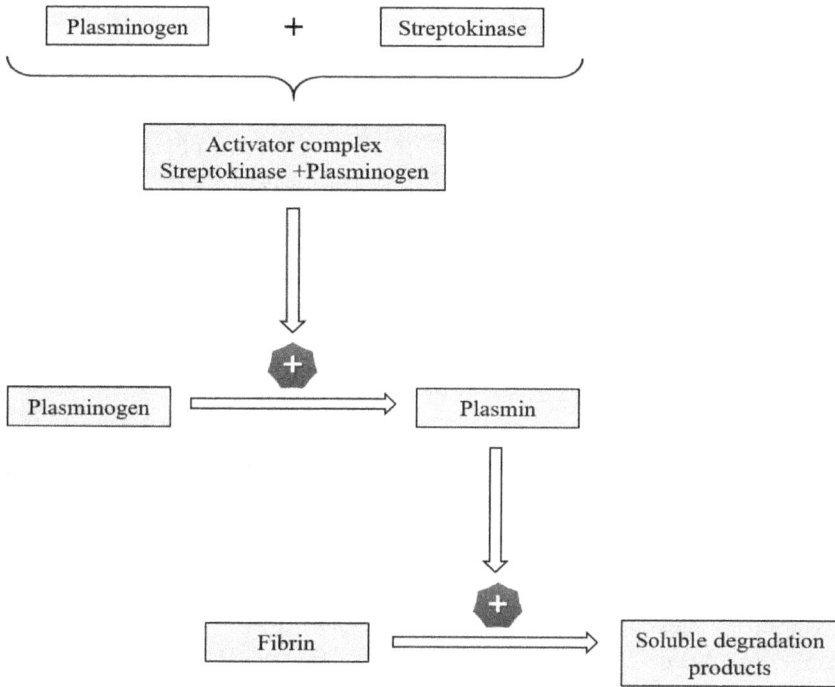

FIGURE 14.3 Mechanism action of streptokinase.

A to O on the basis of serological analysis (Couture-Cossette et al., 2018). It has been demonstrated that beta-hemolytic stretptococci of Lancefield groups A, C, and G produce the streptokinase enzyme (Akbar et al., 2020). Due to less erythrogenic toxin, group C is mostly used in the production of streptokinase, but group B causes infection in pregnant females and neonates (Joubert et al., 2021). Similarly, streptokinase isolated from the human source when compared to the streptokinase isolated from procaine showed significant functional and structural differences (Huish et al., 2017).

14.8   STREPTOKINASE MECHANISM AND STRUCTURE

Multiple domains (alpha, beta, and gamma) are present in the structure of streptokinase protein, performing their different function by scanning calorimeter. Streptokinase consists of two carboxyl terminals and one amino terminal (Sawhney et al., 2016). One of the 59 amino acids of the N terminal has the ability to activate low plasminogen and also to participate in the thrombolytic activity (Huish et al., 2017). If the N terminal region is isolated from the streptokinase, then the structure of streptokinase becomes normal. Due to the absence of 59 amino acids out of 414 amino acids, the thrombolytic activity of streptokinase is essentially reduced (Mohanty and Khasa, 2018). Various domains of streptokinase have been investigated to show solid-like properties in high temperatures (Ayinuola et al., 2020).

In a normal hemostatic system, blood components, fats cells, and cholesterol in blood circulation play a vital role in the prevention of clots in the circulatory system, and if the cholesterol, fat cells, and other blood component are deposited in blood circulation, they may cause blood clots or plaques, causing myocardial infarction and other heart diseases (Huang et al., 2020). This formation of plaque may result in deep-vein thrombosis, pulmonary embolism, and stroke. A clot is formed by the fibrinogen, and this clot is degraded by the PLS, which is triggered by the PLG via t-plasminogen activators. Fibrin is accumulated in the internal lining of blood vessels and assists in thrombus formation, causing heart disease and myocardial infarction (Zeng et al., 2020). In a disturbed state, fibrin embolus is not dissolved and results in a pathophysiological disease and boosts thrombus formation and various other heart diseases (Benjamin et al., 2018).

Rapid vaccination of fibrinolytic agents is the treatment of myocardial infarction and assists in the restoration of blood flow without clotting. In aged people, this obstruction is many times greater. In this type of disease, blockage in the blood flow occurs due to blood coagulation (Hamadeh et al., 2020). Coagulated blood is dissolved by plasmin with the help of fibrin to remove the hemostasis. There will be no breakdown of fibrin in case of less plasmin, causing thrombosis (Rosenberg and Parikh, 2011). The blood clot consists of fibrin and erythrocytes. There are two types of clots. The first is the arterial clot or red clot, consisting of dumped erythrocytes that increase due to reduced shear pressure on internal blood vessel walls. The second is the platelet-rich arterial clots that are formed by great shear pressure, causing atherosclerosis or the internal covering of blood vessels. To substitute the clots in blood vessels, anticoagulant and antiplatelet drugs are used (Mackman, 2012).

14.9 DISCOVERY OF STREPTOKINASE

Acute myocardial infarction (AMI) was generally held as a medical curiosity in the early 19th century. The sound foundation for the use of streptokinase as a thrombolytic agent in the treatment of AMI was laid after a serendipitous discovery by William Smith Tillett in 1933 (Adivitiya and Khasa, 2017). Initial clinical trials of the drug were found in combating fibrinous pleural exudates, tuberculous meningitis, and hemothorax. Streptokinase was used by Sherry and others in 1958 in patients with AMI, and the focus of treatment was changed from palliation to cure. Conflicting results were produced using streptokinase in initial trials (Kunamneni et al. 2018). In 1933, the history of thrombolytic therapy was started after discovering the fact that fibrin clots could be dissolved by broth culture of streptococcus strains formed when calcium or thrombin is added to human plasma. This reaction is termed fibrinolysis, and the active agent in the culture is fibrinolysin (Cone et al., 2020). Fibrinolysin has been found by other investigators to be associated most consistently with hemolytic streptococci of Lancefield groups A, C, and G (Cherkaoui et al., 2011). It is frequently encountered in others species of bacteria while not peculiar to those streptococci. Tillet and Garner observed that besides human plasma and human plasminogen, thrombin clots are vulnerable to action fibrinolysin (Zhou, Yang et al., 2017).

This observation has been confirmed by other investigators, and studies have been extended to understanding fibrin in additional animal species. Plasma from rabbits is resistant (Leach et al., 2003). Significant evidence about proteolysis during fibrinolysis was not detected by Garner and Tillet. Garner and Tillet successfully isolated streptokinase in a stable form and demonstrated that it is a protein, while they had compared the action of streptococcal fibrinolysin and other enzymes on fibrin and substrates (Man et al., 2011). Hence, it was suggested by them that the fibrinolysin was an expected enzyme or a catalyst having elevated specificity for human fibrin. Le Mar and Gunderson (1940) conducted a wide-ranging study of the vulnerability of fibrin from various species, and it was concluded that animal fibrin is usually partially or completely resistant when compared to human fibrin, which is most susceptible. Fibrin does not have resistance property itself (Litvinov et al., 2019). Due to the presence of lysin factors as containment in both these reagents, human fibrinogen thrombin clots are prone to fibrinolysin; rabbits lack a suitable lysin factor due to which plasma clots are resistant and termed as a lytic factor (Man et al., 2011). Proteinase precursors are converted to an active enzyme after being activated by streptococcal substances (Klein et al., 2015).

The term fibrinolysin was replaced by streptokinase in 1945 (Tucker and Idell, 2013) for the inactive enzyme, plasminogen, and for the active enzyme, plasmin. The fibrinolytic properties of *Streptococcus equisimilis* are the probable source of streptokinase, which was reported by Evans in 1944. Christensen (1945) reported that erythrogenic toxins are not produced by the strain *S. equisimilis* H46 A, and on semi-synthetic media, they are less fastidious in growth, and there is a possibility for them to act as a commercial source, producing the bulk of streptokinase (Akbar et al., 2020). The prominence of this discovery is underlined by the fact that to date, streptokinase used for thrombolytic therapy is derived from *S. equisimilis* commercially (Lancefield group C) (Sameni et al., 2017).

Initial clinical applications of streptokinase are found in contesting hemothorax, fibrinous tuberculous meningitis, and pleural exudates. Streptokinase was first used in 1958 in patients with AMI, and this changed the focus of treatment. Streptokinase was efficaciously used to dissolve intravascular clots artificially induced in the marginal ear veins of rabbits (Hosseini et al., 2018). After the successful use of streptokinase in the suspension of intra-arterial thrombi, the idea was highly attractive to use fibrinolytic agents in treating coronary thrombosis, particularly since no drug was available that could actually favor expanding the prognosis after myocardial infarction (Thygesen et al., 2018). Streptokinase was used for dissolving intracoronary clots for the first time. A key hindrance to the role of streptokinase as a thrombolytic agent was evidenced (Sezer et al., 2018), and its imminent use in treating myocardial infarction until 1980 in the clinical setting was not extensive. In 1985, a huge number of papers were published, including small trials mostly, in a struggle to establish a standard protocol (Helal et al., 2018) for streptokinase use in AMI. Within 1.5 to 3 hours of administration of streptokinase, reperfusion rates as high as 90% were achieved. With delayed treatment, the prognosis became deteriorated (Tourani et al., 2018).

14.10   COMMONLY USED FIBRINOLYTIC AGENTS

Streptokinase is a type of extracellular enzyme synthesized by different groups of cocci. The structure of streptokinase comprises two regions: the first is the amino end, while the second is the carboxyl end. The carboxyl end functions to activate the plasminogen in the complex, and the amino acid end functions to identify the substrate (Chou et al., 2012). Groups A, C, and G and various groups of beta-hemolytic streptococci most widely produce it. It has been internationally utilized in therapy, mostly in myocardial infarction, because it shows strong activity to dissolve fibrin (Ząbczyk et al., 2019). In the circulatory system, the inactive precursor of plasmin, also known as plasminogen, activates the plasmin. Plasminogen is activated by streptokinase and is converted into the plasmin by making a complex in response to interaction with serum protein. Plasminogen, in the presence of streptokinase, with the help of that complex plasmin, is formed its inactive plasminogen. The fibrin protein is broken by the plasmin that is present in the blood. Streptokinase possesses a half-life of 23 minutes. Hence, because of its short half-life, it is not apt for therapy. The summary of the plasminogen activator is given in Table 14.2. To enhance its therapeutic potential, the half-life of streptokinase is increased by the researchers by modifying its structure (Mannully et al., 2019).

To acquire maximum perks of thrombolytic therapy in reinstating blood flow, minimizing destruction of heart muscles, and stabilizing the functions of heart, early treatment is predominantly important (Horlocker et al., 2019). Many centers, such as ISIS-3 (Third International Study of Infract Survival), GISSI (Gruppo Italiano per lo Studio della Sopravvienza nell'Infarto Miocardico), and GUSTO (Global Utilization of Streptokinase and Tissue Plasminogen Activator for Occluded Coronary Arteries), compared urokinase, tissue plasminogen activator (tPA), with streptokinase (Natsuaki et al., 2019).

14.11   STRUCTURE AND PROPERTIES OF STREPTOKINASE

Several studies have been conducted for reporting the chemical and physical properties of streptokinase. It is a monomer protein that has an extinction coefficient of 9.4 (Sevostyanov et al., 2020). Streptokinase is an efficacious and economical drug,

TABLE 14.2
Summary of Plasminogen Activator

Streptokinase (SK)	Tissue Plasminogen Activators (tPA)	Urokinase-Plasminogen Activators (uPA)
Non-enzymatic protein	Serine protease	Serine protease
Molecular weight 47 kDa	Molecular weight 70 kDa	Molecular weight 55 kDa
Non-fibrin-specific activators	Fibrin-specific activators	Non-fibrin-specific activators
Produced by fermentation	Produced by the vesicles of endothelium cells	Produced by the kidney
(Zia, 2020)	(Baharifar et al., 2020)	(Mei et al., 2020)

widely used clinically to cure and treat pulmonary embolism, myocardial infarction, thromboembolic strokes, and deep-vein thrombosis (Zayet et al., 2020). Streptokinase is a protein isolated from pathogenic strains of the *Streptococcus* family, having a molecular weight of 47 kDa and 414 amino acids in the polypeptide chain (Iseppi et al., 2020). It is a bacterial protein deficient in typical or anticipated chemical or biological action besides lacking structural homologs (Baharifar et al., 2020).

The only drawback of streptokinase is that it may cause allergy. However, streptokinase is the best medication to treat heart-related diseases (Mamede et al., 2020). Antigenic protein production and hypersensitivity in cells result in allergy, and there are four types of allergy responses observed in allergic patients (Mennini et al., 2020). As streptokinase is obtained from non-human sources, the immune system is activated. Neutralizing antibodies are the main source of massive allergic reactions, which reduce the antigenicity of streptokinase, and the mutated form of streptokinase is produced through genetic manipulation (Prabhakar, 2020). The presence of bacteria inside the body is the main cause of the four types of allergies. According to the type of hypersensitivity responses, symptoms may vary. Anaphylaxis, rash, itchy skin, and hives are the main symptoms of type 4 hypersensitivity (Prasitdumrong et al., 2020). During thrombolytic therapy, antibodies are produced, which shows that there are some genes for the antigenic region that can be cloned from selected *Streptococcus* pyogenes, and the product of genes has the same antigenicity and substrate specificity, and expression is checked in the *Escherichia coli*. In this way, genes control transcription and translation. Likewise, the deletion of 42 amino acids from the C terminal of the streptokinase reduces the antigenicity (Zia, 2020). Adverse reactions can be seen in the form of the complex immune mechanism, antigenicity, and serum sickness reactions. Scientists use different methodologies and tools to reduce the adverse effects of streptokinase by mutating the sequence of streptokinase, which is immunogenic streptokinase, normally contain three domains. It is confirmed from a previous study that immunity is induced in the host cell for immunogenicity by the third number domain (Ayinuola et al., 2020).

PCR is very effective with site-directed mutagenesis. In the host cell, the altered gene possesses both anti-immunogenic and clot-dissolving effects (Hasanpour et al., 2020). Increased levels of anti-streptokinase and perivascular complexes besides immunogenic proteins (IgA, IgG, C3, and IgM) are seen by the minimal dose of streptokinase, and fibrinogen shows some kind of allergic effects (Zia, 2020).

Both the short- and long-term benefits have been demonstrated by these agents in many large-scale clinical trials to save lives (Locke et al., 2020). Early treatment is particularly important to gain maximum benefits of thrombolytic therapy conducted to restore blood flow, control the damage to heart muscles, and preserve heart functions (Dridi et al., 2020). Blood cells occluded by a fibrin matrix constrict the blood flow, forming a blood clot or thrombus. Thrombolysis or fibrinolysis is an enzyme-mediated suspension of the fibrin clot. The enzyme responsible for fibrinolysis in mammalian circulation is plasmin, which is a trypsin-like serine protease (Kojima et al., 2020). Inactive protein plasminogen forms fibrinolytically active plasmin, which is present in circulation. A limited proteolytic cleavage is involved in the conversion of the inactive plasminogen to fibrinolytic plasmin, and various plasminogen activators intercede this process. tPA and uPA are two plasminogen activators that occur naturally in the blood (Mei et al., 2020).

Inhibitors of plasminogen activators (e.g., plasminogen activator inhibitor-1, PAI-1, a fast-acting inhibitor of tPA and uPA) and plasmin (e.g., al-antiplasmin, a2 macroglobulin) modulate the fibrinolytic activity in circulation (Moore and Moore, 2020). In clinical intrusion, recombinant forms of normal human plasminogen activators tPA and uPA are used. Streptokinase (sPA) is another plasminogen activator frequently used. Streptokinase is a bacterial protein that is not present in human circulation naturally (Whyte and Mutch, 2020).

Direct fibrinolytic activity is not associated with streptokinase, tPA, and uPA. The activity of the thrombolytic drug is given in Figure 14.4. They manifest their therapeutic action by activating the blood plasminogen to plasmin which is responsible for dissolving the clot. In contrast to tPA and uPA, which are proteases, streptokinase does not possess its own enzymatic activity (Lakshmanan et al., 2020). High-affinity 1:1 stoichiometric complex (i.e., the streptokinase-plasminogen activator complex) results, which is a protease with elevated specificity that proteolytically activates other plasminogen molecules to plasmin (Hassan et al., 2020).

Thus, the plasminogen activating action incorporated by streptokinase profoundly differs from the proteolytic activation resulting from tPA and uPA. To treat AMI, streptokinase is as good as recombinant tPA (Aghaeepoor et al., 2019). Additionally, it is unquestionably more cost-effective than the currently available, competing recombinant tPA streptokinase (Tran et al., 2018). In spite of this fact, the research conducted on streptokinase continues, and it is truly a viable therapy used worldwide in proper health care schemes (Mohanty and Khasa, 2018).

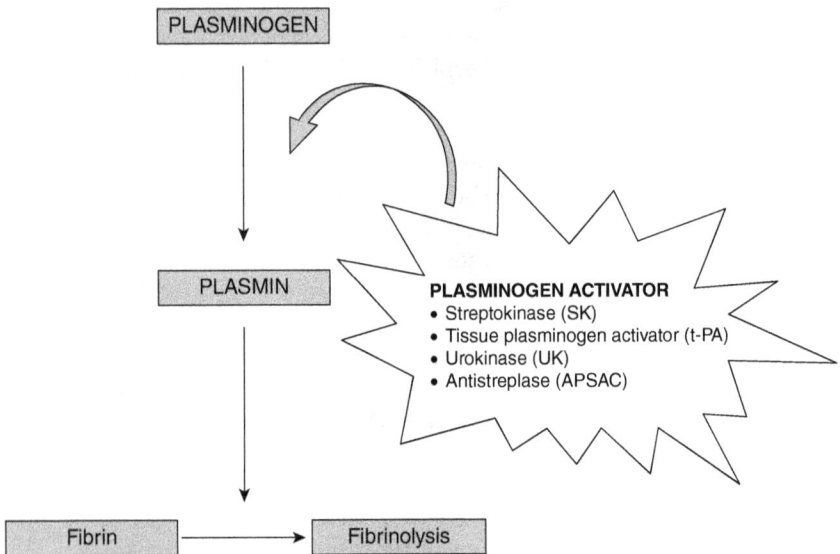

FIGURE 14.4 Mechanism action of thrombolytic agents.

14.12 PRODUCTION, PURIFICATION, AND CHARACTERIZATION OF STREPTOKINASE

The production of streptokinase ensures high benefits. With this highly thrombolytic agent, different investigators are applying genetic engineering techniques to understand the function and structure relationship. Enhancing the production of streptokinase is extremely significant to characterize the gene products and effectively relate their kinetic properties to native streptokinase proteins. The administration of streptokinase as a therapeutic agent is increasing exponentially, which highlights both its qualitative development and quantitative augmentation (Tran et al., 2018). Therefore, along with the selection of a potent microbial culture, opting for medium components is vital in the same token for streptokinase production since optimum cultural conditions are required by microorganisms for growth and metabolite synthesis (Fan et al., 2018). One factor at a time is involved in the traditional media optimization strategy, which is simple in its approach. However, the region of optimal response cannot be assessed since the effects of all factors are not considered at a time. In contrast to this, statistically determined experiments show more precision as compared to classical ones. The administration of streptokinase as a therapeutic agent is exponentially increasing, and this phenomenon lays emphasis on its qualitative improvement and quantitative enhancement (Arshad et al., 2019).

The growth of microorganisms is always influenced by process optimization, which can improve production considerably and reduce production costs significantly. In the enzyme extraction process, even little improvement in culture condition can have a significant effect on commercially successful production (Cai et al., 2019). Regarding the growth of streptokinase, fed-batch culture is used, but continuous culture shows more production two to three times more elevated than fed-batch culture. Unceasing production has been more successfully considered for industrial-scale production. By providing an energy source (in the form of glucose), the biomass yield may increase significantly (Aghaeepoor et al., 2019). Streptokinase is produced from *Streptococcus mutans* by liquid-state fermentation, and optimization studies were performed using rise polishing, sugarcane bagasse, steep corn liquor, and molasses as substrates in varying concentrations ranging from 0.1% to 0.8% (Babu and Devi, 2020). Isolated streptococcus from a throat swab of acute tonsilitis patients was subjected to produce streptokinase by liquid-state fermentation in previous studies. Streptokinase has been purified from ammonium sulfate precipitation, ion-exchange chromatography, and gel filtration chromatography (Ali and Bavisetty, 2020). Hydrophobic interaction chromatography (HIC) was exploited for the purification of recombinant streptokinase by (Guo et al., 2018). Anti-streptokinase antibodies can be separated by immune-affinity chromatography (Zhou et al., 2017; Meng et al., 2017). Kinetic and thermodynamic parameters have been studied for the advantageous applications of streptokinase (Guo et al., 2017). Therefore, along with opting for a potent microbial culture, the choice of medium components is correspondingly vital for streptokinase production since optimum cultural conditions for growth and metabolite synthesis are required by microorganisms (Babu et al., 2015). The conventional strategy of media optimization, which involves one factor at a time, is simple in its approach. However, the region of optimal response cannot assess since the effects of all factors are not considered at a time. In contrast to this, statistically determined experiments show more precision as compared to the classical media optimization strategy as they are time-saving and economical. Furthermore, in the latter case, several variables are appraised at the same time (Verma, 2015).

14.13   APPLICATIONS OF STREPTOKINASE

Streptokinase has been widely used in developing countries due to its low cost and high effectiveness. Native streptokinase can be produced inexpensively by fermentation and is useful in thrombolytic therapy. It is given to the patient who is suffering from myocardial infarction, and intracoronary therapy is initiated within six hours of AMI, and the mortality rate is reduced (Jing et al., 2020). Some studies suggest that streptokinase is an efficient treatment like tPA (De Rosa et al., 2020). Mutated streptokinase produces three times fewer anti-streptokinase antibodies than natural streptokinase. Some kind of allergic reaction is provoked due to streptokinase, In developed countries like the USA, tPA is used as a thrombolytic agent, while developing countries are still using streptokinase just because of the expensiveness of tPA for the treatment of myocardial infarction. Furthermore, the price of tPA is ten times greater than that of streptokinase (Gramegna et al., 2020).

According to an estimate, about 65–75% of the 4 billion people in developing countries rely on conventional medicines, predominantly drugs of plants origin, to meet their primary health care needs. Modern pharmacopeia has derived 30–40% of the drugs from the plants. Hence, medicinal plants increase in importance in developing countries as well as in developed countries due to the following reasons: they are nontoxic in nature, have non-noxious effects, and are easily available at an inexpensive rate. About 600 plants species have great importance in allopathic and homeopathic medicine. About 300–400 species are similar to that which have already been used in Greek medicine and traditional medicine (Lascarrou et al., 2020).

Heart diseases have become a global problem. Many factors are involved in the progression of heart diseases. Dyslipidemia, hypertension, smoking, and diabetes, among others, are associated with the production of reactive oxidative species (ROS). Myocardial damage induce by ischemia reperfusion is due to the generation of ROS (Li et al., 2020). The heart is kept protected by plants from cardiovascular diseases by their cardioprotective action in nature by providing nutritional substances with cohesive structure, mainly phytoconstituents that aid to maintain and restore the heart activity and balance the body's systems. It might be remarkable and possibly rewarding with this basic information to study the native medicinal plants to ascertain their cardioprotective potential in available in animals (rabbits) (Wang et al., 2020).

Plasminogen is directly activated by both uPA and tPA having the M/W of 55 and 70 kDa, respectively, and both are glycoproteins responsible for the production of uPA and excreted through urine. Alternatively, vascular cells cause the production of tPA (Moore et al., 2020; Whyte and Mutch, 2021). The therapeutic use of tPA is achieved from the culture of recombinant animals cells (Hoffmann et al., 2020). Tissue plasminogen activators activate clot-bound plasminogen 100 times more effectively compared to circulatory plasminogen. AMI is usually recognized as a heart attack. In these conditions, enough blood is not supplied to the heart, causing damage to the heart muscles. Mostly, myocardial infarction occurs because a thrombus or blood clot formed in coronary arteries, causing the blockage of arteries (Silverman et al., 2020). The factors responsible for coronary artery disease are high cholesterol level in the blood, high blood pressure, deficiency of exercise, obesity, diabetes, smoking, deprived diet, and too much drinking (Akbar et al., 2020).

The mechanism of myocardial infarction consists of the rupture of an atherosclerosis plaque that causes complete blockage of coronary arteries in underdeveloped countries. The rate of mortality rises to 80% in CVD. However, with deaths, males and females are almost equivalent. From past 30 years, streptokinase is utilized in therapy of thrombolytic counting and cure of AMI. To associate the medical effectiveness of recombinant PA and SK, a lot of trials have been conducted. Unfortunately, these trials have not exposed a clear preference for either type of drug. But streptokinase is more effective as compared to recombinant tissue PA in cure of AMI (Rafipour et al., 2020). But streptokinase utilization is dangerous and it is indeed more cost-effective. Investigations on streptokinase carry on, particularly in the inferior healthcare systems of the world, and it remains the most important affordable therapeutic agent (Sheli et al., 2018).

In the medical field, the applications of streptokinase have been fundamentally increased, so the organism's demand to produce the enzyme has been efficiently higher. Due to these higher demands, the application of streptokinase must be improved qualitatively and quantitatively. It has been observed that usually, wild bacterial species secretes a reduced quantity of streptokinase. Nowadays, streptokinase is industrially produced by fungi and bacteria. A large number of countries produce streptokinase in the whole world, and each year, 0.4–0.5 million patients are being cured with streptokinase successfully. Hence, for the synthesis of streptokinase at a large scale under optimized media and strain development, different techniques are used (Rahimi et al., 2020). To treat the clots in arteries, streptokinase is preferred. Recently, this enzyme is observed to heal patients successfully who suffered from AMI (Sevostyanov et al., 2020).

14.14 CONCLUSION

In the human body, blood circulation is affected by the formation of blood clots, causing serious heart-related diseases and heart-related problems, such as AMI. There are many thrombolytic agents, such as streptokinase, tPa, and urokinase. Streptokinase is the most useful agent among them all, even in underdeveloped countries, because of its low cost and easy availability. This agent is naturally synthesized by using different strains of streptococci from groups C, A, and G, and its main function is to convert plasminogen into the plasmin by activating it and breaking the fibrin, which is the main cause of blood clots deposited in the veins. Streptokinase therapy is most useful in heart attack and stroke. Heart diseases have become a global problem. Many factors are involved in the progression and causation of heart diseases. This review describes the efficacy of streptokinase as a therapeutic agent against myocardial infarction as blood circulation is affected by the formation of blood clots that cause serious heart diseases. Streptokinase is the most useful agent among them all and is used for treating heart diseases because of its low cost and easy availability. This agent is naturally synthesized by using different strains of streptococci. It is given to a patient who is suffering from myocardial infarction, and intracoronary therapy is initiated. This review is helpful in describing the clinical uses of streptokinase for the treatment of heart diseases and for describing the action mechanism of streptokinase.

14.15 ACKNOWLEDGMENTS

Consejo Nacional de Ciencia y Tecnología (MX) is thankfully acknowledged for partially supporting this work under Sistema Nacional de Investigadores (SNI) program awarded to Hafiz M. N. Iqbal (CVU: 735340).

14.16 CONFLICT OF INTERESTS

The author(s) declare no conflicting interests.

REFERENCES

Adivitiya and Y. P. Khasa. 2017. The evolution of recombinant thrombolytics: Current status and future directions. *Bioengineered*, 8: 331–358.

Aghaeepoor, M., A. Akbarzadeh, F. Kobarfard, A. A. Shabani, E. Dehnavi, S. J. Aval and M. R. A. Eidgahi. 2019. Optimization and high level production of recombinant synthetic Streptokinase in E. coli using Response Surface Methodology. *Iranian Journal of Pharmaceutical Research: IJPR,* 18: 961.

Akbar, G., M. Zia, A. Ahmad, N. Arooj and S. Nusrat. 2020. Review on streptokinase with its antigenic determinants and perspectives to develop its recombinant enzyme with minimum immunogenicity. *Journal of Innovative Sciences*, 6: 17–23.

Ali, A. M. M. and S. C. B. Bavisetty. 2020. Purification, physicochemical properties, and statistical optimization of fibrinolytic enzymes especially from fermented foods: A comprehensive review. *International Journal of Biological Macromolecules*, 163, 1498–1517.

Ali, Q., I. Dhande, P. Samuel and T. Hussain. 2016. Angiotensin type 2 receptor null mice express reduced levels of renal angiotensin converting enzyme-2/angiotensin (1–7)/ Mas receptor and exhibit greater high-fat diet-induced kidney injury. *Journal of the Renin-Angiotensin-Aldosterone System,* 17: 1470320316661871.

Angelucci, F., K. Čechová, R. Průša and J. Hort. 2019. Amyloid beta soluble forms and plasminogen activation system in Alzheimer's disease: Consequences on extracellular maturation of brain-derived neurotrophic factor and therapeutic implications. *CNS Neuroscience & Therapeutics*, 25: 303–313.

Arshad, A., M. A. Zia, M. Asghar and F. A. Joyia. 2019. Enhanced production of streptokinase by chemical mutagenesis of streptococcus agalactiae EBL-20. *Brazilian Archives of Biology and Technology,* 62.

Ayinuola, Y. A., T. Brito-Robinson, O. Ayinuola, J. E. Beck, D. Cruz-Topete, S. W. Lee, V. A. Ploplis and F. J. Castellino. 2020. Streptococcus co-opts a conformational lock in human plasminogen to facilitate streptokinase cleavage and bacterial virulence. *Journal of Biological Chemistry: JBC*. RA120.016262.

Babu, M., T. Durga Devi, P. Mäkinen, M. Kaikkonen, H. P. Lesch, S. Junttila, A. Laiho, B. Ghimire, A. Gyenesei and S. Ylä-Herttuala. 2015. Differential promoter methylation of macrophage genes is associated with impaired vascular growth in ischemic muscles of hyperlipidemic and type 2 diabetic mice: Genome-wide promoter methylation study. *Circulation Research,* 117: 289–299.

Babu, V. and C. S. Devi. 2020. A statistical application for the enhanced production of streptokinase from a mutant strain Streptococcus equinus VIT_VB2. *National Academy Science Letters*: 1–7.

Baharifar, H. and A. Amani. 2016. Cytotoxicity of chitosan/streptokinase nanoparticles as a function of size: An artificial neural networks study. *Nanomedicine: Nanotechnology, Biology and Medicine*, 12: 171–180.

Baharifar, H., M. Khoobi, S. A. Bidgoli and A. Amani. 2020. Preparation of PEG-grafted chitosan/streptokinase nanoparticles to improve biological half-life and reduce immunogenicity of the enzyme. *International Journal of Biological Macromolecules,* 143: 181–189.

Bansal, M. (2020). Cardiovascular disease and COVID-19. *Diabetes & Metabolic Syndrome: Clinical Research & Reviews*, 14(3): 247–250.

Basak, S., P. Singh and M. Rajurkar. 2016. Multidrug resistant and extensively drug resistant bacteria: A study. *Journal of Pathogens*. Article ID 4065603. 2016, 1–5.

Benjamin, Emelia J., Salim S. Virani, Clifton W. Callaway, Alanna M. Chamberlain, Alexander R. Chang, Susan Cheng, Stephanie E. Chiuve et al. 2018. Heart disease and stroke statistics—2018 update: A report from the American Heart Association. *Circulation*, 137(12): e67–e492.

Cai, D., Y. Rao, Y. Zhan, Q. Wang and S. Chen. 2019. Engineering Bacillus for efficient production of heterologous protein: Current progress, challenge and prospect. *Journal of Applied Microbiology,* 126: 1632–1642.

Capak, P., H. Aussel, M. Ajiki, H. McCracken, B. Mobasher, N. Scoville, P. Shopbell, Y. Taniguchi, D. Thompson and S. Tribiano. 2007. The first release COSMOS optical and near-IR data and catalog. *The Astrophysical Journal Supplement Series,* 172: 99.

Chapurina, Y. E., A. S. Drozdov, I. Popov, V. V. Vinogradov, I. P. Dudanov and V. V. Vinogradov. 2016. Streptokinase@ alumina nanoparticles as a promising thrombolytic colloid with prolonged action. *Journal of Materials Chemistry B*, 4: 5921–5928.

Cherkaoui, A., S. Emonet, J. Fernandez, D. Schorderet and J. Schrenzel. 2011. Evaluation of matrix-assisted laser desorption ionization-time of flight mass spectrometry for rapid identification of beta-hemolytic streptococci. *Journal of Clinical Microbiology*, 49: 3004–3005.

Chou, S. S., M. De, J. Luo, V. M. Rotello, J. Huang and V. P. Dravid. 2012. Nanoscale graphene oxide (nGO) as artificial receptors: Implications for biomolecular interactions and sensing. *Journal of the American Chemical Society*, 134: 16725–16733.

Christensen, L. R. 1945. Streptococcal fibrinolysis: a proteolytic reaction due to a serum enzyme activated by streptococcal fibrinolysin. *The Journal of General Physiology*, 28: 363–383.

Chu, A. J. 2011. Tissue factor, blood coagulation, and beyond: An overview. *International Journal of Inflammation*. Article ID 367284. 2011, 1–30.

Cone, S. J., A. T. Fuquay, J. M. Litofsky, T. C. Dement, C. A. Carolan and N. E. Hudson. 2020. Inherent fibrin fiber tension propels mechanisms of network clearance during fibrinolysis. *Acta biomaterialia*, 107, 164–177.

Cortes-Canteli, M., J. Paul, E. H. Norris, R. Bronstein, H. J. Ahn, D. Zamolodchikov, S. Bhuvanendran, K. M. Fenz and S. Strickland. 2010. Fibrinogen and β-amyloid association alters thrombosis and fibrinolysis: A possible contributing factor to Alzheimer's disease. *Neuron*, 66: 695–709.

Couture-Cossette, A., A. Carignan, A. Mercier, C. Desruisseaux, L. Valiquette and J. Pepin. 2018. Secular trends in incidence of invasive beta-hemolytic streptococci and efficacy of adjunctive therapy in Quebec, Canada, 1996–2016. *PloS One*, 13: e0206289.

De Leonardis, E., B. Lutz, S. Cocco, R. Monasson, H. Szurmant, M. Weigt and A. Schug. 2015. Protein and RNA structure prediction by integration of co-evolutionary information into molecular simulation. *Biophysical Journal,* 108: 13a–14a.

De Rosa, S., C. Spaccarotella, C. Basso, M. P. Calabrò, A. Curcio, P. P. Filardi, M. Mancone, G. Mercuro, S. Muscoli and S. Nodari. 2020. Reduction of hospitalizations for myocardial infarction in Italy in the COVID-19 era. *European Heart Journal*, 41: 2083–2088.

Dridi, H., A. Kushnir, R. Zalk, Q. Yuan, Z. Melville and A. R. Marks. 2020. Intracellular calcium leak in heart failure and atrial fibrillation: A unifying mechanism and therapeutic target. *Nature Reviews Cardiology*: 1–16.

Elmi, S., G. Sahu, K. Malavade and T. Jacob. 2019. Role of tissue plasminogen activator and plasminogen activator inhibitor as potential biomarkers in psychosis. *Asian Journal of Psychiatry*, 43: 105–110.

Fan, Y., J. M. Miozzi, S. D. Stimple, T.-C. Han and D. W. Wood. 2018. Column-free purification methods for recombinant proteins using self-cleaving aggregating tags. *Polymers*, 10: 468.

Gramegna, M., L. Baldetti, A. Beneduce, L. Pannone, G. Falasconi, F. Calvo, V. Pazzanese, S. Sacchi, M. Pagnesi and F. Moroni. 2020. ST-segment—elevation myocardial infarction during COVID-19 pandemic: Insights from a regional public service healthcare hub. *Circulation: Cardiovascular Interventions*, 13: e009413.

Guo, D. S., X. J. Ji, L. J. Ren, G. L. Li and H. Huang. 2017. Improving docosahexaenoic acid production by Schizochytrium sp. using a newly designed high-oxygen-supply bioreactor. *AIChE Journal*, 63: 4278–4286.

Guo, D.-S., X.-J. Ji, L.-J. Ren, G.-L. Li, X.-M. Sun, K.-Q. Chen, S. Gao and H. Huang. 2018. Development of a scale-up strategy for fermentative production of docosahexaenoic acid by Schizochytrium sp. *Chemical Engineering Science*, 176: 600–608.

Hamadeh, A., A. Aldujeli, K. Briedis, K. Tecson, J. S. Sanchez, M. Al Dujeili,... P. McCullough. 2020. TCT CONNECT-213 clinical characteristics and outcomes of patients with COVID-19 and STEMI treated with fibrinolytic therapy. *Journal of the American College of Cardiology*, 76(17 Supplement S): B89–B90.

Hasanpour, A., F. Esmaeili, H. Hosseini and A. Amani. 2020. Use of mPEG-PLGA nanoparticles to improve bioactivity and hemocompatibility of streptokinase: In-vitro and in-vivo studies. *Materials Science and Engineering: C*, 118: 111427.

Hasanpour, A., F. Esmaeili, H. Hosseini and A. Amani. 2021. Use of mPEG-PLGA nanoparticles to improve bioactivity and hemocompatibility of streptokinase: In-vitro and in-vivo studies. *Materials Science and Engineering: C*, 118: 111427.

Hassan, M. M., S. Sharmin, H.-J. Kim and S.-T. Hong. 2020. Identification and characterization of plasmin-independent thrombolytic enzymes. *Circulation Research*, 128(3): 386–400.

Helal, A. M., S. M. Shaheen, W. A. Elhammady, M. I. Ahmed, A. S. Abdel-Hakim and L. E. Allam. 2018. Primary PCI versus pharmacoinvasive strategy for ST elevation myocardial infarction. *IJC Heart & Vasculature*, 21: 87–93.

Hoffmann, M., H. Kleine-Weber, S. Schroeder, N. Krüger, T. Herrler, S. Erichsen, T. S. Schiergens, G. Herrler, N.-H. Wu and A. Nitsche. 2020. SARS-CoV-2 cell entry depends on ACE2 and TMPRSS2 and is blocked by a clinically proven protease inhibitor. *Cell*, 181(2): 271–280.

Horlocker, T. T., E. Vandermeulen, S. L. Kopp, W. Gogarten, L. R. Leffert and H. T. Benzon. 2019. Regional anesthesia in the patient receiving antithrombotic or thrombolytic therapy: American Society of Regional Anesthesia and Pain Medicine Evidence-Based Guidelines. *Obstetric Anesthesia Digest*, 39: 28–29.

Hosseini, A., S. Akhavan, M. Menshaei and A. Feizi. 2018. Effects of streptokinase and normal saline on the incidence of intra-abdominal adhesion 1 week and 1 month after laparotomy in rats. *Advanced Biomedical Research*, 7.

Huang, G., C. Ding, Y. Li, T. Zhang and X. Wang. 2020. Selenium enhances iron plaque formation by elevating the radial oxygen loss of roots to reduce cadmium accumulation in rice (Oryza sativa L.). *Journal of Hazardous Materials*, 398: 122860.

Huish, S., C. Thelwell and C. Longstaff. 2017. Activity regulation by fibrinogen and fibrin of streptokinase from Streptococcus pyogenes. *PloS One,* 12: e0170936.

Iseppi, R., R. Tardugno, V. Brighenti, S. Benvenuti, C. Sabia, F. Pellati and P. Messi. 2020. Phytochemical composition and in vitro antimicrobial activity of essential oils from the Lamiaceae Family against Streptococcus agalactiae and Candida albicans Biofilms. *Antibiotics,* 9: 592.

Jing, Z.-C., H.-D. Zhu, X.-W. Yan, W.-Z. Chai and S. Zhang. 2020. Recommendations from the Peking Union Medical College Hospital for the management of acute myocardial infarction during the COVID-19 outbreak. *European Heart Journal.* 1791–1794.

Joubert, J., S. M. Meiring, C. Conradie, S. Lamprecht and W. J. van Rensburg. 2021. The effects of streptokinase in a Chacma baboon (Papio ursinus) model of acquired thrombotic thrombocytopenic purpura. *Clinical and Experimental Medicine,* 1–12.

Karnabatidis, D., S. Spiliopoulos, D. Tsetis and D. Siablis. 2011. Quality improvement guidelines for percutaneous catheter-directed intra-arterial thrombolysis and mechanical thrombectomy for acute lower-limb ischemia. *Cardiovascular and Interventional Radiology,* 34: 1123–1136.

Keramati, M., M. M. Aslani and F. Roohvand. 2020. In silico design and in vitro validation of a novel PCR-RFLP assay for determination of phylogenetic clusters of streptokinase gene alleles in streptococci groups. *Microbial Pathogenesis,* 139: 103862.

Kikkert, W. J., N. van Geloven, M. H. van der Laan, M. M. Vis, J. Baan, K. T. Koch, R. J. Peters, R. J. de Winter, J. J. Piek and J. G. Tijssen. 2014. The prognostic value of bleeding academic research consortium (BARC)-defined bleeding complications in ST-segment elevation myocardial infarction: A comparison with the TIMI (Thrombolysis In Myocardial Infarction), GUSTO (Global Utilization of Streptokinase and Tissue Plasminogen Activator for Occluded Coronary Arteries), and ISTH (International Society on Thrombosis and Haemostasis) bleeding classifications. *Journal of the American College of Cardiology,* 63: 1866–1875.

Klein, M. I., G. Hwang, P. H. Santos, O. H. Campanella and H. Koo. 2015. Streptococcus mutans-derived extracellular matrix in cariogenic oral biofilms. *Frontiers in Cellular and Infection Microbiology,* 5: 10.

Kojima, Y., Y. Machida, S. Palani, T. R. Caulfield, E. S. Radisky, S. H. Kaufmann and Y. J. Machida. 2020. FAM111A protects replication forks from protein obstacles via its trypsin-like domain. *Nature Communications,* 11: 1–14.

Kunamneni, A., A. Ballesteros, F. J. Plou and M. Alcalde. 2007. Fungal laccase—a versatile enzyme for biotechnological applications. *Communicating Current Research and Educational Topics and Trends in Applied Microbiology,* 1: 233–245.

Kunamneni, A., Ogaugwu, C. and Goli, D. 2018. Enzymes as therapeutic agents. In: *Enzymes in Human and Animal Nutrition.* Academic Press: 301–312.

Lakshmanan, A., Z. Jin, S. P. Nety, D. P. Sawyer, A. Lee-Gosselin, D. Malounda, M. B. Swift, D. Maresca and M. G. Shapiro. 2020. Acoustic biosensors for ultrasound imaging of enzyme activity. *Nature Chemical Biology,* 16: 988–996.

Lascarrou, J. B., S. Ehrmann, P. Potier, J. Reignier, E. Canet and C. des enseignants de Medecine Intensive. 2020. Effect of brief encouragement to use twitter on knowledge of the critical-care literature by ICU residents: The randomized controlled IMKREASE trial. *Journal of Critical Care,* 60: 69–71.

Le Mar, J. D. and M. F. Gunderson. 1940. Studies on streptococcal fibrinolysis. *Journal of Bacteriology,* 39: 717–725.

Leach, J. K., A. Edgar, E. Patterson, Y. Miao and A. E. Johnson. 2003. Accelerated thrombolysis in a rabbit model of carotid artery thrombosis with liposome-encapsulated and microencapsulated streptokinase. *Thrombosis and Haemostasis,* 90: 64–70.

Li, H., B. Xia, W. Chen, Y. Zhang, X. Gao, A. Chinnathambi, S. A. Alharbi and Y. Zhao. 2020. Nimbolide prevents myocardial damage by regulating cardiac biomarkers, antioxidant level, and apoptosis signaling against doxorubicin-induced cardiotoxicity in rats. *Journal of Biochemical and Molecular Toxicology,* 34: e22543.

Litvinov, R. I., R. M. Nabiullina, L. D. Zubairova, M. A. Shakurova, I. A. Andrianova and J. W. Weisel. 2019. Lytic susceptibility, structure, and mechanical properties of fibrin in systemic lupus erythematosus. *Frontiers in Immunology*, 10: 1626.

Locke, M., P. Rigsby, C. Longstaff and F. subcommittee. 2020. An international collaborative study to establish the WHO 4th International Standard for Streptokinase: Communication from the SSC of the ISTH. *Journal of Thrombosis and Haemostasis,* 18(06): 1501–1505.

Mackman, N. 2012. New insights into the mechanisms of venous thrombosis. *The Journal of Clinical Investigation*, 122: 2331–2336.

Mahara, G., C. Wang, K. Yang, S. Chen, J. Guo, Q. Gao, W. Wang, Q. Wang and X. Guo. 2016. The association between environmental factors and scarlet fever incidence in Beijing region: Using GIS and spatial regression models. *International Journal of Environmental Research and Public Health,* 13: 1083.

Mahmoud, A. A. A., H. E. Mahmoud, M. A. Mahran and M. Khaled. 2020. Streptokinase versus unfractionated heparin nebulization in patients with severe acute respiratory distress syndrome (ARDS): A randomized controlled trial with observational controls. *Journal of Cardiothoracic and Vascular Anesthesia,* 34: 436–443.

Mamede, L. D., K. G. de Paula, B. de Oliveira, J. S. C. Dos Santos, L. M. Cunha, M. C. Junior, L. R. C. Jung, A. G. Taranto, D. de Oliveira Lopes and S. Y. Leclercq. 2020. Reverse and structural vaccinology approach to design a highly immunogenic multi-epitope subunit vaccine against Streptococcus pneumoniae infection. *Infection, Genetics and Evolution,* 85: 104473.

Man, A. J., H. E. Davis, A. Itoh, J. K. Leach and P. Bannerman. 2011. Neurite outgrowth in fibrin gels is regulated by substrate stiffness. *Tissue Engineering Part A*, 17: 2931–2942.

Mannully, S. T., C. Shanthi and K. K. Pulicherla. 2019. Lipid modification of staphylokinase and its implications on stability and activity. *International Journal of Biological Macromolecules,* 121: 1037–1045.

McMahon, B. J. and H. C. Kwaan. 2015. Components of the plasminogen-plasmin system as biologic markers for cancer. In: *Advances in Cancer Biomarkers*. Springer: 145–156.

Mehra, M. R., S. S. Desai, S. Kuy, T. D. Henry and A. N. Patel. 2020. Cardiovascular disease, drug therapy, and mortality in Covid-19. *New England Journal of Medicine*, 382(25): e102.

Mei, T., B. Shashni, H. Maeda and Y. Nagasaki. 2020. Fibrinolytic tissue plasminogen activator installed redox-active nanoparticles (t-PA@ iRNP) for cancer therapy. *Biomaterials*, 259: 120290.

Mennini, M., V. Fierro, G. Di Nardo, V. Pecora and A. Fiocchi. 2020. Microbiota in non-IgE-mediated food allergy. *Current Opinion in Allergy and Clinical Immunology*, 20: 323–328.

Mohanty, S. and Y. P. Khasa. 2018. Engineering of deglycosylated and plasmin resistant variants of recombinant streptokinase in Pichia pastoris. *Applied Microbiology and Biotechnology,* 102: 10561–10577.

Moore, H. B., C. D. Barrett, E. E. Moore, R. C. McIntyre, P. K. Moore, D. S. Talmor, F. A. Moore and M. B. Yaffe. 2020. Is there a role for tissue plasminogen activator as a novel treatment for refractory COVID-19 associated acute respiratory distress syndrome? *The Journal of Trauma and Acute Care Surgery*, 88: 1.

Moore, H. B. and E. E. Moore. 2020. Temporal changes in fibrinolysis following injury. In: *Seminars in Thrombosis and Hemostasis*. Thieme Medical Publishers: 189–198.

Natsuaki, M., T. Morimoto, H. Shiomi, K. Yamaji, H. Watanabe, S. Shizuta, T. Kato, K. Ando, Y. Nakagawa and Y. Furukawa. 2019. Application of the Academic Research Consortium High Bleeding Risk Criteria in an all-comers registry of percutaneous coronary intervention. *Circulation: Cardiovascular Interventions*, 12: e008307.

Nedaeinia, R., H. Faraji, S. H. Javanmard, G. A. Ferns, M. Ghayour-Mobarhan, M. Goli, . . . M. Ranjbar. 2020. Bacterial staphylokinase as a promising third-generation drug in the treatment for vascular occlusion. *Molecular Biology Reports*, 47(1), 819–841.

Prabhakar, P. K. 2020. Bioreactor strategies to increase the engineered protein production in Lactococcus lactis. *Plant Archives*, 20: 3183–3191.

Prasitdumrong, H., K. Duangmee, P. Boonmuang, W. Santimaleeworagun, Y. Oppamayun, C. Sonsupap and T. Nakkaratniyom. 2020. Incidence of urticaria, angioedema, and type I hypersensitivity reactions associated with fibrinolytic agents in Thailand using the database of the health product vigilance center. *Asian Pacific Journal of Allergy and Immunology*. DOI: 10.12932/AP-181119-0694

Raee, M. J., A. Ghasemian, S. Maghami, M. B. Ghoshoon and Y. Ghasemi. 2017. Cloning, purification and enzymatic assay of streptokinase gene from Streptococcus pyogenes in Escherichia coli. *Minerva Biotecnologica*, 29: 8–13.

Rafipour, M., M. Keramati, M. M. Aslani, A. Arashkia and F. Roohvand. 2020. Contribution of streptokinase-domains from groups G and A (SK2a) streptococci in amidolytic/proteolytic activities and fibrin-dependent plasminogen activation: A domain-exchange study. *Iranian Biomedical Journal*, 24: 15.

Rahimi, N., M. Alinezhad Chamazketi, A. Yaghoubi Nezhad and F. Talaeizadeh. 2020. Optimization of streptokinase mutant protein purification method using affinity chromatography technique. *Chemical Methodologies*, 4: 671–678.

Rizzoli, R., S. Boonen, M.-L. Brandi, O. Bruyère, C. Cooper, J. A. Kanis, J.-M. Kaufman, J. Ringe, G. Weryha and J.-Y. Reginster. 2013. Vitamin D supplementation in elderly or postmenopausal women: A 2013 update of the 2008 recommendations from the European Society for Clinical and Economic Aspects of Osteoporosis and Osteoarthritis (ESCEO). *Current Medical Research and Opinion*, 29: 305–313.

Rosenberg, J. D. and S. R. Parikh. 2011. Anticoagulation therapy as a supplement to recanalization for the treatment of sigmoid sinus thrombosis: a case report. *Ear, Nose & Throat Journal*, 90: 418–422.

Ruf, W., J. Disse, T. C. Carneiro-Lobo, N. Yokota and F. Schaffner. 2011. Tissue factor and cell signalling in cancer progression and thrombosis. *Journal of Thrombosis and Haemostasis*, 9: 306–315.

Sameni, M., M. Gholipourmalekabadi, M. Bandehpour, M. Hashemi, F. Sahebjam, V. Tohidi and B. Kazemi. 2017. Evaluation of In vivo bioactivity of a mutated streptokinase. *Novelty in Biomedicine,* 5: 71–77.

Sawhney, P., S. Kumar, N. Maheshwari, S. Singh Guleria, N. Dhar, R. Kashyap and G. Sahni. 2016. Site-Specific Thiol-mediated PEGylation of streptokinase leads to improved properties with clinical potential. *Current Pharmaceutical Design*, 22: 5868–5878.

Sevostyanov, M., A. Baikin, K. Sergienko, L. Shatova, A. Kirsankin, I. Baymler, A. Shkirin and S. Gudkov. 2020. Biodegradable stent coatings on the basis of PLGA polymers of different molecular mass, sustaining a steady release of the thrombolityc enzyme streptokinase. *Reactive and Functional Polymers*: 104550.

Sezer, M., N. van Royen, B. Umman, Z. Bugra, H. Bulluck, D. J. Hausenloy and S. Umman. 2018. Coronary microvascular injury in reperfused acute myocardial infarction: A view from an integrative perspective. *Journal of the American Heart Association*, 7: e009949.

Sheli, K. B., M. Ghorbani, A. Hekmat, B. Soltanian, A. Mohammadian and R. Jalalirad. 2018. Structural characterization of recombinant streptokinase following recovery from inclusion bodies using different chemical solubilization treatments. *Biotechnology Reports*, 19: e00259.

Shibata, K., T. Hashimoto, T. Miyazaki, A. Miyazaki and K. Nobe. 2019. Thrombolytic therapy for acute ischemic stroke: Past and future. *Current Pharmaceutical Design,* 25(3): 242–250.

Silverman, A. D., A. S. Karim and M. C. Jewett. 2020. Cell-free gene expression: An expanded repertoire of applications. *Nature Reviews Genetics*, 21: 151–170.

Tadayon, A., R. Jamshidi and A. Esmaeili. 2015. Delivery of tissue plasminogen activator and streptokinase magnetic nanoparticles to target vascular diseases. *International Journal of Pharmaceutics*, 495: 428–438.

Thygesen, K., J. S. Alpert, A. S. Jaffe, B. R. Chaitman, J. J. Bax, D. A. Morrow, H. D. White and E. G. o. b. o. t. J. E. S. o. C. A. C. o. C. A. H. A. W. H. F. T. F. f. t. U. D. o. M. Infarction. 2018. Fourth universal definition of myocardial infarction (2018). *Journal of the American College of Cardiology*, 72: 2231–2264.

Tourani, S., S. Bashzar, S. Nikfar, H. Ravaghi and M. Sadeghi. 2018. Effectiveness of tenecteplase versus streptokinase in treatment of acute myocardial infarction: A meta-analysis. *Tehran University Medical Journal TUMS Publications*, 76: 380–387.

Tran, K., C. Gurramkonda, M. A. Cooper, M. Pilli, J. E. Taris, N. Selock, T. C. Han, M. Tolosa, A. Zuber and C. Peñalber-Johnstone. 2018. Cell-free production of a therapeutic protein: Expression, purification, and characterization of recombinant streptokinase using a CHO lysate. *Biotechnology and Bioengineering*, 115: 92–102.

Tucker, T. and S. Idell. 2013. Plasminogen—plasmin system in the pathogenesis and treatment of lung and pleural injury. In: *Seminars in Thrombosis and Hemostasis*. Thieme Medical Publishers: 373–381.

Verma, M. K. 2015. *Fundamentals of Carbon Dioxide-enhanced Oil Recovery (CO2-EOR): A Supporting Document of the Assessment Methodology for Hydrocarbon Recovery Using CO2-EOR Associated with Carbon Sequestration*. US Department of the Interior, US Geological Survey. Washington, DC.

Virani, S. S., A. Alonso, E. J. Benjamin, M. S. Bittencourt, C. W. Callaway, A. P. Carson, . . . American Heart Association Council on Epidemiology and Prevention Statistics Committee and Stroke Statistics Subcommittee. (2020). Heart disease and stroke statistics—2020 update: A report from the American Heart Association. *Circulation*, 141(9): e139–e596.

Vishnu, V. Y. and M. P. Srivastava. 2019. Innovations in acute stroke reperfusion strategies. *Annals of Indian Academy of Neurology*, 22(1): 6.

Wang, J., S. Toan, R. Li and H. Zhou. 2020. Melatonin fine-tunes intracellular calcium signals and eliminates myocardial damage through the IP3R/MCU pathways in cardiorenal syndrome type 3. *Biochemical Pharmacology*, 174: 113832.

Whyte, C. S. and N. J. Mutch. 2021. uPA-mediated plasminogen activation is enhanced by polyphosphate. *Haematologica*, 106: 522.

Wunderlich, K., P. Smittenaar and R. J. Dolan. 2012. Dopamine enhances model-based over model-free choice behavior. *Neuron*, 75: 418–424.

Yamashita, T., T. Kamiya, K. Deguchi, T. Inaba, H. Zhang, J. Shang, K. Miyazaki, A. Ohtsuka, Y. Katayama and K. Abe. 2009. Dissociation and protection of the neurovascular unit after thrombolysis and reperfusion in ischemic rat brain. *Journal of Cerebral Blood Flow & Metabolism*, 29: 715–725.

Yepes, M., B. D. Roussel, C. Ali and D. Vivien. 2009. Tissue-type plasminogen activator in the ischemic brain: More than a thrombolytic. *Trends in Neurosciences*, 32: 48–55.

Yusuf, S., P. Joseph, A. Dans, P. Gao, K. Teo, D. Xavier, . . . P. Pais. 2021. Polypill with or without aspirin in persons without cardiovascular disease. *New England Journal of Medicine*, 384(3): 216–228.

Ząbczyk, M., G. Królczyk, G. Czyżewicz, K. Plens, S. Prior, S. Butenas and A. Undas. 2019. Altered fibrin clot properties in advanced lung cancer: Strong impact of cigarette smoking. *Medical Oncology*, 36: 37.

Zayet, S., T. Klopfenstein, R. Kovåcs, S. Stancescu and B. Hagenkötter. 2020. Acute cerebral stroke with multiple infarctions and COVID-19, France, 2020. *Emerging Infectious Diseases*, 26: 2258–2260.

Zeng, J., J. Huang and L. Pan. 2020. How to balance acute myocardial infarction and COVID-19: The protocols from Sichuan Provincial People's Hospital. *Intensive Care Medicine*, 46(6): 1111–1113.

Zhou, C., W. Qi, E. N. Lewis and J. F. Carpenter. 2016. Characterization of sizes of aggregates of insulin analogs and the conformations of the constituent protein molecules: A concomitant dynamic light scattering and Raman spectroscopy study. *Journal of Pharmaceutical Sciences,* 105: 551–558.

Zhou, P.-P., J. Meng and J. Bao. 2017. Fermentative production of high titer citric acid from corn stover feedstock after dry dilute acid pretreatment and biodetoxification. *Bioresource Technology,* 224: 563–572.

Zhou, X.-l., Y.-c. Yang, L. Liu and Y. Zhang. 2017. Research on screening and fermentation properties of fibrinolysin producing strains from fermented soya beans. *China Condiment:* 9.

Zia, M. A. 2020. Streptokinase: An efficient enzyme in cardiac medicine. *Protein and Peptide Letters*, 27: 111–119.

Zuo, Y., M. Warnock, A. Harbaugh, S. Yalavarthi, K. Gockman, M. Zuo, . . . D. A. Lawrence. 2021. Plasma tissue plasminogen activator and plasminogen activator inhibitor-1 in hospitalized COVID-19 patients. *Scientific Reports*, 11: 1–9.

15 Laccase-Assisted Biocatalytic Removal of Lignin from Lignocellulosic Biomass

Sadia Noreen, Sara Rehman, Memoona Asif, Muhammad Bilal, and Hafiz M. N. Iqbal

ABSTRACT

Unprecedented developments in industrialization and the growing worldwide population have caused a dramatic rise in energy demand, resulting in a rapid diminution of fossil assets. This scenario has led to the exploration of potential substitute energy sources that are environmentally friendly, cheaper, and bio-renewable. The utilization of lignocellulosic waste biomass for the production of advanced biofuels, bioenergy, and high-value chemicals has been appraised as a prospective strategy because it does not compete with the food supply. Nevertheless, the lignin constituent of this biomass interferes with its effective exploitation by acting as a natural blockade. Lignocellulose pretreatment using chemical and physical means for efficient lignin decomposition and improving the level of fermentable sugars is unsafe and costly. Microbe-based biological approaches and their enzymatic armory are potentially capable of disrupting lignin polymers to expose carbohydrate moieties for producing sugars. Laccases are a prodigious class of biocatalysts with a high potential for biomass delignification. In this chapter, we spotlight the laccase-assisted biocatalytic removal of lignin from lignocellulosic biomass and its catalytic conversion to bioethanol.

Keywords: laccase, lignocellulosic biomass, lignin, lignin-degrading fungi, biofuel

CONTENTS

15.1 Introduction 296
15.2 Lignocellulosic Waste 301
15.3 Lignin-Degrading Fungi and Enzymes 302
15.4 Catalytic Cycle of Laccase 304
15.5 Catalytic Conversion of Lignocellulosic Biomass to Bioethanol 306
15.6 Natural and Synthetic Polymers Are Used as Supports 307
15.7 Delignification of Plant Biomass 310

DOI: 10.1201/9781003202998-15

15.8   Conclusion and Future Perspectives 311
15.9   Acknowledgments... 312
15.10 Conflict of Interests 312
References... 312

15.1   INTRODUCTION

The enzyme laccase was discovered in 1883 from the extract of the tree *Rhus vernicifera*. It belongs to a family of multicopper oxidases and oxidizes specific substrates by utilizing its four coppers and reducing oxygen and water in this reaction (An et al., 2015; Paliwal et al., 2019). The reaction mechanism of laccase reduces the oxygen instead of hydrogen peroxides, which makes the conditions of a chemical reaction milder. That's why this enzyme gained the attention of scientists for the purpose of detoxification and delignification of lignocellulosic biomass. Commercially available laccases are most commonly produced by polypore basidiomycete named *T. versicolor* due to its stability at a wide range of pH and temperatures—pH 5 to 8 and 10 to 45°C, respectively (Stevens and Shi, 2019).

Moreover, laccases have also been isolated from thermophilic microorganisms, which are stable and functional at a higher temperature range (50 to 279°C). Thermophilic laccase that was expressed in *Trichoderma reesei* from *Melanocarpus albomyces* was capable of oxidizing a number of phenolic substrates: syringaldazine, guaiacol and 2,6-dimethoxyphenol (DMP). (Kiiskinen et al., 2004). Thermophilic laccase (PPLCC2) was expressed by An et al. from fungal sources (e.g., *Trichoderma reesei* from *Melanocarpus albomyces*) and oxidized substrates DMP and 2,2'-azino-bis(3-ethylbenzthioazoline-6-sulphonic acid) or ABTS at 60°C (An et al., 2015). Thermophilic laccase has also been isolated from bacterial and archaeal sources by scientists for the delignification of lignin-containing biomass. These thermostable laccase-like enzymes have been extracted from *B. subtilis*, *Thermus thermophilus* (TtL), and *Haloferax volcanii* (from the Dead Sea), which oxidized ABTS substrate up to 75°C and 92°C and syringaldazine and DMP accompanying nonphenolic substrate ABTS at 50°C, respectively (Hullo et al., 2001, Uthandi et al., 2010). In spite of the temperature, laccases also require optimum pH to work efficiently. For example, PPLCC2 work oxidize ABTS substrate at pH 3.5–5, while TtL optimally oxidizes ABTS at pH 4.5 and syringaldazine at 5.5 (Chio et al., 2019; Miyazaki, 2005).

Lignocellulosic biomass is primarily composed of three main polysaccharides: hemicelluloses, lignin, and celluloses. The dry weight of carbohydrates in lignocellulosic biomass consists of five-carbon sugar (pentose sugar) and six-carbon sugars (hexose sugars). Physically, cellulose is an amorphous or crystalline solid, and structurally, it has repeating subunits of D-glucose linked together by β(1,4) glycosidic linkages. Multiple cellulose chains join to form a special structure called cellulose fibrils. These fibrils are held together by van der Waals forces and hydrogen bonds (Pérez et al., 2002). Due to this intensive hydrogen bonding between the cellulose fibrils in lignocellulosic biomass, conventional hydrolytic enzymes are unable to hydrolyze it (Nishiyama et al., 2002). Polysaccharide hemicellulose is a heterogeneous compound that contains different pentose sugars, including xylose and arabinose, as well as different hexose sugars, like glucose, mannose, and galactose. Moreover, it

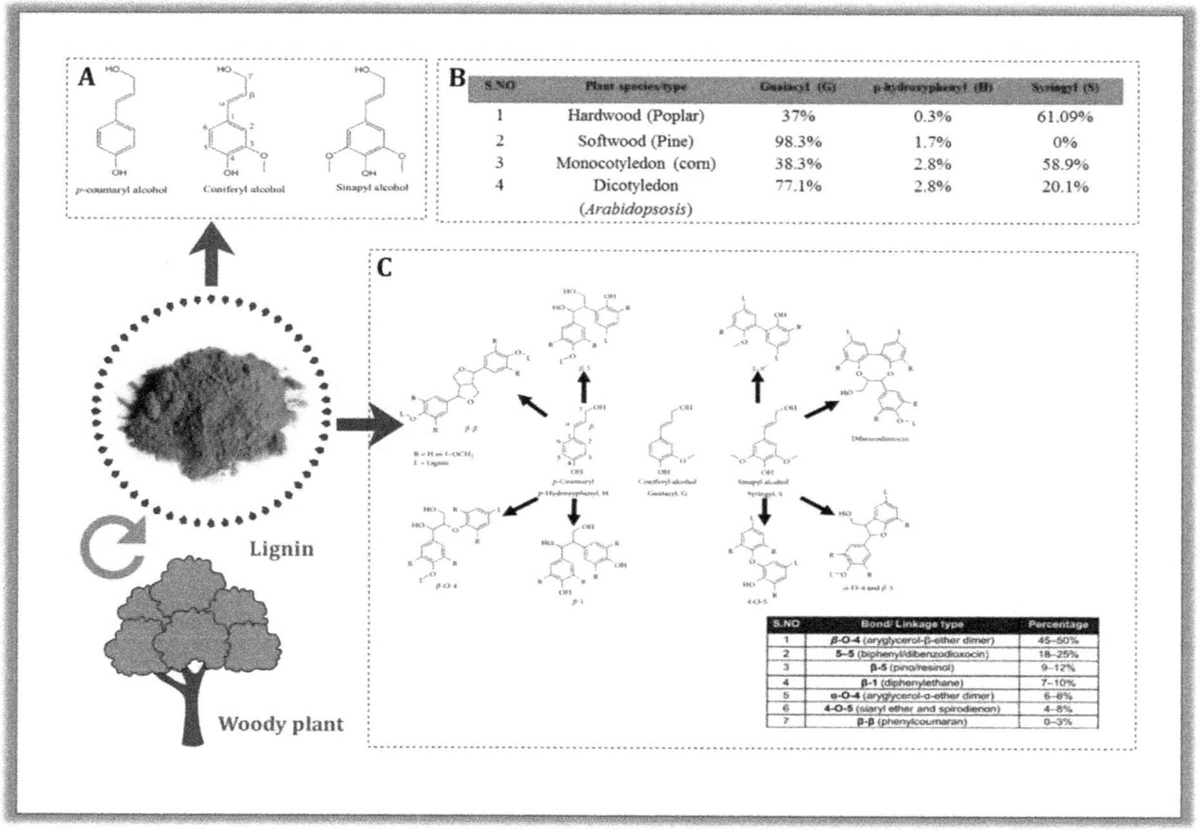

FIGURE 15.1 Graphical representation of lignin and its different aspects. A: 2D depiction of three primary lignin polymer-forming unit (monomer; G; H; S units). B: Percentage of monomer units in a different type of lignin based on plant origin. C: Lignin polymer including respective percentage and constructive linkages. Reprinted from Singh et al. (2021), an open access article distributed under the terms of the Creative Commons CC-BY license.

also has sugar acids. Hemicellulose serves as an intermediate between cellulose and lignin. Hemicellulose, as compared to cellulose is less amorphous, crystalline, and heavier with short monomeric chains. The branching of hemicelluloses has $\beta(1,4)$ glycosidic linkages and occasionally $\beta(1,3)$ glycosidic linkages (Kumar et al., 2009). Representation of lignin and its different aspects are shown in Figure 15.1 (Singh et al., 2021).

Lignin, the third and main constituent of lignocellulosic biomass, is the aromatic polymer composed of methoxy, phenyl, and propane groups. This structural characteristic of lignin makes it resistant to hydrolyze by conventional hydrolytic enzymes (Kuila, 2019). Lignin, being the second most abundant organic material on the earth (the first one is cellulose), makes up 20–30% of wood's dry weight. Lignin interlocks the cellulose and hemicellulose, which provides physical strength to lignocellulosic material and makes it unavailable for enzymatic action (Grelska and Magdalena, 2020). Lignin is a non-carbohydrate polymer made up of three principal monolignols— p-coumaryl alcohol, coniferyl alcohol, and sinapyl alcohol (Longe et al., 2018). These monolignols are held together by C-C and ether bonds, which make lignin resistant to harsh environmental conditions as well as conventional hydrolytic reactions by enzymes (Bugg et al., 2011). The classification of lignin is based on the substituent group attached to the phenylpropanoid backbone: guaiacyl-lignin and guaiacyl-syringyl. The guaiacyl lignin has a substituent methoxy group at the third carbon position, while guaiacyl-syringyl lignin has a methoxy group at the third and fifth carbon position.

Occurrence percentages of lignin depend on the type of plant. These types of plants include hardwood, softwood, and grass, and their occurrence percentages of lignin are 20 to 25%, 28 to 32%, and 17 to 24%, respectively (Singh et al., 2021). For bioethanol production in refineries, constituent sugars of cellulose and hemicellulose are of core importance as the monomeric sugars can be converted biotechnologically into bioethanol. But the main restriction to use the lignocellulosic material in bioethanol production is that it has lignin in it which makes the cellulose and hemicellulose unavailable to the enzymatic conversion of these polymers into simple monomeric sugars and ultimately production of bioethanol badly affected (Cabral and Venancio,

---

TABLE 15.1

Laccase for Lignin Removal in Various Lignocellulosic Biomasses

Fungal strain	Lignocellulosic Biomass	Fermentation Conditions	Lignin Removal (%)	References
Cerrena sp. B.Md.T.A.1	Corncob	Different concentrations of corncob (5%, 10%, and 15% [w/v]) were treated with laccase by incubating in a shaker at 40°C, 100 rpm, for 24 hours.	15%	Muryanto et al. (2021)

Fungal strain	Lignocellulosic Biomass	Fermentation Conditions	Lignin Removal (%)	References
Cerrena unicolor	Organosolv lignin via corncob	—Laccase + violuric acid + 50 mg lignin + 0.5 mL isopropanol —Laccase + 2,2'-azino-bis (3-ethylbenzothiazoline-6-sulphonic acid) + 50 mg lignin + 0.5 mL isopropanol —Laccase + 4-tert-butyl2,6-di-methylphenol + 50 mg lignin + 0.5 mL isopropanol —Laccase + 1-hydroxy benzotriazole + 50 mg lignin + 0.5 mL isopropanol	73% 49% 43% 39%	Longe et al. (2018)
Trametes versicolor, *Ganoderma lucidum*, and *Pleurotus ostreatus*	Wheat straw	15 g feedstock + laccase + sodium malonate buffer (0.5 mM) having pH 4.5 up to the final volume of the 100 mL flask	58.5%	Asgher et al. (2017)
Shizophyllum commune	—Bagasse —Banana stalk —Corncobs —Sugarcane —Wheat straw	30 g lignocellulosic material + laccase + 50 mM Na-malonate buffer (pH 5) up to the final volume of 200 mL for 48 hours at 35°C	47.5% 61.7% 58.6% 63.6%	Asgher et al. (2016a)
Pleurotus eryngii WC 888	—Banana stalk —Rice straw —Corn stove	Delignification process performed at room temperature for 48 hours	39.15% 43.45% 49.52%	Asgher et al. (2016b)
Pleurotus sapidus	—Sugarcane bagasse —Rice straw —Wheat straw —Corncobs	15.5 mL laccase + 20 g substrate in each flask, + 0.5 mM sodium malonate buffer having pH 4.5 for 48 hours at 35 °C	51.08% 53.10% 45.6% 61.16%	Asgher et al. (2016c)
Lentinus squarrosulus MR13	Bambusa bamboos	Enzyme (laccase) + substrate (incubation for 8 hours)	73.9%	Mukhopadhyay and Banerjee (2015)

(Continued)

TABLE 15.1 (*Continued*)

Fungal strain	Lignocellulosic Biomass	Fermentation Conditions	Lignin Removal (%)	References
Ganoderma lucidum	Wheat straw	30 g substrate used with different doses of enzymes + sodium malonate buffer (0.5 mM) having pH 4.5 initially at 35°C for 48 hours	39.6%	Asgher et al. (2014)
Trametes villosa	—*Eucalyptus globulus* —*Pennisetum purpureum*	Substrate + laccase (50 U g⁻¹) + 2.5 HBT was kept in a thermostatic shaker at 50°C and 170 rpm, using the 2 g (dry weight) samples at 6% consistency (w/w) in sodium tartrate buffer 50 mM (pH 4) in O₂ (atmosphere) for the 24 hours.	48% 32%	Gutiérrez et al. (2012)
Trametes villosa	Pulp of kraft	Incubation of substrate together with 2.5 mmol aryl substituted N-hydroxy phthalimide (mediator) and 2,000 units of laccase in sodium citrate buffer 0.1 M (pH 5) for 20 hours at 50°C.	42.5%	Annunziatini et al. (2005)
Trametes versicolor	*Eucalyptus globulus* kraft pulp	25 g substrate (pulp) + mediator violuric acid at pH 4.5 was heated at 45°C. 20 IU/g of laccase was entered in 3.8 L capacity reactor in well-stirred conditions	54.2%	Oudia et al. (2008)

2020). To overcome this problem, scientists used the idea of pretreatment of ligno-cellulosic biomass before biotechnological procedures. Pretreatment delignifies the lignin that has interlocked the celluloses and hemicelluloses. This treatment makes the cellulose and hemicellulose available for the hydrolytic enzymes that can maximally convert the lignocellulosic pretreated feedstock into biofuel or bioethanol by fermentation process (Asgher et al., 2013).

15.2 LIGNOCELLULOSIC WASTE

The lignocellulosic material of plants consists of three main elements: lignin, hemicellulose, and cellulose. The lignin is a crucial constituent of the plant cell wall, giving rigidity and protection to the cellulose, which is readily degradable by the attack of pathogens. It is synthesized from precursor phenyl propanoid by polymerization in higher plants. The precursors of the lignin, like coumaryl p-alcohol, sinapyl alcohol, and coniferyl alcohol, are in an aromatic ring and a three-carbon atom side chain. Because of its complex structure and non-hydrolyzable linkages, lignin is more difficult to break than cellulose or hemicellulose. Lignin is the amplest aromatic compound on the planet. It consists of approximately 20–30% of higher plant cell walls. It makes a matrix adjoining the cellulose and hemicellulose (Nunes and Kunamneni, 2018).

In this way, it strengthens the wood and also protects against biological and chemical degradation. Background molecules of lignin are the p-coumaryl, sinapyl, and coniferyl alcohol. The aromatic alcohols also make the lignin molecule (Huber et al., 2018). Lignin is made up of phenylpropanoid units, covalently linked with a number of bonds, hence making a mash-like three-dimensional structure. Plants synthesize lignin through the oxidation of para- hydroxycinnamyl alcohols by using peroxidases. The lignocellulosic biomass from plants is a renewable source of chemicals, energy, and food (Wang et al., 2020).

The lignin is an intricate polyphenolic aromatic complex compound and is able to enormously hamper the catalytic activity of enzymes for lignocellulosic material degradation. Because the yields of biofuel or perhaps biochemicals are especially determined by the production of monomeric sugars by the process of fermentation of lignocellulosic plant biomass, a pretreatment reaction of distracting the structures of lignocellulosic recalcitrant is necessary for the exposure of cellulose and hemicellulose fibers required for the enzymatic or saccharification reactions (Kamimura et al., 2019) Lignocellulosic biomass is the major part of renewable biomass on Earth. The lignocellulosic biomass is the best choice for the raw materials employed in manufacturing chemicals, polymers, and biofuels that give remarkable economic value and eco-friendliness. The agricultural and agro-industrial activities produce tons of it annually. Its use does not hinder food supplies as it has a lower cost than crude oil or any other agriculturally important feed stock (Chukwuma et al., 2020).

The lignocellulosic biomass carries lignin and the polysaccharides like cellulose, hemicellulose, ash, minerals, salts, and pectin. It contains nearly 15 to 20% lignin, 40 to 50% cellulose, and 25 to 30% hemicellulose. Cellulose consists of D-glucose monomers linked together by the linear β(1,4) linkage, making ~50% of the plant biomass. Hemicellulose is made up of multiple polymers of the polysaccharides having small chains. They consist of five different sugar monomers: L-arabinose, D-mannose, D-glucose, D-xylose, and D-galactose. The lignin is a plenteous complex polymer found naturally as it is less biodegradable because of multiplex chemical linkage joining its monomers leading to an unmanageable compound of the lignocellulosic biomass (Kumar and Ram, 2020).

Lignin has an amorphous and aromatic nature. Lignin, being the hydrophobic polymer, acts like a barrier averse to water penetration. It makes up 20–35% of the cell wall of plants based on the source of biomass. It consists of phenyl propane units—mono lignols/lignol precursors—made by the radical polymerization of units of syringyl, p-hydroxyphenyl and guaiacyl from the sinapyl, p-coumaryl alcohol, and coniferyl precursors. The lignin comprises an aromatic polymer with excessive molecular weight, having many biologically constant esters or ether links (Tsegaye et al., 2019).

The microfibrils of cellulose embedded in the complex matrix of lignin and hemicellulose hinder the way to hemicellulases and cellulases. Degradation of the lignocellulosic material is a great challenge because of its complexity. The elimination of lignin from lignocellulosic biomass magnifies the efficiency of hemicellulose and cellulose hydrolysis. It eases the utilization of biomass carbohydrate portions to produce cellulosic ethanol and many biofuels (Kumar and Ram, 2020).

15.3   LIGNIN-DEGRADING FUNGI AND ENZYMES

Fungi are the most effective microorganisms to break down the lignocellulosic materials into three remarkable polymers. Lignin-degrading fungal species include the most common filamentous fungi that are ubiquitously found in lignocellulosic biomass, soil, and plants. Brown and white rot fungi are effective in lignocellulosic biomass degradation like wheat straw, Bermuda grass, soft woods, and wood chips. White rot fungi can degrade lignin, hemicellulose, and cellulose, while the brown rot fungi, being limited to cellulose and hemicellulose, do not affect lignin (Chukwuma et al., 2020). White rot fungi have a powerful capability of degrading lignin because of their high laccase activity and well-established hyphal organization, which can effectively penetrate plant cell walls (Christopher et al., 2014). The lignocellulolytic fungi like *Penicillium*, *Aspergillus*, *Schizophyllum*, *Trichoderma*, *Sclerotium*, and *Phanerochaete* are capable of producing large quantities of enzymes extracellularly (Chukwuma et al., 2020).

Most microorganisms have the ability to degrade lignin anaerobically more efficiently than aerobically because aromatic rings are present in lignin. Wood-rotting basidiomycetous fungi in nature are the most efficient lignin degraders that cause white rot in wood. So they are probably nature's significant mediators for reutilizing the organic molecules of wood. Lignin degradation occurs majorly by the efficient ligninolytic enzymes that are secreted by white rot fungi that belong to the basidiomycetes (Mayolo-Deloisa et al., 2020). Recently, there are studies on the biological pretreatments involving lignocellulosic biomass by using the fermentation in solid state (SSF) by microorganisms. Organisms are capable of degrading lignin exclusively to water (H_2O) and carbon dioxide (CO_2) as end products by exposing celluloses and hemicelluloses as hydrolyzable nutrients in the medium. The white rot species of basidiomycetes are the only ones that are known to synthesize complex enzyme machinery to degrade the lignin, consisting of extracellular polyphenol oxidases, such as manganese peroxidase or MnP, the laccase or LAC, and the lignin peroxidase LiP, being very effective in lignin breakdown. The ligninases are extracellular and nonspecific, participating in different reactions or oxidizing the aromatic structure of the lignin and breaking the links between the base units. Fungi can break down lignin aerobically by using the extracellular enzymes that are collectively called lign-inases. The two families of the ligninolytic enzymes are broadly considered to play an important part in degradation (enzymatic): phenol oxidase (laccase).

Oxidoreductases enzymes catalyze oxidative conversions of lignin derivatives, phenols, aromatic acids, alcohols, aldehydes, dyes, amines, anilines, poly-aromatic hydrocarbons (PAHs), quinones, substituted phenols, thiol residues, and non-aromatic alcohols. The lignocellulosic biomass degradation in fungi employs mechanisms that are majorly classified as hydrolytic and oxidative. The oxidative mechanism of lignin degradation produces free radicals of the active oxygen species, largely hydroxyls. Many fungi make hydrogen peroxides by the enzymatic action of aryl alcohol oxi-dase, glyoxaline oxidase, and pyranose-2-oxidase. The reaction of iron and hydro-gen peroxide produces hydroxyl radicals that carry out the degradation of lignin to products with less molecular weight. The hydrolytic mechanism of degradation of lignin involves the breakdown of glycosidic linkages by hydrolytic enzymes of fungi (Mayolo-Deloisa et al., 2020).

Laccases and MnP catalyze the oxidative mechanism of the degradation of lignin. In the case of multi-copper oxidase laccase, it is mediated by free radicals. MnP car-ries out hydrolysis of hydrogen peroxide by oxidizing Mn^{2+} to Mn^{3+} (Tsegaye et al., 2019). Regardless of the greater potential of fungi for degradation of lignin, the need for long residence time leads them impractical for production at a large scale. It requires a large space and enhances the production costs. Biological pretreatment is a cheap alternative due to eco-friendliness, mild temperature operation, and no chem-ical requirements (Chukwuma et al., 2020). The best-known lignocellulosic bio-mass-degrading enzymes: laccase (EC 1.10.3.2), versatile peroxidase (EC 1.11.1.16), lignin peroxidase (EC 1.11.1.14), and manganese peroxidase (EC 1.11.1.13). These enzymes occur in a variety of molecular weights based on the sequence of amino acids and composition of cofactors (Kumar and Ram, 2020).

Laccase (EC 1.10.3.2) or oxygen oxidoreductase; benzenediol is an N-glycosylated extracellular multi-copper enzyme that oxidizes different phenolic and nonphenolic compounds through a single electron transfer along with reduction of the dioxygen to water. It contains monomeric, dimeric, and tetrameric glycoproteins. It is majorly found in bacteria, insects, higher plants, and white rot fungi. The physicochemical properties like molecular size, isoelectric point, activity, and stability are based on their source. The fungal laccases are widely used in bioremediation or bio-refineries due to their high redox potential (Tocco et al., 2021). The laccase consists of nearly 500 amino acids (residues) and three consecutive domains with β barrel topology that is circulated in a single molecule. The first domain has 150 amino acids, the second one has 150 to 300 residues, and the third one has 300 to 500 amino acids. Three types of copper atoms maintain amino acids' active site. T1 Cu, known as substrate-reducing site, exhibits diverse triangle combinations. The active site con-sists of a sequence of amino acids like a cysteine and the two histidines, the equato-rial ligands. Other multi-copper enzymes (oxidases) have an additional ligand with methionine. T2 Cu shows a combination of the two histidines and the H_2O molecule. T3 Cu has two copper molecules together with the six histidines found in the two groups of the three active sites for sharing oxygen, which is reduced to H_2O and split out. Laccases are the ideal green catalysts owing to their oxidizing property in a huge diversity of compounds by utilizing O_2 and releasing H_2O (only by-product) (Sankar et al., 2020).

The laccase is mainly divided into three types based on its functions: (1) degradation of biopolymers, (2) ring cleavage of organic compounds, and (3) cross-linking structure of monomers, depending on the dispersion of copper atoms between three changed binding sites. The copper atoms play an important part in the catalytic mechanisms of enzymes. Owing to the low substrate specificity of laccases, they can degrade numerous compounds having phenolic structures, including lignin (Kumar and Ram, 2020).

Laccases have diverse functions as these are important in lignin degradation and biosynthesis, plant pathogenesis, fungal virulence factors, iron metabolism, pigment formation in fungal spores, and plants kernel browning. The fungal laccases are important in detoxification, sporulation, lignin degradation, fructification, and phyto-pathogenicity. The ability of laccases to survive extreme conditions, such as high temperatures and a wide pH range, makes them useful for biotechnological purposes (Tsegaye et al., 2019; Christopher et al., 2014).

Laccases are extracellular enzymes consisting of monomeric, dimeric, and tetrameric glycoproteins having nearly five hundred amino acid residues. Laccase is the high potential oxidative enzyme, having broad substrate specificity for aromatic compounds, making it a promising choice for xenobiotic (containing hydroxyl and amine groups) degradation. Laccases catalyze three major types of reactions—cleave rings in organic compounds, cross-link monomers, or degrade biopolymers—based on the dispersion of copper atoms in three changed binding states (Kumar and Ram, 2020). The unbound laccase can't tolerate extreme conditions and inhibitors and is not much stable as it is highly soluble. It is immobilized in order to increase stability and reutilization by preventing proteolysis, achieve enzyme structure rigidification using covalent attachment, and establish beneficial micro-environments (Daronch et al., 2020). The activity of a biocatalyst is also improved in both aqueous and organic phases for large time periods. The support permits substrate diffusion to the active site. Enzyme recovery by sedimentation, centrifugation, and other methods of physical separation and its continuous reuse is also made possible. (Chong-Cerda et al., 2020). Laccase oxidative biocatalysis arises as an effective alternative in advanced oxidation processes' development for xenobiotics removal.

15.4 CATALYTIC CYCLE OF LACCASE

Nanobiocatalysts' catalytic activity can be evaluated in terms of the biotransformation potential of the phenol. After the approval of successful laccase immobilization, scientists aim to examine the application of life cycle principles that are useful for the reformulation of production schemes of the most suitable support. The degradation of lignin is done in a three-stage pathway in which firstly the lignin oxidation reduces copper, the movement of electrons occurs from copper to two groups of copper atoms, and oxygen reduction to water occurs at centers of copper type 2 and type 3. In laccase oxidation, harmful effects of intermediate compounds to cells are avoided as the early phase of delignification utilizes O_2 instead of H_2O_2. Laccases oxidize

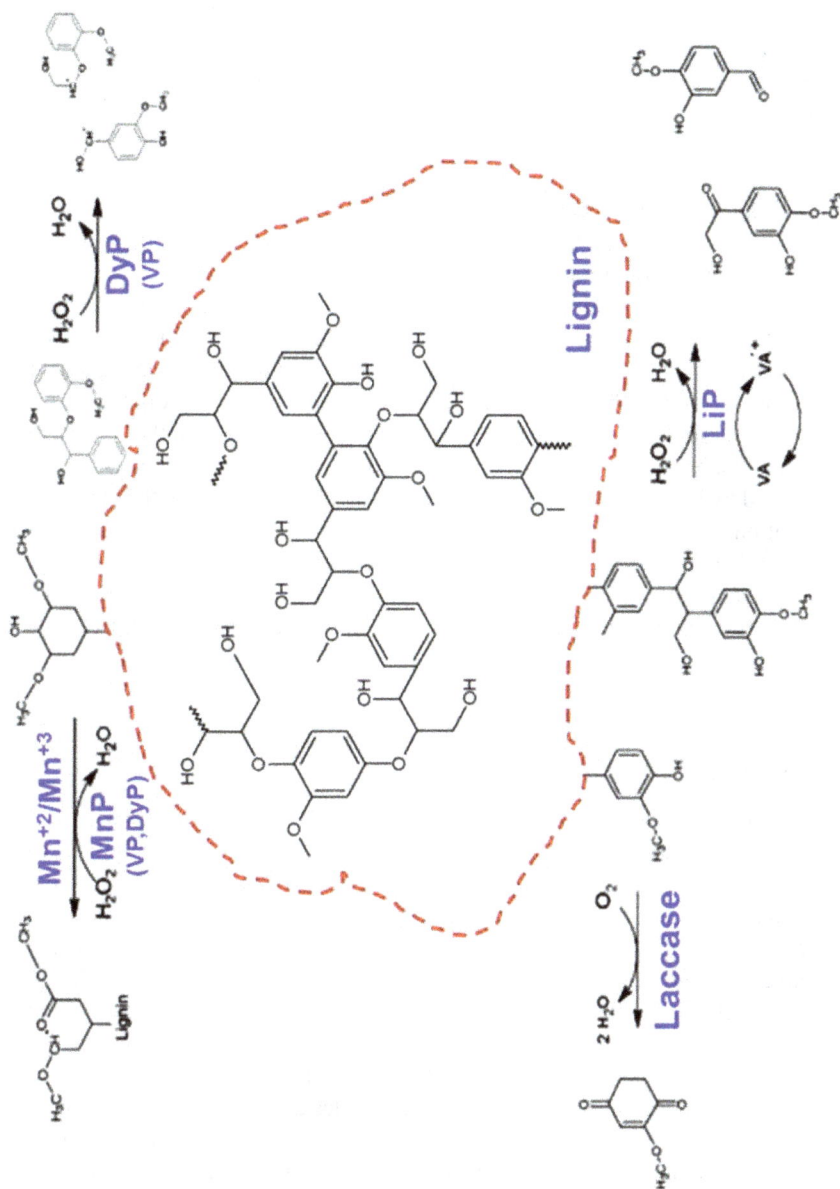

FIGURE 15.2 A comparative schematic lignin degradation reaction representation of lignin-modifying enzymes; reprinted from Singh et al., 2021, an open-access article distributed under the terms of the Creative Commons CC-BY license.

a huge number of substrates and aromatic diamines, polyphenols, and methoxy-substituted phenols by breaking the alkyl-aryl and Cα-Cβ bonds by oxidation (Tsegaye et al., 2019).

Laccase acts on lignocellulosic materials having phenolic and nonphenolic compounds for degradation and detoxification. Biodegradation of lignin with a laccase-mediated system involves starting an oxidative attack on the lignin phenolic moiety (N-OH, pyrazolone, and phenothiazine type) and reduces the lignin G-units and the aromatic structures. Phenolic degradation (like β-1 lignin model) occurs by phenoxy radicals' generation that takes toward an aromatic ring cleavage, Cα oxidation, alkyl-aryl cleavage, and Cα-Cβ cleavage. Oxidation with laccase-HBT destroys the Cα-Cβ linkage, creating carboxylic acid groups (in modified pulp lignin) (Kumar and Ram 2020). Figure 15.2 illustrates a comparative schematic lignin degradation reaction representation of lignin modifying enzymes (Singh et al., 2021).

Nonphenolic (like β-O-4 linked) oxidation by the laccase HBT opens aromatic ring (oxidation of π electron), oxidizes Cα and cleaves the Cα-Cβ and β-ether bonds for producing the carboxylic acids and aromatic carbonyl compounds. The beta-aryl radical (cation) or the Cα (benzylic) radical intermediates generates oxidation reactions of the substrate through an electron transfer mechanism. Beta-ether cleavage of β-O-4 lignin substructure results from the reaction with Cα-peroxy intermediate radical (formed by benzylic radical). If the reaction involves cleavage of the aromatic ring, aryl cation radicals are formed from the degraded products. The typical products of lignin degradation are 4-ethyl-2,6-dimethoxybenzaldehyde, 2,6-dimethoxy-4-((E)-prop-1-enyl) benzaldehyde, and 2, 6-dimethoxy-4-methylbenzaldehyde. Lignin molecular weight and the phenolic content influence pathways catalyzed by laccases (Christopher et al., 2014).

15.5 CATALYTIC CONVERSION OF LIGNOCELLULOSIC BIOMASS TO BIOETHANOL

In order to produce bioethanol optimally from the lignocellulosic material or any other sugar source, the content of sugar as a carbon source for the fermentation process should be optimal (Paliwal et al., 2019). There are two ways for obtaining the highest content of sugars from lignocellulosic material. These methods include acidolysis or enzymatic hydrolysis. Enzymatic hydrolysis is a much more significant, cost-effective, productive, and environment-friendly method as compared to acidolysis (De Bhowmick et al., 2018). The main problem of obtaining sugar monomers from lignocellulosic material is that it is reinforced by lignin molecule, which is reluctant to the enzymatic reaction, so non-efficient hydrolysis of lignocellulosic material takes place, which in turn gives less output of sugar molecules and consequently produces less bioethanol after fermentation (Christopher et al., 2014). So there is a need for pretreatment of lignocellulosic biomass in an effective way to avoid the wastage of sugar monomers and formation of by-production (conventional pretreatments).

Effective pretreatment of lignocellulosic material must delignify the lignin and make the cellulose and hemicellulose available to hydrolytic enzymes for their conversion into simple monomeric sugars. There are a number of methods that are being used for the pretreatment of lignocellulosic material. These methods include physical, chemical, physicochemical, and biological methods (Nishiyama et al., 2002). The choice of pretreatment technology is directly depending on the feedstock being used in the process (Kumar P et al., 2009). Pretreatments, including chemical and physico-chemical, are considered to be most effective. The use of acid and bases for pretreat-ment of lignocellulosic material is included in the chemical method, while ammonia fiber explosion or expansion, steam explosion, wet oxidation, liquid hot water, and so on are included in the physicochemical method. Both of these methods are cost-effective, but the physicochemical method has less environmental effects as compared to the chemical method. The main purpose for all pretreatment technologies is only to dis-solve lignin and hemicellulose and make cellulose free for enzymatic reaction to be converted into simple constituent sugars for the sake of bioethanol production by the fermentation process. Other pretreatment methodology, including milling, organo-solv, and ionic liquids (ILs), are also in use, but they have limitations of requiring specific operational techniques and cost for them (Tuck et al., 2012).

Pretreatment is a step of breaking down the lignocellulosic material into its con-stituent polymeric cellulose, hemicellulose, and lignin. These polymers are then depolymerized or hydrolyzed completely into simple sugars by specific enzymes—cellulose, hemicellulolase, and ligninase (Kuila, 2019). Other cellulases, including endoglucanases, cellobiohydrolases, and β-glucosidases, aid in cellulose hydroly-sis to the glucose subunits. Hemicellulose and lignin are hydrolyzed by different hemicelulases (β-xilosidases, esterases, α-L-arabinofuranosidases, and xylanases) and the ligninases (oxidases producing H_2O_2, laccases, reductases, and peroxidases), respectively. For the effective hydrolysis of the constituents of lignocellulosic bio-mass, it is necessary to provide the mixture of all the abovementioned enzymes, which requires cost. Moreover, enzymes also need optimization conditions for the reaction. These are the drawbacks of this method. To overcome the cost problems, low-cost substrates are preferentially being used (Abdel-Raheem and Shearer 2002; Grelska and Magdalena, 2020). After cellulose and hemicellulose are hydrolyzed to sugars, the fermentation (microbial) is processed, which utilizes these sugars as a car-bon source and converts them to their products and most commonly into bioethanol.

Three types of fermentation can be processed for bioethanol production. These processes are the consolidating bioprocessing (CBP), the separate hydrolysis and fer-mentation (SHF), and the simultaneous saccharification and (co)fermentation (SSF/SSCF) (Longe et al., 2018). Different microorganisms have been used for bioethanol production. These are fungi, bacteria, and yeasts, and among all microorganisms, the *Saccharomyces cerevisiae* (yeast) is well-known and the most utilized microorgan-ism in the alcohol industries for bioethanol production. This is because it can utilize all kinds of hexoses as a carbon source and convert them to ethanol. But it is unable to convert pentose sugars into ethanol. That's why other microorganisms that could convert all types of sugars into ethanol are getting more attention (Bugg et al., 2011).

15.6   NATURAL AND SYNTHETIC POLYMERS ARE USED AS SUPPORTS

The ideal properties of support for enzyme immobilization include physical resistance to compression, inertness toward derivatization, hydrophilicity, biocompatibility, availability at low cost, and resistance to attack of microbes. The natural polymers used as support materials include alginate, carrageenan, cellulose, chitin, chitosan, collagen, pectin, sepharose, and starch. In addition, many synthetic polymers that have good mechanical stability and are easily modified are used as supports. Numerous inorganic supports are also used, like alumina, mesoporous silica, silica, and zeo-lites. The silica-based supports are the best matrices in the industrial production of enzyme products and research purposes. The carriers with large surface areas allow better immobilization efficiency. Different types of magnetic or nonmagnetic nano-materials are used for laccase immobilization, like gold (Au), silver (Ag), zinc oxide (ZnO), carbon nanoparticles, titanium dioxide (TiO_2), and chitosan-coated magnetic nanoparticles. These are also useful in medicine, agriculture, food, and electronics (Nile et al., 2020). Immobilizations of these ligninolytic enzymes by using different immobilization strategies make them economically valuable at an industrial scale because immobilization significantly improves the thermo-stability, reusability, stor-age ability, and catalytic activities of these novel enzymes.

Biocatalysts are immobilized by chemical or physical interactions to link enzymes and supports. The physical approach involves nonspecific interactions by hydrogen bonds and hydrophobic and ionic interactions. Entrapment and adsorption techniques do not need functionalized supports (Bilal and Iqbal, 2019). There are two basic methods for immobilization. Physical coupling methods include entrap-ment of enzyme within a 3D matrix, its encapsulation in an organic or inorganic polymer, and its adsorption to support surface by ionic exchange, whereas covalent bonding makes sure the irreversible binding of the enzyme to the matrix of support. Laccase immobilization occurs on different nanoparticles like silver and gold, chi-tosan-coated magnetic particles, and carbon nanotubes.

Two different immobilization techniques are used: (1) ionic exchange between enzyme and nanoparticle and (2) covalent bonding of protein to nanoparticle sur-face using glutaraldehyde or carbodiimide as the cross-linkers. Glutaraldehyde is a bifunctional and versatile agent, reacts with a number of enzyme moieties, and is majorly involved in proteins' primary amino groups, although it can eventually react with other groups (e.g., thiols, phenols, and imidazoles), whereas the carbodiimide makes amide linkage between carboxylates and amino terminal groups of enzymes.

Laccase can be efficiently immobilized on a number of supports like activated carbon porous glass, TiO_2, membranes, mesoporous materials, and microsphere. Fe_3O_4 magnetic nanoparticles are regarded as promising supports due to their large surface area, lower mass transfer hindrance, and very well-defined surface proper-ties. Moreover, enzymes immobilized on Fe_3O_4 magnetic nanoparticles are not only capable of maintaining unique performance but can also be easily separated or recy-cled. Because of these benefits, the potential applications of magnetic nanoparticles are widely investigated for enzyme immobilization, and enzyme stability can be improved (Zhiguo et al., 2020).

Entrapment is done by inorganic or organic polymeric matrices, encapsulation, gel/ fiber entrapment, embedding in metal-organic frameworks, or adsorption to support the surface via ionic exchange. Chemical techniques involve covalent bond formation. In covalent bonding, the enzyme is irreversibly bound to the support matrix (Liang et al., 2020). Nanoparticles give a homogenous core-shell structure that is functionalized to react with enzymes' nucleophilic group. The structure of nanoparticles (porous or non-porous), ionic strength, pH, additives, and protein concentration affect biocatalyst and covalent bond effectiveness between support and enzyme (Shao et al., 2019).

Many new carriers and technology have been implemented to improve the immobilization of the traditional enzymes, which intend to improve the enzyme activity; loading efficiency and constancy to reduce the enzyme represent a cost in industrial biotechnology. These include CLEAs (cross-linked enzyme aggregates), click chemistry of technology, mesoporous support, and single-enzyme nanoparticles. CLEAs methodology is basically a grouping of enzyme immobilization and purification in one step. This method is applied to the synthesis of combi-CLEAs, which consist of two or more enzymes for utilization in one or several steps in industries. In recent times, the metal-organic framework (MOF) is considered an attractive supporting matrix for the enzyme's immobilization. MOF possesses a well-ordered porous structures and surface area, which can be modified with numerous functional groups (Chong-Cerda et al., 2020). For immobilization of enzymes or biomolecules, these are useful nano-structured materials because of their distinctive characteristics, such as large accessible surface area, high pore volume, excellent biocompatibility, enhanced biodegradability, superior morphology, oxygenated functional groups' affluence and enhanced functionalization potential with the nanoparticles, and outstanding enzyme loading capability when used as support and superb chemo-stability and thermo-stability. The appearance of different functional groups, like carboxylic, epoxide, or hydroxyl, leads to the development of strong biocatalyst-matrix interactions without modifying graphene surface or using the cross-linking agents. These characteristics make graphene-based nanomaterials ideal matrices for enzyme immobilization by covalent binding, entrapment, or adsorption (Rouhani et al., 2020).

Covalent immobilization involves the use of suitable cross-linkers based on functional groups on the nanomaterial support surface. Glutaraldehyde and carbodiimide cross-linkers are generally used. The typical cross-linker glutaraldehyde is used for graphene-based nanoparticles having amine functional groups. GLU, being a bifunctional and versatile agent, reacts with various enzyme moieties mainly involves the protein's primary amino groups, may react with phenols, thiols, imidazoles, or other groups. The graphene surface can be activated with BSA for providing free amine groups to cross-link with enzymes. EDC (1-ethyl-3,3-dimethylaminopropyl carbodiimide), like carbodiimides, attacks the carboxyl group of nanomaterial forming O-acylisourea that is highly reactive and directly used to form amide linkage with an amino group of enzymes (Adeel et al., 2018). Graphene oxide-laccase nano-assemblies are formed by settling several fGO-laccase layers until desired thickness is achieved. The graphene oxide (amino-functionalized) is good support for immobilization and is used with a number of enzymes for various applications (Wenqing et al., 2020). Graphene immobilized nano-biocatalysts allow manipulation of a nano-scale environment of enzymes, increase operational stability, and improve electron transfer, loading of enzymes, catalytic efficiency, and potential of biocatalysts. They can expand the practical utility of the biocatalyst. Graphene immobilized nano-biocatalysts are exploited in different biotechnological sectors like biocatalytic transformation, biosensor and biofuel cell development, pollutant degradation, and microchip bioreactors (Adeel et al., 2018).

15.7 DELIGNIFICATION OF PLANT BIOMASS

The delignification is required for the separation of cellulose and hemicellulose from lignin. In the pretreatment, the lignin modification or partial separation occurs, and its dissolution or removal from the cellulose and hemicellulose enhances their accessibility to hydrolytic enzymes. Schematic representation of ligninolysis and lignin deconstruction potential of ligninolytic enzymes are shown in Figure 15.3 (Bilal and Iqbal, 2020). Various delignification techniques are used like mechanical (extrusion, microwave, or ultrasound), physicochemical (steam explosion, hot water, NH_3 fiber explosion or CO_2 explosion), chemical (alkali hydrolysis or acid hydrolysis, selective oxidation, reductive catalytic or enzymatic fractionation, or ionic liquids), and biological (redox enzymes or microorganisms). Redox enzymes are important to overcome drawbacks of other methods, such as enzyme inhibitors and wastes produced (as a result of harsh conditions like high pressure, temperature, solvents, or extreme pH) due to partial biomass degradation and also prevent furfural and phenolics formation (Wang et al., 2019).

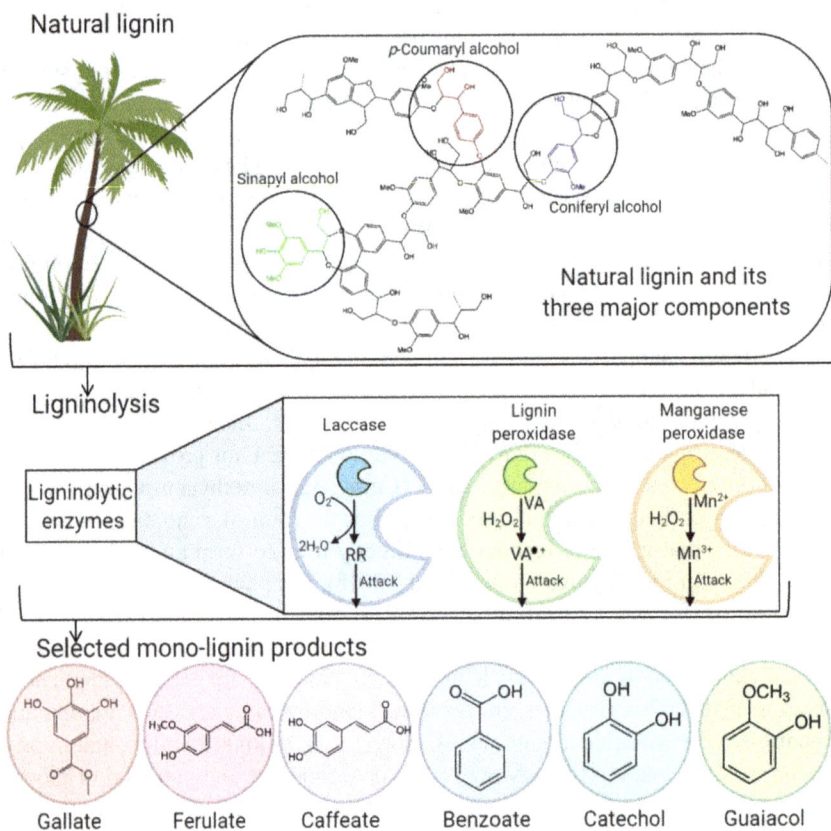

FIGURE 15.3 Schematic representation of ligninolysis and lignin deconstruction potential of ligninolytic enzymes; reprinted from Bilal and Iqbal, 2020.

Laccase enzyme gets extensive attention for their capacity of specifically lignin degradation. Although the biological pretreatment has the merit of being risk-free and low-cost, consuming less energy, and being more respectful of the environment, it also has the disadvantage of being a time-consuming process, generally a number of days to several months, with regard to accomplishing cellulose restoration comparable to those achieved with the physicochemical pretreatments. A number of strategies are proposed to overcome the major limitations of biofuel production employing the lignocellulosic biomass, like the use of enzyme cocktails for the biomass pretreatment. The engineering of white rot fungi is another way for improving lignocellulose biomass degradation, enhancing the lignocellulolytic activity, and leading to the synergistic combination between catalysts to increase degradation of the sugar (Ameta et al., 2020). The consumption of holocellulose carbon source by microorganisms for their expansion and metabolic requirements results in loss of plant biomass and leads to bioconversion. Research revealed that 33% of the cellulose in the wheat straw had been converted into glucose after a month of pretreatment with laccase. Hemicellulose damage of up to 27% had been attendant from corn stover treated with *C. subvermispora* for a one-and-a-half-month incubation period (Wan and Li, 2010). In addition, more industrial cellulolytic enzyme regarding saccharification after SSF method becomes necessary, which causes expense raise within bio-energy production.

To encounter the demerits of micro-flora of pretreatments, there is another effective strategy—the treatment of lignocellulosic material with the ligninolytic enzyme. Previously, utilization of these enzymes in expressions of LiP, laccase, and MnP for the degradation of lignin on a large scale has received much attention because these enzymes act specifically on the aromatic or polyphenolic linkages of lignin, resulting in lignin depolymerization, and reduce hindrance of side products (Mukhopadhyay and Banerjee, 2015). Laccases are unable to oxidize the nonphenolic compounds (having high redox potential, making >80% of lignin) because of relatively lower redox potential (0.5 to 0.8 V). These are also unable to enter small pores of the cell wall in plants. Laccase mediator systems with higher redox potential ($E° > 1.1$ V) have the ability to oxidize lignin efficiently (Zerva et al., 2019). The free ligninolytic enzymes can cause a number of limitations like low operational stability in extreme pH and temperature conditions, reuse, and difficult recovery. Most problems are overcome by the immobilization of enzymes on solid supports like xerogels, clay, sand, nanoparticles, or nano-fibrous polymers. In the case of LBM (the cross-linked molecular network of lignin-based on monolignols), lignin insolubility is overcome by using two methods; by using ILs or DESs that solubilize lignin to some extent so that it can be effectively oxidized by enzyme, or by using appropriate mediators as the molecular shuttles that permit effective electrons' flux to laccase through the lignin. For enzyme immobilization, the choice of support and method largely affects the enzyme stability and activity (Tocco et al., 2021).

15.8 CONCLUSION AND FUTURE PERSPECTIVES

Considering the circular economy and biorefinery themes, lignocellulosic biomass has gained popularity as a meaningful feedstock for producing an array of platform chemicals and biofuel production. Nevertheless, recalcitrancy hinders its utilization from exposing carbohydrates for cellulose hydrolysis, necessitating a pretreatment process to modify and remove lignin polymers. In this context, laccase has been found an exciting lignin-oxidizing biocatalyst with significant potential to remove lignin and other inhibiting agents produced during the treatment process. Its mediated lignin oxidation may facilitate the delignification of biomass in a cost-effective, greener, and eco-friendlier way. Using laccases is speculated to enhance the energy efficacy of bioprocess. However, intensive research efforts are needed for the large-scale and real-time deployment of laccase-based biocatalytic systems for lignin removal.

15.9   ACKNOWLEDGMENTS

Consejo Nacional de Ciencia y Tecnología (CONACYT) Mexico is thankfully acknowledged for partially supporting this work under Sistema Nacional de Investigadores (SNI) program awarded to Hafiz M.N. Iqbal (CVU: 735340).

15.10   CONFLICT OF INTERESTS

The listed author(s) declare no conflicting interests.

REFERENCES

Abdel-Raheem, A., and Shearer, C. A. (2002). Extracellular enzyme production by freshwater ascomycet. *Fungal Divers,* 11, 1–19.

Adeel, M., Bilal, M., Rasheed, T., Sharma, A., and Iqbal, H. M. (2018). Graphene and graphene oxide: Functionalization and nano-bio-catalytic system for enzyme immobilization and biotechnological perspective. *International Journal of Biological Macromolecules*, 120, 1430–1440.

Ameta, S. K., Avinash, K. R., Divya, H., Rakshit, A., and Suresh, C. A. (2020). Use of nano-materials in food science. biogenic nanoparticles and their use in agro-ecosystems. doi: 10.1007/978-981-15-2985-6_24.

An, H., Xiao, T., Fan, H., and Wei, D. (2015). Molecular characterization of a novel thermosta-ble laccase PPLCC2 from the brown rot fungus *Postia placenta* MAD-698-R. *Electronic Journal of Biotechnology,* 18(6), 451–458.

Annunziatini, C., Baiocco, P., Gerini, M. F., Lanzalunga, O., and Sjögren, B. (2005). Aryl sub-stituted N-hydroxyphthalimides as mediators in the laccase-catalysed oxidation of lignin model compounds and delignification of wood pulp. *Journal of Molecular Catalysis B: Enzymatic*, 32(3), 89–96.

Asgher, M., Ahmad, Z., and Iqbal, H. M. N. (2013). Alkali and enzymatic delignification of sugarcane bagasse to expose cellulose polymers for saccharification and bio-ethanol pro-duction. *Industrial Crops and Products*, 44, 488–495.

Asgher, M., Ahmad, Z., and Iqbal, H. M. N. (2017). Bacterial cellulose-assisted de-lignified wheat straw-PVA based bio-composites with novel characteristics. *Carbohydrate Poly-mers*, 161, 244–252.

Asgher, M., Bashir, F., and Iqbal, H. M. N. (2014). A comprehensive ligninolytic pre-treatment approach from lignocellulose green biotechnology to produce bioethanol. *Chemical Engineering Research and Design*, 92(8), 1571–1578.

Asgher, M., Ijaz, A., and Bilal, M. (2016c). Lignocellulose-degrading enzyme production by Pleurotus sapidus WC 529 and its application in lignin degradation. *Turkish Journal of Biochemistry*, 41(1), 26–36.

Asgher, M., Khan, S. W., and Bilal, M. (2016b). Optimization of lignocellulolytic enzyme production by *Pleurotus eryngii* WC 888 utilizing agro-industrial residues and bio-ethanol production. *Romanian Biotechnological Letters*, 21(1), 11133.

Asgher, M., Wahab, A., Bilal, M., and Iqbal, H. M. N. (2016a). Lignocellulose degradation and production of lignin modifying enzymes by *Schizophyllum commune* IBL-06 in solid-state fermentation. *Biocatalysis and Agricultural Biotechnology*, 6, 195–201.

Bilal, M., and Iqbal, H. M. (2019). Chemical, physical, and biological coordination: An interplay between materials and enzymes as potential platforms for immobilization. *Coordination Chemistry Reviews*, 388, 1–23.

Bilal, M., and Iqbal, H. M. (2020). Ligninolytic enzymes mediated ligninolysis: An untapped biocatalytic potential to deconstruct lignocellulosic molecules in a sustainable manner. *Catalysis Letters*, 150(2), 524–543.

Bugg, T. D., Ahmad, M., Hardiman, E. M., and Rahmanpour, R. (2011). Pathways for degradation of lignin in bacteria and fungi. *Natural Product Reports*, 28(12), 1883–1896.

Cabral Silva, A. C., and Venancio, A. (2020). Application of laccase for mycotoxin decontamination. *World Mycotoxin Journal*, 2021, 14(1), 61–73.

Chio, C., Sain, M., and Qin, W. (2019). Lignin utilization: A review of lignin depolymerization from various aspects. *Renewable and Sustainable Energy Reviews,* 107, 232–249.

Chong-Cerda, R., Laura, L., Rocio, C. R., Carlos Eduardo, H. L., Azucena, G. H., Guadalupe, G. S., and Abelardo, C. M. (2020). Nanoencapsulated laccases obtained by double-emulsion technique. Effects on enzyme activity pH-dependence and stability. *Catalysts*, 10(9), 1085.

Christopher, L. P., Bin, Y. and Yun, J. (2014). Lignin biodegradation with laccase-mediator systems. *Frontiers in Energy Research*, 2, 12.

Chukwuma, O. B., Mohd, R., Husnul Azan, T., and Norli, I. (2020). Lignocellulolytic enzymes in biotechnological and industrial processes: A review. *Sustainability*, 12, 7282.

Daronch, N. A., Kelbert, M., Pereira, C. S., de Araujo, P. H. H., and de Oliveira, D. (2020). Elucidating the choice for a precise matrix for laccase immobilization: A review. *International Journal of Chemical Engineering*, 307, 125506.

De Bhowmick, G., Sarmah, A. K., and Sen, R. (2018). Lignocellulosic biorefinery as a model for sustainable development of biofuels and value-added products. *Bioresource Technology*, 247, 1144–1154.

Grelska, A., and Magdalena, N. (2020). White rot fungi can be a promising tool for removal of bisphenol A., bisphenol S., and nonylphenol from wastewater. *Environmental Science and Pollution Research*, 27, 39958–39976.

Gutiérrez, A., Rencoret, J., Cadena, E. M., Rico, A., Barth, D., José, C., and Martínez, Á. T. (2012). Demonstration of laccase-based removal of lignin from wood and non-wood plant feedstocks. *Bioresource Technology*, 119, 114–122.

Nile, S. H., Venkidasamy, B., Dhivya, S., Arti, N., Jianbo, X., and Guoyin, K. (2020). Nanotechnologies in food science: Applications, recent trends, and future perspectives. *Nano-Micro Letters,* 12, 45.

Huber, D., Bleymaier, K., Pellis, A., Vielnascher, R., Daxbacher, A., Greimel, K. J., et al. (2018). Laccase catalyzed elimination of morphine from aqueous systems. *New Biotechnology*. 42, 19–25.

Hullo, M.-F., Moszer, I., Danchin, A., Martin-Verstraete, I., (2001). CotA of *Bacillus subtilis* is a copper dependent laccase. *Journal of Bacteriology*, 183(18), 5426–5430.

Kamimura, N., Sakamoto, S., Mitsuda, N., Masai, E., and Kajita, S. (2019). Advances in microbial lignin degradation and its applications. *Current Opinion in Biotechnology*, 56, 179–186.

Kiiskinen, L-L., Kruus, K., Bailey, M., Ylosmaki, E., Siika-aho, M., and Saloheimo, M. (2004). Expression of *Melanocarpus albomyces* laccase in *Trichoderma reesei* and characterization of the purified enzyme. *Microbiology*, 150(9), 3065–3074.

Kuila, A. (ed.) (2019). *Sustainable Biofuel and Biomass: Advances and Impacts*. Boca Raton: CRC Press.

Kumar, A., and Ram, C. (2020). Ligninolytic enzymes and its mechanisms for degradation of lignocellulosic waste in environment. *Heliyon*, 6, 2.

Kumar, P., Barrett, D. M., Delwiche, M. J., and Stroeve, P. (2009). Methods for pretreatment of lignocellulosic biomass for efficient hydrolysis and biofuel production. *Industrial & Engineering Chemistry Research*, 48, 3713–3729.

Liang, S., Wu, X. L., Xiong, J., Zong, M. H., and Lou, W.Y. (2020). Metal-organic frameworks as novel matrices for efficient enzyme immobilization: An update review. *Coordination Chemistry Reviews*, 406, 213149.

Longe, L. F., Couvreur, J., Leriche Grandchamp, M., Garnier, G., Allais, F., and Saito, K. (2018). Importance of mediators for lignin degradation by fungal laccase. *ACS Sustainable Chemistry & Engineering*, 6(8), 10097–10107.

Mayolo-Deloisa, K., Gonzalez-Gonzalez, M., and Rito-Palomares, M. (2020). Laccases in food industry: Bioprocessing, potential industrial and biotechnological applications. *Frontiers in Bioengineering and Biotechnology*, 8, 222.

Miyazaki, K. (2005). A hyperthermophilic laccase from Thermus thermophilus HB27. *Extremophiles*, 9(6), 415–425.

Mukhopadhyay, M., and Banerjee, R. (2015). Purification and biochemical characterization of a newly produced yellow laccase from *Lentinus squarrosulus* MR13. *3 Biotech*, 5(3), 227–236.

Muryanto, M., Muharramah, R., Falah, S., and Hidayat, A. (2021, January). Enzymatic biodelignification of Corncob by Laccase (Lac) from Cerrena Sp. B. Md. TA 1. In *IOP Conference Series: Materials Science and Engineering* (Vol. 1011, No. 1, p. 012030). Bristol, UK: IOP Publishing.

Nishiyama, Y., Langan, P., and Chanzy, H. (2002). Crystal structure and hydrogen-bonding system in cellulose Iβ from synchrotron X-ray and neutron fiber diffraction. *Journal of the American Chemical Society*, 124(31), 9074–9082.

Nunes, C. S., and Kunamneni, A. (2018). Chapter 7 laccases; properties and applications. In *Enzymes in Human and Animal Nutrition*, eds. C. S. Nunes, and V. Kumar. Cambridge, MA: Academic Press, 133–161.

Oudia, A., Queiroz, J., and Simões, R. (2008). The influence of operating parameters on the biodelignification of Eucalyptus globulus kraft pulps in a laccase—violuric acid system. *Applied Biochemistry and Biotechnology*, 149(1), 23–32.

Paliwal, R., Giri, K., and Rai, J. P. N. (2019). Microbial ligninolysis: Avenue for natural ecosystem management. In *Biotechnology: Concepts, Methodologies, Tools, and Applications* (pp. 1399–1423). Hershey, PA: IGI Global.

Pérez, J., Munoz-Dorado, J., De la Rubia, T. D. L. R., and Martinez, J. (2002). Biodegradation and biological treatments of cellulose, hemicellulose and lignin: An overview. *International Microbiology*, 5(2), 53–63.

Rouhani, S., Shohreh, A., Rose, W. K., Bhekie, B. M., and Titus, A. M. M. (2020). Laccase immobilized Fe_3O_4-graphene oxide nanobiocatalyst improves stability and immobilization efficiency in the green preparation of sulfa drugs. *Catalysts*, 10, 459.

Sankar Muthuvelu, K., Ravikumar, R., Roselin, N. S., Vignesh, B., and Rajendren. (2020, June 1). A novel method for improving laccase activity by immobilization onto copper ferrite nanoparticles for lignin degradation. *International Journal of Biological Macromolecules*, 152, 1098–1107.

Shao, B., Liu, Z., Zeng, G., Liu, Y., Yang, X., Zhou, C., Chen, M., Liu, Y., Jiang, J., and Yan, M. (2019). Immobilization of laccase on hollow mesoporous carbon nanospheres: Noteworthy immobilization, excellent stability and efficacious for antibiotic contaminants removal. *Journal of Hazardous Materials*, 362, 318–326.

Singh, A. K., Bilal, M., Iqbal, H. M., Meyer, A. S., and Raj, A. (2021). Bioremediation of lignin derivatives and phenolics in wastewater with lignin modifying enzymes: Status, opportunities and challenges. *Science of The Total Environment*, 777, 145988.

Stevens, J. C., and Shi, J. (2019). Biocatalysis in ionic liquids for lignin valorization: Opportunities and recent developments. *Biotechnology Advances*, 37(8), 107418.

Tocco, D., Cristina, C., Maura, M., Andrea, S., and Enrico, S. (2021). Recent developments in the delignification and exploitation of grass lignocellulosic biomass. *ACS Sustainable Chem. Eng.*, 9, 2412–2432.

Tsegaye, B., Chandrajit, B., and Partha, R. (2019). Microbial delignification and hydrolysis of lignocellulosic biomass to enhance biofuel production: An overview and future prospect. *Bulletin of the National Research Centre*, 43(1), 1–16.

Tuck, C. O., Pérez, E., Horváth, I. T., Sheldon, R. A., and Poliakoff, M. (2012). Valorization of biomass: Deriving more value from waste. *Science*, 337(6095), 695–699.

Uthandi, S., Saad, B., Humbard, M. A., and Maupin-Furlow, J. A. (2010). LccA, an archaeal laccase secreted as a highly stable glycoprotein into the extracellular medium by Haloferax volcanii. *Applied and Environmental Microbiology*, 76(3), 733–743.

Wan, C., and Li, Y. (2010). Microbial pretreatment of corn stover with *Ceriporiopsis subvermispora* for enzymatic hydrolysis and ethanol production. *Bioresource Technology*, 101(16), 6398–6403.

Wang, F., Ling, Xu, Liting, Z., Zhongyang, D., Haile, Ma, and Norman, T. (2019, December 9). Fungal laccase production from lignocellulosic agricultural wastes by solid-state fermentation: A review. *Microorganisms*, 7(12), 665.

Wang, F., Owusu-Fordjour, M., Xu, L., Ding, Z., and Gu, Z. (2020). Immobilization of laccase on magnetic chelator nanoparticles for apple juice clarification in magnetically stabilized fluidized bed. *Frontiers in Bioengineering and Biotechnology*, 8, 589.

Wenqing, L., Lin, B., Zhang, W., Yan, J., Liu, H., and Xi, Z. (2020). The amino-functionalized graphene oxide nanosheet preparation for enzyme covalent immobilization. *Journal of Nanoscience and Nanotechnology*. Pub date; July 27, 2020.

Zerva, A., Stefan, S., Evangelos, T., and Jasmina, N. R. (2019). Applications of microbial laccases: Patent review of the past decade (2009–2019). *Catalysts*, 9(12), 1023.

Zhiguo, Li., Zhiming, C., Qingpeng, Z., Jiaojiao, S., Song, Li., and Xinhua, Liu (2020, November). Improved performance of immobilized laccase on Fe_3O_4@C-Cu^{2+} nanoparticles and its application for biodegradation of dyes. *Journal of Hazardous Materials*, 399(15), 123088.

16 Omics Approaches for the Production of the Microbial Enzymes and Applications

Heena Parveen, Anuj Chaudhary,
Parul Chaudhary, Rabiya Sultana, Govind Kumar,
Priyanka Khati, Meenakshi Rana, and Pankaj Bhatt

ABSTRACT

Pollution loads due to excessive industrialization have resulted in the deterioration of the environment quality. Recalcitrant chemical compounds in the ecosystem pose a serious threat to living organisms. The biological methodology is a promising and eco-friendly approach that gains a significant substitute in degrading pollutants without affecting the quality of the environment. Microbes are ubiquitously distributed in the divergent environment, and their enzymatic machinery can be exploited to transform toxic chemicals into nontoxic and degradable forms. Nowadays, various omics methods have been used to study the diversity of microbes in their natural niche and also monitor the mechanism of microbial activity in the bioremediation process. The present chapter focuses on the importance of microbial enzymes in the detoxification of xenobiotics and the efficacy of modern multi-omics techniques in exploring the microbial communities to know the structural and functional characteristics of the bioremediation process.

CONTENTS

16.1   Introduction 318
16.2   Bioremediation Using Microbial Enzymes 319
16.3   Dehydrogenase 320
16.4   Genomics for Bioremediation 321
16.5   Metagenomics to Study the Enzymes in Bioremediation 322
16.6   Proteomics in Bioremediation 323
16.7   Transcriptomics in Bioremediation 324
16.8   Metabolomics to Study Enzymes and Bioremediation 325
16.9   Conclusion 326
References 326

DOI: 10.1201/9781003202998-16

16.1   INTRODUCTION

The total condition of the environment is intrinsically tied to the quality of life on earth. Unfortunately, as science, technology, and industry advance, a vast quantity of trash, ranging from sewage to radioactive waste, is released into the ecosystem, causing a serious threat to mankind's survival on earth. Contamination of the environment with a composite of xenobiotics has developed a foremost ecological hazard around the world. Because of their high toxicity, lengthy persistence, and low biodegradability many xenobiotic chemicals have a negative influence on the environment [1]. Synthetic substances produced in vast quantities for industrial, agricultural, and residential usage are referred to as xenobiotics in the context of environmental contaminants. Polycyclic aromatic hydrocarbons (PAHs), phenolic composites, special care products, and pesticides are examples of environmental xenobiotics. Concerns about their possible negative impacts have grown as they have become more common in diverse environmental compartments. As a result of their toxicity, they pose unprecedented health concerns and threats to the environment's safety and security [2–4]. Physical decomposition (e.g., filtration, incineration, coagulation) and chemical disintegration (UV oxidation, chemical precipitation, dichlorination) have progressed into new waste disposal systems, but these methods are complex and uneconomical, cause generations of toxic by-products, and lack public acceptance. An alternative and eco-friendly approach that has been well accepted is that diverse species, such as bacteria, fungi, algae, and plants, can be efficiently used to remediate contaminants [5–6].

As public awareness has grown, a slew of new ways based on cutting-edge scientific technologies has emerged to assess and address this difficult worldwide problem. Bioremediation is a generally established technology for cleaning a polluted site in an environmentally benign and supportable manner. Bioremediation is a fascinating method for eradicating pollutants from the atmosphere. Microbes, such as fungi and bacteria, have long been thought to be superior organisms for pollutant decontamination. It ensures a low-cost, straightforward, and environmentally friendly cleanup approach [7–9]. The ubiquitous presence of the microorganisms in natural and extreme habitats allows them to exploit these recalcitrant composites as a carbon and nitrogen source during their metabolic activities and causes mineralization into less toxic molecules [10–11]. Various microbes are involved in the degradation process—*Pseudomonas, Micrococcus, Rhodococcus*, and *Flavobacterium*—and fungi such as *Penicillium, Trichoderma, Aspergillus* and *Fusarium*, are isolated from contaminated soil and water niche. Fungi are usually better at decomposing complex and big macromolecules such as lignin, while bacteria are better at degrading mononuclear aromatic chemicals [12–13]. Xenobiotic degradation is facilitated by a wide range of enzymatic machinery, including hydroxylating, dehydrogenating, and hydrolyzing systems as well as full cleavage systems [14].

Microbial bioconversion solutions rely on a wide group of creatures that are native to the contaminated area and have enormous metabolic power. Isolating and purifying the native microbes provides insight into microbial metabolites and degradation pathways [15]. Identification of varied microbial populations from contaminated sites via a culturable approach is a difficult process that is restricted to rapidly developing microbial variety. Thus, advanced culture-independent molecular approaches (genomics, proteomics, metabolomics, fluxomics etc.) offer a viable way to uncover the functional dynamics of microbial populations in the realm of bioremediation. Hence, the chapter will address how to use omics technology for environmental monitoring and bioremediation in a modern and practical manner.

16.2 BIOREMEDIATION USING MICROBIAL ENZYMES

Microbial metabolism is thought to be the enticing process for pesticide breakdown into nonhazardous form under soil habitat, and it is the base for all bioremediation and bioaugmentation techniques [16]. The immense diversity of microorganisms under an environment where they act as natural recyclers, which transforms xenobiotic pollutants into harmless compounds, generally CO_2 water, nitrogen compounds and methane [17]. Due to the extensive genetic diversity and functioning, microbial populations have the ability to remediate any contaminated environment. Microorganisms releases metabolites with respect to their environment, which mediates the transformation process. For the remediation of resistive and organopollutants, fungi and bacteria depend on the engagement of distinct extracellular and intracellular enzymes. Algae, in addition to bacteria and fungus, are used in pesticide remediation, however to a lower level. Several green and blue-green algae have been discovered to break down organophosphorus insecticides when isolated from soil or water [18]. Some of the essential microbial-based enzymes are discussed below in respect to their hydrolysis mechanism toward different xenobiotic compounds.

Laccase: Laccases come under the category of multi-copper oxidases, present in a variety of microbes and plants that have received a lot of attention due to their ability to oxidize aromatic and aliphatic amines, phenols, polyphenols, hydroxy indoles, and other nonphenolic chemicals (carbohydrates). Laccase catalytic action in xenobiotic biodegradation with lignin-like structures has picked attention, and its biodegradative properties on a variety of pollutants have been thoroughly investigated. Fungal laccases have been extensively studied in the decolorization and detoxification of effluents. **Balcázar-López et al. [19]** reported a modified strain of *Trichoderma atroviridae* that expresses an extracellular laccase enzyme was employed in the degradation of phenolic compounds from wastewater, showing its potential for xenobiotic compound degradation. Industries are the main sources of phenolic or other compounds that release their effluents into water bodies, resulting in the accumulation of these recalcitrant compounds. Furthermore, lignin peroxidase (LiP) and manganese peroxidase (MnP) are both also involved in the catalysis of lignin organic pollutants. White rot fungus has gained a major position in expressing these enzymes. White rot species, such as *Trametes versicolor*, is involved in pesticide degradation in sandy soil [20]. Other reported laccase-producing basidiomycetes include *Polyporus brumalis, Ganoderma lucidum, Cryptococcus neoformans Agaricus bisporus, Coprinopsis cinerea, Fomes fomentarius, Schizophyllum commune, Panus rudis,* and various *Pleurotus* sp. and *Trametes* sp. [21]. The well-known pollutants polyaromatic hydrocarbons (PAHs) are ubiquitously distributed in the environment, and due to their persistency, toxicity, and carcinogenic character, they become a great concern for the quality of the ecosystem. Microbial laccase enzymes subsequently degrade PAHs into CO_2 and other less hazardous compounds [22]. Despite the fungal system, bacterial species are also catalyzing the xenobiotic compounds by utilizing their laccase enzyme complex. But in contrast to fungus, bacterial laccase is not versatile in its activity. In a report by **Lu et al. [23]**, a recombinant strain of *Bacillus licheniformis* was able to degrade carmine and reactive black synthetic dyes through alkaline laccase within 1 hour of activity. In another recent report, a recombinant laccase from *Bacillus vallismortis* was demonstrated in the bioremediation of aquaculture. Hence, laccase has a lot of promise for treating wastewater, including phenolic and nonphenolic chemicals, PHAs, synthetic dyes, and other developing contaminants [24].

Dehalogenase: Microbial dehalogenases is involved in the breaking of carbon-halogen bonds, which is a crucial step in the aerobic mineralization of numerous halogenated chemicals found in the environment. Dehalogenase enzyme cleaves carbon and halogen bonds using hydrolytic and oxygenolytic process, which might execute dehalogenation by replacing the halogen with a hydroxyl group and hydrogen atom from water and H_2 [25–26]. The inducible enzyme system that produces dehalogenases has sparked a rising interest in the application of microbial processes in the removal and recovery of halogen-contaminated environments. The usefulness of these microbes in pollution degradation in complicated and volatile halogenated settings stems from their exceptional survival skills [27]. Zu et al. (2012) reported *Bacillus* strain isolated from sludge sample was involved in debromination of 2,4,6-tribromophenol (TBP) employing methylated and reductive debromination mechanisms. TBP is generally used as a flame retardant (BFRs) and wood preservative, is ubiquitously found in soil and aquatic environments, and causes developmental fetotoxicity, neurotoxicity, and embryotoxicity. The process of dehalogenation occurs via different reaction mechanisms: reductive, hydrolytic, and others [28]. **Liang et al. [29]** also found tetra-bromobisphenol-A (TBBPA) degrading gene (*tbbpaA*) through whole-genome sequencing in *Ochrobactrum* sp. T, which is capable of degrading TBBPA, a flame retardant. Because of their carcinogenic, mutagenic, and cytotoxic qualities, organohalides are extremely harmful to living things. The identification, isolation, and characterization of new dehalogenation microbial species might be viewed as a significant means of mobilizing adaptability in microorganisms for organohalide pollution biodegradation. Efforts to make the prospecting and discovery of novel dehalogenases with potential for broader application easier have paid off, with numerous technologies, such as protein engineering quickening the discovery of novel dehalogenases with better or modified activity.

16.3 DEHYDROGENASE

Dehydrogenases are oxidoreductases that can be found in bacteria, yeast, plants, and animals [30–34]. Dehydrogenases use a coenzyme like $NAD^+/NADP^+$ as an electron acceptor to catalyze the oxidation-reduction reaction. Alcohol dehydrogenase enzymes convert alcohol into aldehydes or ketones. Other aromatic dehydrogenases reported include naphthalene dihydrodiol dehydrogenase, polyethylene glycol dehydrogenase, benzyl alcohol dehydrogenase, and others [35]. Previously, bacterial cell-free extracts digest the industrially generated xenobiotics of various molecular weights, and the polyethylene glycol dehydrogenase activity was detected [36]. Similarly, another dehydrogenase, dye-linked polypropene glycol dehydrogenase in the periplasm or membrane of *Stenotrophomonas maltophilia*, is energetic in high-molecular-weight PPG degradation, whereas a cytoplasm-located enzyme was found to be active in hydrolyzing low-molecular-weight composite [37]. In a recent study, a novel dehydrogenase, 17β-hydroxysteroid dehydrogenase (17β-HSDx) present in *Rhodococcus* sp. P14, showed better activity in steroid bioremediation [38]. In previous studies, *Rhodococcus* sp. was originally involved in the degradation of various polycyclic aromatic hydrocarbons.

Other hydrolytic enzymes, such as lipases, esterases, cellulases, amylases, proteases, and peroxidases, can be used to manage waste generated during the processing of food, to degrade plastics and insecticides, to treat biofilm mass, and to treat oil-contaminated soils, among other applications. Hydrolytic enzymes have a wide range of possible applications, including feed additives, biological sciences, and chemical industries. These esterases, amidases, and proteases may break down into ester, amide, and peptide bonds, resulting in compounds with low toxicity. Carbamate/parathion hydrolases from *Pseudomonas*, *Nocardia*, *Flavobacterium*, and *Bacillus cereus* have been effectively employed in the hydrolysis of pollutants like carbaryl, parathion, and coumaphos [39]. Hence, microbial enzymes are used throughout the biodegradation of harmful contaminants in an eco-friendly and cost-effective manner for recovering degraded soil's biological and physicochemical qualities [40].

16.4   GENOMICS FOR BIOREMEDIATION

Microbial degradation of recalcitrant compounds is a community-based strategy that engages multiple species that act synergistically. To study the complex profile and comprehend the bioremediation mechanism, various genomic approaches have evolved, which provide in-depth knowledge of microbial activity. The understanding of community dynamics and the plethora of bacteria actively engaging in bioremediation has been expanded due to cultivation-independent analysis of microorganisms from contaminated areas. Hence, the development of genomic technology has contributed to the long-term remediation of polluted surroundings [41]. The advancement of molecular, biotechnological and bioinformatic technologies for bioremediation challenges has resulted in understanding microbial gene-level systems in bioremediation (**Fig. 16.1**). Furthermore, the direct analysis of microorganisms in a polluted site opened up new insights in the scientific community for sharing uncultured microbial knowledge with the applications like metagenomics, proteomics, metabolomics, and transcriptomics [42–44]. The genomic study of *Gordonia* sp. and *Bacillus megaterium* showed that clusters of alkaline hydrolase genes present in bacteria degrade the naphthalene and xenobiotic compounds. The genomic study of *Cycloclasticus* sp. revealed that six ring hydroxylating dioxygenase enzymes are involved in pyrene degradation [45–47].

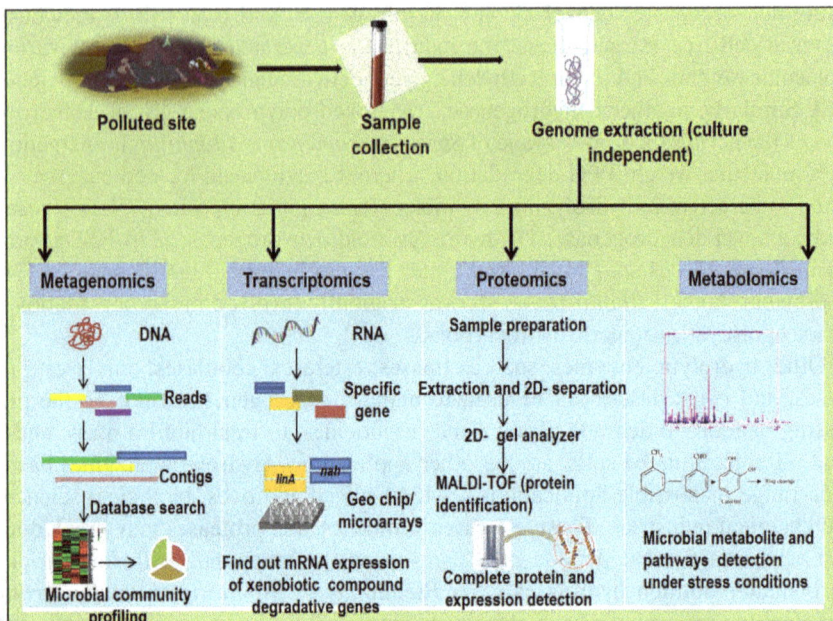

FIGURE 16.1 Different *omics* platforms for monitoring microbial dynamics in bioremediation.

16.5   METAGENOMICS TO STUDY THE ENZYMES IN BIOREMEDIATION

Metagenomic approaches are predominantly involved in the detection of microbial diversity of environmental samples. It centers on the unearthing of novel microbial products from the isolated DNA samples of the varied environmental niches. The metagenome (i.e., the pooled DNA tank obtained from environmental samples) acts as the genetic source used for the analysis of new species identification and for the discovery of the unexplored realms of biochemical pathways [48]. Moreover, microorganisms were difficult to culture as pure isolates could be processed easily with the aid of the techniques of metagenomics [49]. It plays an indispensable role in pollution monitoring in soil, water, and air. Microbial metagenomics aid in the selection of efficient genes and microbial communities that serve as biomarkers for pollution [50]. Previous reports stated the metagenomic analysis of data obtained from hydrocarbon-polluted areas enabled the application of biomarkers specific for the polluted hydrocarbon site. Specific software like MetaBoot assist in the comparison and identification of pollution biomarkers from contaminated environments [51].

Besides the aforementioned field, metagenomics plays an essential role in the identification of enzymes and bioactive compounds. Different enzymes having bioremediation capability, have been detected by the use of metagenomic methods. These enzymes originated from sources like the activated sludge, wastewater, soil, compost, the rumen of cows, polluted artificial soils, oil-contaminated areas, and environmental effluents. The enzymes obtained possessed properties for degradation of alkane, polyaromatic hydrocarbon, phenol, hexadecane, and trichlorophenol together, with monooxygenases, esterases, dioxygenases, laccases, and carboxylesterases. Altogether, the enzymes also help in the degradation of plastics, dyes, fumigants, insecticides, diesel, and pesticides [52]. Bioremediation plays an essential role in maintaining the bio-economy. A stable global environment is created by bioremediation technologies by maintaining clean water and healthy soil by eliminating the pollutants from natural sources. Pesticide-contaminated sites have been studied previously using metagenomics. There are various taxa found in soil that are able to degrade organophosphate pesticides—*Koribacter*, *Hypomicrobium*, *Bradyrhizobium*, and *Burkholderia* are the most common soil bacteria, contain degrading genes, and play important roles in the carbon and nitrogen cycle [53]. Alkane-degrading (alkane 1-monooxygenase) and LadA genes utilize dioxygen involved in the oxidation of alkane compounds by produced by *Nocardiodaceae* and *Mycobacteriaceeae* family [54]. **Zhang et al. [45]** found that the biostimulation process significantly increased the population of *Proteobacteria* and functional genes involved in carbon metabolic pathway and aromatic/alkene compounds degradation using the KEGG database. *Sphongomonas* and *Sphingobium* degrade aromatic hydrocarbons, which contain pbp and xyl gene cluster, determined by whole-genome sequencing [55]. Atrazine-degrading bacteria such as *Arthrobacter* sp. and *Sphingobium fuliginis* (organophosphate degrading), were analyzed using a sequencing platform to unravel their degradation potential [56]. **Negi and Lal [57]** reported that *Actinobacteria*, *Planctomycetes*, and *Acidobacteria* present in lindane-contaminated sites and showed their degradation with degradative enzymes, such as lyases, isomerases, and oxidoreductases. A bacterial consortium of *Rhodococcus*, *Achromobacter*, *Varivorex*, and *Pseudomonas* involved in biophenyl degradation is revealed by metagenomics [58]. Bacterial taxon such as *Actinobacteria* was the main degrader under aerobic conditions, while *Acidobacter* and *Bacteriodetes* are involved under anaerobic conditions. Members of *Peptococcaceae* and *Geobacteraceae* found in jet fuel contaminated sites and showed the metabolic potential to mitigate the toluene and benzene [59].

16.6   PROTEOMICS IN BIOREMEDIATION

The study of proteomics deals with the global expression of proteins. In 1995, the terms "proteome" and "proteomics" were introduced, which emerged as a key post-genomic trait out of the large complex datasets of genome sequencing. In proteomic analysis, the observed phenotype is the direct outcome of the activity of proteins, and hence the field of proteomics is regarded as vital as compared to the other fields [60]. Conventionally, the two-dimensional polyacrylamide gel electrophoresis (2-DE) was used for the proteomics study in relation to bioinformatics and mass spectrometry (MS). In bioremediation, especially in the case of PAH degradation, the proteome of membrane proteins is of utmost interest, wherein alterations in any site of the bacterium would lead to changes in cell surface receptors and proteins [61].

MS has transformed the sensitivity in the identification of proteins by many folds in less amount of time together with the analysis of peptides [62]. Altogether, the technique of liquid-chromatography MS (LC-MS) has paved a new direction for the detection and identification of potential water contaminants. Moreover, to evaluate the fate of the organic contaminants like surfactants, pesticides, algal toxins, and so on, the degradation products along with the metabolites were focused primarily during the treatment process of water [63]. Environmental conditions play an important role in the protein expression of an organism. The presence of toxic substances in the environment could stimulate the organism to adapt to different physiological changes in contrast to the normal behavioral response. The dawn of proteomics thus allows an extensive inspection of the abundance and composition of proteins together with the detection of the major proteins engaged in the organism's response in the predefined physiological condition [64–65]. Varied reports have depicted the fold changes of the protein sets with respect to the specific pollutants present [66]. The impact of trichloroethylene on protein expression by the microbial community was checked in anaerobic conditions. **Bastida et al. [47]** observed that catechol dioxygenase and dihydrodiol dehydrogenase was present in *Sphingomonadales* and involved in compost bioremediation. In the bioremediation process, metaproteome analysis is used to recognition of differentially expressed proteins and their genes, protein structure, and function characterization. Fluctuations in proteins were observed in *Arthrobacter phenanthrenivorans* cells when grown with phenanthrene and different carbon sources using LC-MS. Similarly, **Gregson et al. [67]** used LC/MS shotgun proteomics to study the differences in the proteome of *Oleispira antarctica* involved in hydrocarbon degradation when grown with n-alkanes. A total of 1,094 proteins were identified using proteomic analysis in pyrene degrading *Achromobacter xylosoxidans* [68]. Upregulation of p450 monooxygenase and O-methyltransferase in *P.chryosporium* is involved in the biotransformation of TBBPA through oxidative hydroxylation [69].

16.7   TRANSCRIPTOMICS IN BIOREMEDIATION

Transcriptome includes the transcribed subset of genes of an organism, which interlinks the cellular phenotype, the genome, and the proteome. Gene expression regulation acts as the prime key for adaptation to changes in the environment leading to survival. Transcriptomics portrays the method in a genome-wide range. In transcriptomics, one of the powerful platforms that facilitate the expression of mRNA level determination of each gene of an organism is the DNA microarrays [70–71]. The crucial phase in microarray experiments is data elucidation. A large amount of data may evolve in a specific stress condition due to the upregulation and downregulation of a wide range of genes. Hence, it becomes tremendously complex in statistical matters in addition to systemic and random errors.

Earlier studies depicted the use of DNA microarrays for the evaluation of physiology of pure environmental cultures [72]. In addition to it, DNA microarrays were also engaged in the case of mixed microbial cultures in order to display the gene expression profile. The whole-genome DNA microarray technology contributes to the bioremediation process by analyzing the expression of all the genes in the entire genome under varied environmental situations [73]. The results obtained aid in the identification of the regulatory circuits in the specific organisms. Comparative transcriptomics was used to determine the degradative pathways for phenol and to improve tolerance and consumption by *Rhodococcus opacus* [74]. Transcriptome examines the role of *Pseudomonas aeruginosa* in crude oil degradation by differentially expressed genes [75]. *Achromobacter* sp. is involved in hydrocarbon degradation by upregulation of monooxygenase and dehydrogenases enzymes [76]. Multiple enzymes, like oxygenases, reductases, isomerases, and kinases, are involved in different pathways of aromatic, phenols, and other xenobiotics [77]. *Sphingobacterium multivorum* is involved in hexaconazole degradation and generate different metabolites and upregulated the monooxygenases and dehydrogenases revealed by transcriptome analysis [78]. Transcriptomic analysis revealed that *Dehalococcoides mccartyi* contains rRNA-encoding genes and reductive dehalogenases (tceA, hup, vhu, and vcrA) involved in trichloroethene degradation [79]. Guo et al. [80] revealed that *Rhodococcus* sp. degrades phenol via the ketoadipate pathway determined using transcriptomics.

16.8 METABOLOMICS TO STUDY ENZYMES AND BIOREMEDIATION

Metabolomics is the study that involves the quantification and characterization of the metabolites produced by an organism in a desired environment [81]. Previously, metabolite studies were limited to specific target analytes, but now metabolomics has expanded its horizon and aims to detect every single metabolite along with its functional role. The analysis of metabolites is complicated due to its dynamic range and diversity. Moreover, the prime concern in metabolomics is to reveal the information and interlink the varied metabolic compositions present in the metabolic pathways of the biological system.

Metabolites are the substances or products yielded from the host organism in response to cellular or environmental changes [82–83]. These metabolites could be further differentiated into primary and secondary ones, which drive essential functions of cells, like signal transduction, energy storage, and production, in addition to apoptosis. Apart from the host organism, the metabolites could also be obtained from exogenous sources, xenobiotics, and other microorganisms. Usually, the metabolites produced are organic compounds of low molecular weight and which actively participate in the metabolic reactions for the normal maintenance of the organism. Certain mixtures of metabolites exist, which have their unique characteristics. Some microorganisms produce metabolites, which act as protectors against environmental stresses obtained in both in vivo and in vitro decontamination [84]. The principle of the metabolite study primarily focuses on the quantification of all the metabolites present in the cellular system. The target of the study could be achieved by technologies dealing with sampling, extraction, sample preparation, and sample analysis. Several spectroscopic methods can be used to depict the metabolites formed by different microbes [85]. Metabolites can be detected using chromatography, such as liquid and gas chromatography, and detection occurs through mass spectrometry, and nuclear magnetic resonance shows spectra that consist of different peaks that can identify/quantify metabolites. Metabolomics and bioinformatics databases have allowed understanding the microbial population and the genes involved in catabolic pathways. Metagenomics and metabolomic analysis were used to study petroleum hydrocarbons degrading microbial communities **[86]**. Metabolomic and proteogenomic tactics were used to detect the pathways and enzymes in *Mycobacterium* sp. to degrade acetyl tributyl citrate and dibutyl phthalate **[87]**. *Metarhizium brunneum* is involved in the metabolic biodegradation pathway for the removal of ametryn, which generates different metabolites, such as diethyl ametryn, 2-hydroxy atrazine, and ethyl hydroxylated ametryn, revealed by LC-MS/MS metabolomic analysis **[88]**. *Burkholderia* sp. produces 196 polar metabolites in the degradation of methylcarbamate **[89]**. *Drechsler* sp. and *Sinorhizobium* sp. produce polar metabolites, polyhydroxyalkanoates, monocerin, and fatty acids, which are the products of phenanthrene degradation **[90]**.

16.9   CONCLUSION

The huge risk posed to the environment is due to human actions, which necessitated innovative approaches to clean up the environment. Omics approaches have eradicated the limits to examine the mechanisms elaborated in several bioremediation pathways. These multi-omics methods provide novel organisms for bioremediation and degradative pathways at the molecular level. The study of functional genes in different metabolic pathways for remediation purposes is crucial when trying to link microbial diversity with detailed ecological functions. In this chapter, better knowledge of the role of microbial enzymes in the bioremediation of xenobiotics compounds is explained by the utilization of omics-based technology, which provides the prospect to measure the mechanisms, pathways, and functions of enzymes.

REFERENCES

1 Mishra S, Lin Z, Pang S, Zhang W, Bhatt P, Chen S (2021). Recent advanced technologies for the characterization of xenobiotic-degrading microorganisms and microbial communities. *Front. Bioeng. Biotechnol.* 9, 632059. doi: 10.3389/fbioe.2021.632059.

2 Kim KH, Jahan SA, Kabir E, Brown RCB (2013). A review of airborne polycyclic aromatic hydrocarbons (PAHs) and their human health effect. *Environ. Int.* 60, 71–80. doi: 10.1016/j.envint.2013.07.019.

3 Dovrak P, Nikel PI, Damborsky J, de Lorenzo V (2017). Bioremediation 3.0: Engineering pollutant-removing bacteria in the times of systemic biology. *Biotechnol. Adv.* 35, 845–866. doi: 10.1016/j.biotechadv.2017.08.001.

4 Ravindra AP, Haq SA (2019). Effects of xenobiotics and their biodegradation in marine life. In *Smart Bioremediation Technologies: Microbial Enzymes*, ed. P. Bhatt. Academic Press, 63–81. doi: 10.1016/B978-0-12-818307-6.00004-4.

5 Chandrakant SK, Rao SS (2011). Role of microbial enzymes in the bioremediation of pollutants: A review. *Enzyme Res.* 2011. Article ID 805187, 11 pages. https://doi.org/10.4061/2011/805187.

6 Kumar A, Sharma A, Chaudhary P, Gangola S (2021). Chlorpyrifos degradation using binary fungal strains isolated from industrial waste soil. *Biologia.* https://doi.org/10.1007/s11756-021-00816-8.

7 Chaudhary P, Sharma A (2019). Response of nanogypsum on the performance of plant growth promotory bacteria recovered from nanocompound infested agriculture field. *Envi. Ecol.* 37(1B), 363–372.

8 Khati P, Chaudhary P, Gangola S, Sharma A (2019b). Influence of nanozeolite on plant growth promotory bacterial isolates recovered from nanocompound infested agriculture field. *Environ. Eco.* 37(2), 521–527.

9 Agri U, Chaudhary P, Sharma A (2021). In vitro compatibility evaluation of agriusable nanochitosan on beneficial plant growth-promoting rhizobacteria and maize plant. *Natl. Acad. Sci. Lett.* https://doi.org/10.1007/s40009-021-01047-w.

10 Siles JA, Margesin R (2018). Insights into microbial communities mediating the bioremediation of hydrocarbon-contaminated soil from Alpine former military site. *Appl. Microbiol. Biotechnol.* 102, 4409–4421. doi: 10.1007/s00253-018-8932-6.

11 Zhan H, Wang H, Liao L, Feng Y, Fan X, Zhang L (2018). Kinetics and novel degradation pathway of cypermethrin in *Acinetobacter baumannii* ZH-14. *Front. Microbiol.* 9, 98. doi: 10.3389/fmicb.2018.00098.

12 Mishra V, Lal R, Srinivasan R (2001). Enzymes and operons mediating xenobiotic degradation in Bacteria. *Crit. Rev. Microbiol.* 27(2), 133–166.

13 Bhatt P, Rene ER, Kumar AJ, Zhang W, Chen S (2020c). Binding interaction of allethrin with esterase: Bioremediation potential and mechanism. *Bioresour. Technol.* 315, 123845. doi: 10.1016/j.biortech.2020.123845.

14 Harayama S, Kok M, Neidle, EL (1992). Functional and evolutionary relationships among diverse oxygenases. *Annu. Rev. Microbiol.* 46, 565.

15 Chandran H, Meena M, Sharma K (2020). Microbial biodiversity and bioremediation assessment through omics approaches. *Front. Environ. Chem.* 1, 570326. doi: 10.3389/fenvc.2020.570326.

16 Magan N, Fragoeiro S, Bastos C (2010). Environmental factors and bioremediation of xenobiotics using white rot fungi. *Mycobiology.* 38(4), 238–248.

17 Azaizeh H, Castro PM, Kidd P (2011). Biodegradation of organic Xenobiotic pollutants in the Rhizosphere. In *Org Xenob Plants.* Springer, Dordrecht, pp. 191–215.

18 Mukherjee I, Gopal M, Dhar DW (2004). Dissipation of chlorpyrifos with *Chlorella vulgaris. Bull. Environ. Contam. Toxiclo.* 73, 358–363.

19 Balcázar-López E, Méndez-Lorenzo LH, Batista-García RA, Esquivel-Naranjo U, Ayala M, Kumar VV (2016). Xenobiotic compounds degradation by heterologous expression of a Trametes sanguineus Laccase in Trichoderma atroviride. *PLoS One.* 11(2), e0147997. doi: 10.1371/ journal.pone.0147997.

20 Fragoeiro S, Magan N (2008). Impact of Trametes versicolor and Phanerochaete crysosporium on differential breakdown of pesticide mixtures in soil microcosms at two water potentials and associated respiration and enzyme activity. *Int Biodeterior Biodegrad.* 62, 376–383.

21 Forootanfar H, Faramarzi MA (2015). Insights into laccase producing organisms, fermentation states, purification strategies, and biotechnological applications. *Biotechnol. Prog.* 31, 1443–1463. doi: 10.1002/btpr.2173.

22 Khlifi R, Belbahri L, Woodward S (2010). Decolourization and detoxification of textile industry wastewater by the laccase-mediator system. *J. Hazard. Mater.* 175(1–3), 802–808.

23 Lu L, Wang TN, Xu TF, Wang JY, Wang CL, Zhao M (2013). Cloning and expression of thermo-alkali-stable laccase of *Bacillus licheniformis* in *Pichia pastoris* and its characterization. *Bioresour. Technol.* 134, 81–86.

24 Sun J, Zheng M, Lu Z, Lu F, Zhang C (2017). Heterologous production of a temperature and PH-stable laccase from Bacillus vallismortis fmb-103 in Escherichia coli and its application. *Process Biochem.* 55, 77–84.

25 Wang D, Li A, Han H, Liu T, Yang Q (2018). A potent chitinase from Bacillus subtilis for the efficient bioconversion of chitin-containing wastes. *Int. J. Biol. Macromol.* 116, 863–868.

26 Wang Y, Feng Y, Cao X, Liu Y, Xue S (2018). Insights into the molecular mechanism of dehalogenation catalyzed by D-2-haloacid dehalogenase from crystal structures. *Sci. Rep.* 8(1), 1454.

27 Ang TF, Maiangwa J, Salleh A, Normi Y, Leow T (2018). Dehalogenases: From improved performance to potential microbial dehalogenation applications. *Molecules.* 23(5), 1100.

28 Zu L, Li G, An T, Wong PK (2012). Biodegradation kinetics and mechanism of 2,4,6-tribromophenol by Bacillus sp. GZT: A phenomenon of xenobiotic methylation during debromination. *Bioresour. Technol.* 110, 153–159.

29 Liang Z, Li G, Mai B, Ma H (2019). An application of a novel gene encoding bromophenol dehalogenase from *Ochrobactrum* sp. T in TBBPA degradation. *Chemosphere.* 217, 507–515.

30 Kumari H, Khati P, Gangola S, Chaudhary P, Sharma A (2021). Performance of plant growth promotory rhizobacteria on maize and soil characteristics under the influence of TiO$_2$ nanoparticles. *Pantnagar J. Res.* 19(1), 28–39.

31 Kumari S, Sharma A, Chaudhary P, Khati P (2020). Management of plant vigor and soil health using two agriusable nanocompounds and plant growth promotory rhizobacteria in Fenugreek. *3Biotech.* 10, 461. https://doi.org/10.1007/s13205-020-02448-2.

32 Kukreti B, Sharma A, Chaudhary P, Agri U, Maithani D (2020). Influence of nanosilicon dioxide along with bioinoculants on *Zea mays* and its rhizospheric soil. *3Biotech.* 10, 345. https://doi.org/10.1007/s13205-020-02329-8.

33 Khati P, Bhatt P, Nisha, Kumar R, Sharma A (2018). Effect of nanozeolite and plant growth promoting rhizobacteria on maize. *3Biotech.* 8, 141. https://doi.org/10.1007/s13205-018-1142-1.

34 Khati P, Chaudhary P, Gangola S, Bhatt P, Sharma A (2017). Nanochitosan supports growth of *Zea mays* and also maintains soil health following growth. *3Biotech.* 7, 81. doi: 10.1007/s13205-017-0668-y.

35 Phale PS, Sharma A, Gautam K (2019). Microbial degradation of xenobiotics like aromatic pollutants from the terrestrial environments. In *Pharmaceuticals and Personal Care Products: Waste Management and Treatment Technology.* doi:10.1016/B978-0-12-816189-0.00011-1

36 Kawai F, Yamanaka H (1989). Inducible or constitutive polyethylene glycol dehydrogenase involved in the aerobic metabolism of polyethylene glycol. *J. Ferment. Bioeng.* 67(4), 300–302.

37 Tachibana S, Kawai F, Yasuda M (2002). Heterogeneity of dehydrogenases of *Stenotrophomonas maltophilia* Showing dye-linked activity with polypropylene glycols. *Biosci. Biotechnol. Biochem.* 66(4), 737–742.

38 Ye T, Peng J, Feng L (2019). A novel dehydrogenase 17β- HSDx from *Rhodococcus* sp. P14 with potential application in bioremediation of steroids contaminated environment. *J. Hazard. Mater.* 362, 170–177.

39 Kumar A, Sharma S (2019). Microbes and enzymes in soil health and bioremediation. In: *Microorganisms for Sustainability.* Springer, Berlin, Germany.

40 Khati P, Sharma A, Chaudhary P, Singh AK, Gangola S, Kumar R (2019a). High-throughput sequencing approach to access the impact of nanozeolite treatment on species richness and eveness of soil metagenome. *Biocatalysis and Agri. Biotech.* 20, 101249. https://doi.org/10.1016/j.bcab.2019.101249.

41 Desai C, Pathak H, Madamwar D (2010). Advances in molecular and '-omics' technologies to gauge microbial communities and bioremediation at xenobiotic/anthropogen contaminated sites. *Bioresour. Technol.* 101, 1558–1569. doi: 10.1016/j.biortech.2009.10.080.

42 Chaudhary P, Khati P, Chaudhary A, Maithani D, Kumar G, Sharma A (2021d). Cultivable and metagenomic approach to study the combined impact of nanogypsum and *Pseudomonas taiwanensis* on maize plant health and its rhizospheric microbiome. *PLoS One.* 16(4), e0250574. https://doi.org/10.1371/journal. pone.0250574.

43 Chaudhary P, Sharma A, Chaudhary A, Khati P, Gangola S, Maithani D (2021b). Illumina based high throughput analysis of microbial diversity of rhizospheric soil of maize infested with nanocompounds and *Bacillus* sp. *Applied Soil Ecology.* 159, 103836. http://doi.org/10.1016/j.apsoil.2020.103836.

44 Mishra S, Lin Z, Pang S, Zhang W, Bhatt P, Chen S (2021). Recent advanced technologies for the characterization of xenobiotic-degrading microorganisms and microbial communities. *Front. Bioeng. Biotechnol.* 9, 632059. doi: 10.3389/fbioe.2021.632059.

45 Zhang Z, Wan J, Liu L, Ye M, Jiang X (2021). Metagenomics reveals functional profiling of microbial communities in OCP contaminated sites with rapeseed oil and tartaric acid biostimulation. *J. Environ. Manage.* 289, 112515. doi: 10.1016/j.jenvman.2021.112515.

46 Delegan YA, Valentovich LN, Shafieva SM, Ganbarov KG, Filonov AE, Vainstein, MB (2019). Characterization and genome analysis of highly efficient thermotolerant oil-degrading bacterium *Gordonia* sp. 1D. *Folia Microbiol.* 64, 41–48. doi: 10.1007/s12223-018-0623-2.

47 Bastida F, Jehmlich N, Lima K, Morris BEL, Richnow HH, Hernández T, et al. (2016). The ecological and physiological responses of the microbial community from a semiarid soil to hydrocarbon contamination and its bioremediation using compost amendment. *J. Proteomics.* 135, 162–169. doi: 10.1016/j.jprot.2015.07.023.

48 Alves LDF, Westmann CA, Lovate GL, et al. (2018). Metagenomic approaches for understanding new concepts in microbial science. *Int. J. Genom.* 2312987. https://doi.org/10.1155/2018/2312987

49 Simon C, Daniel R (2011, 2012). Metagenomic analyses: Past and future trends. *Appl. Environ. Microbiol.* 77(4), 1153–1161.

50 Kisand V, Valente A, Lahm A, et al. (2012). Phylogenetic and functional metagenomic profiling for assessing microbial biodiversity in environmental monitoring. *PLoS One.* 7(8), e43630.

51 Wang X, Su X, Cui X, Ning K (2015). MetaBoot: A machine learning framework of taxonomical biomarker discovery for different microbial communities based on metagenomic data. *PeerJ.* 3, e993.

52 Ufarté L, Laville É, Duquesne S, Potocki-Veronese G (2015). Metagenomics for the discovery of pollutant degrading enzymes. *Biotechnol Adv.* 33, 1845–1854.

53 Jeffries TC, Smriti R, Nielsen UN, Lai K, Ijaz A, Nazaries L, Singh BK (2018). Metagenomic functional potential predicts degradation rates of a model organophosphorus xenobiotic in pesticide contaminated soils. *Front. Microbiol.* 9, 147. doi: 10.3389/fmicb.2018.00147.

54 Wang M, Garrido-Sanz D, Sansegundo-Lobato P, Redondo-Nieto M, Conlon R, Martin M, Mali R, Liu X, Dowling DN, Rivilla R, Germaine KJ (2021). Soil microbiome structure and function in ecopiles used to remediate petroleum-contaminated soil. *Front. Environ. Sci.* 9, 39. doi: 10.3389/fenvs.2021.624070.

55 Zhao Q, Yue S, Bilal M, Hu H, Wang W, Zhang X (2017). Comparative genomic analysis of 26 *Sphingomonas* and *Sphingobium* strains: Dissemination of bioremediation capabilities, biodegradation potential and horizontal gene transfer. *Sci. Total Environ.* 609, 1238–1247. doi: 10.1016/j.scitotenv.2017.07.249.

56 Azam S, Parthasarthy S, Singh C, Kumar S, Siddavattam D (2019). Genome organization and adaptive potential of archetypal organophosphate degrading *Sphingobium fuliginis* ATCC 27551. *Genome Biol. Evol.* 11, 2557–2562. doi: 10.1093/gbe/evz189.

57 Negi V, Lal R (2017). Metagenomic analysis of a complex community present in pond sediment. *J. Genomics* 5, 36–47. doi: 10.7150/jgen.16685.

58 Garrido-Sanz D, Manzano J, Martin M, Redondo-Nieto M, Rivillia R (2018). Metagenomic analysis of a biphenyl-degrading soil bacterial consortium reveals the metabolic roles of specific populations. *Front. Microbiol.* 9, 232. doi: 10.3389/fmicb.2018.00232.

59 Hidalgo KJ, Teramoto EH, Soriano AU, Valoni E, Baessa MP, Richnow HH (2020). Taxonomic and functional diversity of the microbiome in a jet fuel contaminated site as revealed by combined application of *in situ* microcosms with metagenomic analysis. *Sci. Total Environ.* 708, 135152. doi: 10.1016/j.scitotenv.2019.135152.

60 Hart EH, Creevey CJ, Hitch T, Kingston-Smith AH (2018). Meta-proteomics of rumen microbiota indicates niche compartmentalisation and functional dominance in a limited number of metabolic pathways between abundant bacteria. *Sci. Rep.* 8, 10504. doi: 10.1038/s41598-018-28827-7.

61 Wang M, Garrido-Sanz D, Sansegundo-Lobato P, Redondo-Nieto M, Conlon R, Martin M, Mali R, Liu X, Dowling DN, Rivilla R, Germaine KJ (2021). Soil microbiome structure and function in ecopiles used to remediate petroleum-contaminated soil. *Front. Environ. Sci.* 9, 39. doi: 10.3389/fenvs.2021.624070.

62 Hart EH, Creevey CJ, Hitch T, Kingston-Smith AH (2018). Meta-proteomics of rumen microbiota indicates niche compartmentalisation and functional dominance in a limited number of metabolic pathways between abundant bacteria. *Sci. Rep.* 8, 10504. doi: 10.1038/s41598-018-28827-7.

63 Sikkema J, deBont JAM, Poolman B (1995). Mechanisms of membrane toxicity of hydrocarbons. *Microbiological Rev.* 59, 201–222.

64 Aebersold R, Mann M (2003). Mass spectrometry-based proteomics. *Nature.* 422, 198–207.

65 Joo WA, Kim CW (2005). Proteomics of Halophilic archaea. *J Chromatogr B Analyt-Technol Biomed Life Sci.* 815, 237–250.

66 Vasseur C, Labadie J, Hebraud M (1999). Differential protein expression by Pseudomonas fragi submitted to various stresses. *Electrophoresis.* 20, 2204–2213.

67 Gregson BH, Metodieva G, Metodiev MV, Golyshin PN, McKew BA (2020). Protein expression in the obligate hydrocarbon-degrading psychrophile *Oleispira antarctica* RB-8 during alkane degradation and cold tolerance. *Environ. Microbiol.* 22, 1870–1883. doi: 10.1111/1462-2920.14956.

68 Nzila A, Ramirez CO, Musa MM, Sankara S, Basheer C, Li QX (2018). Pyrene biodegradation and proteomic analysis in *Achromobacter xylosoxidans*, PY4 strain. *Int. Biodeterior. Biodegrad.* 175, 1294–1305. doi: 10.1016/j.ibiod.2018.03.014.

69 Chen Z, Yin H, Peng H, Lu G, Liu Z, Dang Z (2019). Identification of novel pathways for biotransformation of tertrabromobisphenol A by *Phanerochaete chrysosporium* combined with mechanism analysis at proteome level. *Sci. Total Environ.* 659, 1352–1362. doi: 10.1016/j.scitotenv.2018.12.446.

70 Wilkins JC, Homer KA, Beighton D (2001). Altered protein expression of Streptococcus oralis cultured at low pH revealed by two-dimensional gel electrophoresis. *Appl Environ Microbiol.* 67, 3396–3405.

71 Kim S II, Kim SJ, Nam MH (2002). Proteome analysis of aniline-induced proteins in Acinetobacter lwoffi K24. *Curr Microbiol.* 44, 61–66.

72 Golyshin PN, Martins Dos Santos VA, Kaiser O (2003). Genome sequence completed of Alcanivorax borkumensis, a hydrocarbon-degrading bacterium that plays a global role in oil removal from marine systems. *J Biotechnol.* 106, 215–220.

73 Diaz E (2004). Bacterial degradation of aromatic pollutants: A paradigm of metabolic versatility. *Int Microbiol.* 7, 173–180.

74 Yoneda A, Henson WR, Goldner NK, Park KJ, Forsberg KJ, Kim SJ (2016). Comparative transcriptomics elucidates adaptive phenol tolerance and utilization in lipid-accumulating *Rhodococcus opacus* PD630. *Nucleic Acids Res.* 44, 2240–2254. doi: 10.1093/nar/gkw055.

75 Das D, Mawlong GT, Sarki YN, Singh AK, Chikkaputtaiah C, Boruah HPD (2020). Transcriptome analysis of crude oil degrading *Pseudomonas aeruginosa* strains for identification of potential genes involved in crude oil degradation. *Gene.* 755, 144909. doi: 10.1016/j.gene.2020.144909.

76 Hong YH, Deng MC, Xu XM, Wu CF, Xiao X, Zhu Q (2016). Characterization of the transcriptome of *Achromobacter* sp. HZ01 with the outstanding hydrocarbon-degrading ability. *Gene*. 584, 185–194. doi: 10.1016/j.gene.2016.02.032.

77 Singh A, Chaudhary S, Dubey B, Prasad V (2016). Microbial-mediated management of organic xenobiotic pollutants in agricultural lands. In *Plant Response to Xenobiotics*, eds. A Singh, SM Prasad, RP Singh. Singapore: Springer, 211–230. doi: 10.1007/978-981-10-2860-1_9.

78 An X, Tian C, Xu J, Dong F, Liu X, Wu X (2020). Characterization of hexaconazole-degrading strain *Sphingobacterium multivorum* and analysis of transcriptome for biodegradation mechanism. *Sci. Total Environ*. 722, 37171. doi: 10.1016/j.scitotenv.2020.137171.

79 Mao X, Trembley J, Yu K, Tringe SG, Alvarez-Cohen L (2019). Structural dynamics and transcriptomic analysis of *Dehalococcoides mccartyi* within a TCE-dechlorinating community in a completely mixed flow. *Water Res*. 158, 146–156. doi: 10.1016/j.watres.2019.04.038.

80 Gu Q, Wu Q, Zhang J, Guo W, Ding Y, Wang J (2018). Isolation and transcriptome analysis of phenol-degrading bacterium from carbon-sand filters in full scale drinking water treatment plant. *Front. Microbiol*. 9, 2162. doi: 10.3389/fmicb.2018.02162.

81 Beale DJ, Karpe AV, Ahmed W, Cook S, Morrison PD, Staley C. (2017). A community multi-omics approach towards the assessment of surface water quality in an urban river system. *Int. J. Environ. Res. Public Health*. 14, E303. doi: 10.3390/ijerph14030303.

82 Chaudhary P, Khati P, Gangola S, Kumar A, Kumar R, Sharma A (2021c). Impact of nano-chitosan and *Bacillus* spp. on health, productivity and defence response in *Zea mays* under field condition. *3 Biotech*. 11, 237. http://doi.org/10.1007/s13205-021-02790-z.

83 Chaudhary P, Khati P, Chaudhary A, Gangola S, Kumar R, Sharma A (2021a). Bioinoculation using indigenous *Bacillus* spp. improves growth and yield of *Zea mays* under the influence of nanozeolite. *3Biotech*. 11, 11. https://doi.org/10.1007/s13205-020-02561-2.

84 Malla MA, Dubey A, Yadav S, Kumar A, Hashem A, Abd Allah EF (2018). Understanding and designing the strategies for the microbe-mediated remediation of environmental contaminants using omics approaches. *Front. Microbiol*. 9, 1132. doi: 10.3389/fmicb.2018.01132.

85 Bargiela R, Herbst FA, Martínez-Martínez M, Seifert J, Rojo D, Cappello S, et al. (2015). Metaproteomics and metabolomics analyses of chronically petroleum-polluted sites reveal the importance of general anaerobic processes uncoupled with degradation. *Proteomics*. 15, 3508–3520. doi: 10.1002/pmic.201400614.

86 Lu H, Que Y, Wu X, Guan T, Guo H (2019). Metabolomics deciphered metabolic reprogramming required for biofilm formation. *Sci. Rep*. 9, 13160. doi: 10.1038/s41598-019-49603-1.

87 Wright RJ, Bosch R, Gibson MI, Christie-Oleza JA (2020). Plasticizer degradation by marine bacterial isolates: A proteogenomic and metabolomic characterization. *Environ. Sci. Technol*. 54, 2244–2256. doi: 10.1021/acs.est.9b05228.

88 Szewczyk R, Kusmierska A, Bernat P (2017). Ametryn removal by *Metarhizium brunneum*: Biodegradation pathway proposal and metabolic background revealed. *Chemosphere*. 190, 174–183. doi: 10.1016/j.chemosphere.2017.10.011.

89 Seo J, Keum YS, Li QX (2013). Metabolomic and proteomic insights into carbaryl catabolism by *Burkholderia* sp. C3 and degradation of ten *N*-methylcarbamates. *Biodegradation*. 24, 795–811. doi: 10.1007/s10532-013-9629-2.

90 d'Errico G, Aloj V, Flematti GR, Sivasithamparam K, Worth CM (2020). Metabolites of *Drechslera* sp. endophyte with potential as biocontrol and bioremediation agent. *Nat. Prod. Res*. 11, 1–9. doi: 10.1080/14786419.2020.1737058.

Index

A

acetone, 27
Acinetobacter radioresistens, 12
after breaking the cell wall, 26
agriculture, 1
α-acetolactate decarboxylase, 33
α-amylase, 30
alumina, 259
ammonium sulfate, 27
analytical industry, 5
animal feed, 13
asparaginases, 31
Aspergillus, 2
Aureobasidium pullulans, 10
azo dyes, treatment of, 261

B

Bacillus prodigiosus, 12
Bacillus sp., 10
β-galactosidase, 10
biodegradation and bioremediation, 1
bioprocesses, 2
biotechnological, useful for, 305
blood clots, 5

C

Candida cylindrica, 11
chemical therapies, 182
Cochliobolus sp., 9
combination of lipases, 12
cost-effectiveness, 2

D

dairy industry, 9
debittering enzymes, 8
decolorization, 258
degumming of jute, 12
dehairing is the most important, 11
detergent industry, 12, 13
DNA technology, 5

E

electrophoresis, 28
Endothia parasitica, 9
engineered enzymes, 24
environmental remediation, 198
enzymatic degradation, 179
enzyme cocktails, use of, 311
enzyme committee, 173
enzymes, 5
enzymes concentrated through nanofiltration, 26
Escherichia coli, 23
Est-A and Est-B, 173
esterases, 172
exonucleases, 175
expenses of cultivation, 173
extracellular enzymes, using, 303
extraction of the intracellular enzyme, 26
extremely thermal, 177
extremoenzymes, 2
extremophilic esterases, 178

F

feed industry, 69
fermentation, 208
food industry, 59
frequently used in the food
    industry, 93
freshness, and shelf life of bread, 67
fungal enzymes in bioremediation, 109
fungal laccases, 304
fungi, 103

G

Garner, 278
gaseous pollutants, 46
gel exclusion chromatography, 28
genome mining, including, 22
Geobacillus thermocatenulatus, 44
globalization, 103
glucoamylases, 33
gluconic acid, 8
glucosaminidase, 192
glucose, 181
glucose oxidases, 8
glucosidases, 69
glutaminase, 64
glutathione, 64
gluten, 96
glycol dehydrogenase, 48
grape pomace, 68
growing, 98
growth and development, 66

H

Halobacterium, Halobacillus, and
   Halothermothrix, 23
halogenated organic chemicals, 46
health as in a safety manner, 61
healthcare, 59
hemicellulases, 32
hemp, 11
heteroduplex, 22
high blood pressure, 63
high efficiency, 110
highest produce, 24
highly catalytic, 2
high resolution, 29
high yield, 8
households, 149
human body, 112
human era, 208
hydrocarbons, 45
hydrolase, 48, 50, 151
hydrophobic interaction, 29

I

immobilized form, 25
immobilized metal ion chromatography, 28
immunoadsorption, 28
important and time-consuming process, 11
improper regulation, 5
improvements in bioprocess technology, 19
improves color, 33
improves production, 21
industrial applications, 29
industrial sectors, 1
industrial synthesis, 8
inhibitors, 26
injury, 214
invert sugar, 10
ionic strength changes, 28

J

jamun leaves, 200
Japanese rice wines, 184
Japanese schnapps brewed from rice, 208
jejunum of pigs, 240
jet fuel, 323
join to form a special structure called cellulose, 296
juice processing, 194
juices, 214

K

kappas, 182
key enzymes of ferulic acid biosynthesis,
   183
key function in the hydrolyze reaction, 173

key role in chemically induced, 180
kingdom of plants, 183
knockout mice, 180
known risk factor for atherosclerosis, 182
kraft method utilizes, 182

L

laccase CopA, 47
laccase generated from a recombinant strain, 47
laccase in the processes of dye degradation,
   111
laccases, 42, 46
later evidence, 172
leather industry, 1
ligninolytic enzyme immobilization, 258
like the textile industry, 1
lipases, 172
long process, 42
long shelf life and compatibility, 12
long-term storage, 8

M

maintaining physiological, 5
maltose syrup, 8
market is growing globally, 67
metabolism of purine, 65
metabolize lactose, 66
Microbacterium sp. (ARACC2), 224
microbes, 1
microbes can be modified, 5
microbial enzymes, 1
microbial proteases, 10
microbiome, 222
milk products, 10
modify enzymes, 1
myocardial infarction, 8

N

natural gas, 22
Nectriella pironi, 110
needed to demonstrate, 22
never be, 109
number of reasons, 22
nutraceuticals, 99
nutrition, 108

O

one of their particular, 110
one such improvement comprises, 159
one type of fungus capable of decomposing,
   119
organic compounds, removing, 115
organisms show greater potential, 257

organophospho-hydrolase, 157
other groups of enzymes, 119
oxidase to degrade, 121
oxide systems, 112
oxidoreductase enzymes, 113
oxygen atom, 115
oxygen transmission, 111

P

paper industry utilizes, 61
pentachlorophenol, 48
peroxidases, proteases, 49
pesticides, 47
pharmaceutical applications, 63
pharmaceutical, 61
phenolic and nonphenolic chemicals, 47
phosphorus, 49
place in two methods, 47
polyethylene glycol dehydrogenase, 48
poly-hydroxybutyrate, 50
product formations, 60
protease, 49
Pseudomonas putida F6, 47
pulp and paper industry, 59
purified and crude CotA laccase, 47

Q

quality baking products, 67
quality food, 65
quality of baked products, 67
quality of leather, 73
quantity in milk, 67
quantity of wastes, 70
quicker rehydration, 73
quinon, 110
quoting, 114

R

Rahnella, 152
recently for the remediation, 157
remediation of agrochemicals, 158
remediation of organic compounds, 115
remediation of organophosphates, 154

S

Schyzophyllum commune, 253
sensitivity, 256
steric selectivity, and specificity, 251
structures, 252
supporting materials, 254
surface area, 254
surface immobilization, 254
synthetically, 251

T

termed fibrinolysin, 275
terminal, 275
thrombolytic agents promote the degradation, 273
thrombolytic drugs, 273
thrombus, 272
toxicity test, 261
Triticum aestivum, 261
truncated type of zymogens, 275

U

ubiquitously found in lignocellulosic, 302
ultrasound, 310
unfortunately, as science, 318
urokinase, 276
used like mechanical, 310
using peroxidases, 302
utilization in one or several steps in industries, 310
utilization of biomass, 302

V

variations and hence do not affect, 2
varied chemical and physical conditions, 2
varied environmental, 325
varied metabolic compositions, 325
variety of enzymes, 2
variety of industries, 2
various taxa found in soil, 323
Varivorex, 323
vital as compared, 324
volcanic springs, 2

W

waste, 8
well under extreme, 2
which hydrolysis starts, 231
which microbial enzymes, 2
wide range of applications, 5
wide range of uses, 8
wide range of varied chemical, 2
widely employed in food industries, 8
widely reported for, 9
wine in the baking, 8
with increased efficiency, 5
work efficiently, 2
worldwide attention, 2
worldwide for the production of enzymes, 2
worm gut, 5
wound debridement, 8

X

Xanthomonas, 32
Xanthus, 6

xenobiotics, 151
xenobiotics simultaneously, 157

yeast, 109
yellowing of the samples, 114

Y

Z

years in bioremediation processes, 110

zymogens, 275

For Product Safety Concerns and Information please contact our EU
representative GPSR@taylorandfrancis.com
Taylor & Francis Verlag GmbH, Kaufingerstraße 24, 80331 München, Germany

www.ingramcontent.com/pod-product-compliance
Lightning Source LLC
Chambersburg PA
CBHW060805220326
41598CB00022B/2541